T0296099

LONDON MATHEMATICAL SOCIETY LECTURE NOTE SERIES

Managing Editor: Professor M. Reid, Mathematics Institute,
University of Warwick, Coventry CV4 7AL, United Kingdom

The titles below are available from booksellers, or from Cambridge University Press at
http://www.cambridge.org/mathematics

314 Spectral generalizations of line graphs, D. CVETKOVIĆ, P. ROWLINSON & S. SIMIĆ
315 Structured ring spectra, A. BAKER & B. RICHTER (eds)
316 Linear logic in computer science, T. EHRHARD, P. RUET, J.-Y. GIRARD & P. SCOTT (eds)
317 Advances in elliptic curve cryptography, I.F. BLAKE, G. SEROUSSI & N.P. SMART (eds)
318 Perturbation of the boundary in boundary-value problems of partial differential equations, D. HENRY
319 Double affine Hecke algebras, I. CHEREDNIK
320 L-functions and Galois representations, D. BURNS, K. BUZZARD & J. NEKOVÁŘ (eds)
321 Surveys in modern mathematics, V. PRASOLOV & Y. ILYASHENKO (eds)
322 Recent perspectives in random matrix theory and number theory, F. MEZZADRI & N.C. SNAITH (eds)
323 Poisson geometry, deformation quantisation and group representations, S. GUTT *et al* (eds)
324 Singularities and computer algebra, C. LOSSEN & G. PFISTER (eds)
325 Lectures on the Ricci flow, P. TOPPING
326 Modular representations of finite groups of Lie type, J.E. HUMPHREYS
327 Surveys in combinatorics 2005, B.S. WEBB (ed)
328 Fundamentals of hyperbolic manifolds, R. CANARY, D. EPSTEIN & A. MARDEN (eds)
329 Spaces of Kleinian groups, Y. MINSKY, M. SAKUMA & C. SERIES (eds)
330 Noncommutative localization in algebra and topology, A. RANICKI (ed)
331 Foundations of computational mathematics, Santander 2005, L.M PARDO, A. PINKUS, E. SÜLI & M.J. TODD (eds)
332 Handbook of tilting theory, L. ANGELERI HÜGEL, D. HAPPEL & H. KRAUSE (eds)
333 Synthetic differential geometry (2nd Edition), A. KOCK
334 The Navier–Stokes equations, N. RILEY & P. DRAZIN
335 Lectures on the combinatorics of free probability, A. NICA & R. SPEICHER
336 Integral closure of ideals, rings, and modules, I. SWANSON & C. HUNEKE
337 Methods in Banach space theory, J.M.F. CASTILLO & W.B. JOHNSON (eds)
338 Surveys in geometry and number theory, N. YOUNG (ed)
339 Groups St Andrews 2005 I, C.M. CAMPBELL, M.R. QUICK, E.F. ROBERTSON & G.C. SMITH (eds)
340 Groups St Andrews 2005 II, C.M. CAMPBELL, M.R. QUICK, E.F. ROBERTSON & G.C. SMITH (eds)
341 Ranks of elliptic curves and random matrix theory, J.B. CONREY, D.W. FARMER, F. MEZZADRI & N.C. SNAITH (eds)
342 Elliptic cohomology, H.R. MILLER & D.C. RAVENEL (eds)
343 Algebraic cycles and motives I, J. NAGEL & C. PETERS (eds)
344 Algebraic cycles and motives II, J. NAGEL & C. PETERS (eds)
345 Algebraic and analytic geometry, A. NEEMAN
346 Surveys in combinatorics 2007, A. HILTON & J. TALBOT (eds)
347 Surveys in contemporary mathematics, N. YOUNG & Y. CHOI (eds)
348 Transcendental dynamics and complex analysis, P.J. RIPPON & G.M. STALLARD (eds)
349 Model theory with applications to algebra and analysis I, Z. CHATZIDAKIS, D. MACPHERSON, A. PILLAY & A. WILKIE (eds)
350 Model theory with applications to algebra and analysis II, Z. CHATZIDAKIS, D. MACPHERSON, A. PILLAY & A. WILKIE (eds)
351 Finite von Neumann algebras and masas, A.M. SINCLAIR & R.R. SMITH
352 Number theory and polynomials, J. MCKEE & C. SMYTH (eds)
353 Trends in stochastic analysis, J. BLATH, P. MÖRTERS & M. SCHEUTZOW (eds)
354 Groups and analysis, K. TENT (ed)
355 Non-equilibrium statistical mechanics and turbulence, J. CARDY, G. FALKOVICH & K. GAWEDZKI
356 Elliptic curves and big Galois representations, D. DELBOURGO
357 Algebraic theory of differential equations, M.A.H. MACCALLUM & A.V. MIKHAILOV (eds)
358 Geometric and cohomological methods in group theory, M.R. BRIDSON, P.H. KROPHOLLER & I.J. LEARY (eds)
359 Moduli spaces and vector bundles, L. BRAMBILA-PAZ, S.B. BRADLOW, O. GARCÍA-PRADA & S. RAMANAN (eds)
360 Zariski geometries, B. ZILBER
361 Words: Notes on verbal width in groups, D. SEGAL
362 Differential tensor algebras and their module categories, R. BAUTISTA, L. SALMERÓN & R. ZUAZUA
363 Foundations of computational mathematics, Hong Kong 2008, F. CUCKER, A. PINKUS & M.J. TODD (eds)
364 Partial differential equations and fluid mechanics, J.C. ROBINSON & J.L. RODRIGO (eds)
365 Surveys in combinatorics 2009, S. HUCZYNSKA, J.D. MITCHELL & C.M. RONEY-DOUGAL (eds)
366 Highly oscillatory problems, B. ENGQUIST, A. FOKAS, E. HAIRER & A. ISERLES (eds)
367 Random matrices: High dimensional phenomena, G. BLOWER
368 Geometry of Riemann surfaces, F.P. GARDINER, G. GONZÁLEZ-DIEZ & C. KOUROUNIOTIS (eds)
369 Epidemics and rumours in complex networks, M. DRAIEF & L. MASSOULIÉ
370 Theory of p-adic distributions, S. ALBEVERIO, A.YU. KHRENNIKOV & V.M. SHELKOVICH
371 Conformal fractals, F. PRZYTYCKI & M. URBAŃSKI
372 Moonshine: The first quarter century and beyond, J. LEPOWSKY, J. MCKAY & M.P. TUITE (eds)
373 Smoothness, regularity and complete intersection, J. MAJADAS & A. G. RODICIO
374 Geometric analysis of hyperbolic differential equations: An introduction, S. ALINHAC
375 Triangulated categories, T. HOLM, P. JØRGENSEN & R. ROUQUIER (eds)

376 Permutation patterns, S. LINTON, N. RUŠKUC & V. VATTER (eds)
377 An introduction to Galois cohomology and its applications, G. BERHUY
378 Probability and mathematical genetics, N. H. BINGHAM & C. M. GOLDIE (eds)
379 Finite and algorithmic model theory, J. ESPARZA, C. MICHAUX & C. STEINHORN (eds)
380 Real and complex singularities, M. MANOEL, M.C. ROMERO FUSTER & C.T.C WALL (eds)
381 Symmetries and integrability of difference equations, D. LEVI, P. OLVER, Z. THOMOVA & P. WINTERNITZ (eds)
382 Forcing with random variables and proof complexity, J. KRAJÍČEK
383 Motivic integration and its interactions with model theory and non-Archimedean geometry I, R. CLUCKERS, J. NICAISE & J. SEBAG (eds)
384 Motivic integration and its interactions with model theory and non-Archimedean geometry II, R. CLUCKERS, J. NICAISE & J. SEBAG (eds)
385 Entropy of hidden Markov processes and connections to dynamical systems, B. MARCUS, K. PETERSEN & T. WEISSMAN (eds)
386 Independence-friendly logic, A.L. MANN, G. SANDU & M. SEVENSTER
387 Groups St Andrews 2009 in Bath I, C.M. CAMPBELL *et al* (eds)
388 Groups St Andrews 2009 in Bath II, C.M. CAMPBELL *et al* (eds)
389 Random fields on the sphere, D. MARINUCCI & G. PECCATI
390 Localization in periodic potentials, D.E. PELINOVSKY
391 Fusion systems in algebra and topology, M. ASCHBACHER, R. KESSAR & B. OLIVER
392 Surveys in combinatorics 2011, R. CHAPMAN (ed)
393 Non-abelian fundamental groups and Iwasawa theory, J. COATES *et al* (eds)
394 Variational problems in differential geometry, R. BIELAWSKI, K. HOUSTON & M. SPEIGHT (eds)
395 How groups grow, A. MANN
396 Arithmetic differential operators over the p-adic integers, C.C. RALPH & S.R. SIMANCA
397 Hyperbolic geometry and applications in quantum chaos and cosmology, J. BOLTE & F. STEINER (eds)
398 Mathematical models in contact mechanics, M. SOFONEA & A. MATEI
399 Circuit double cover of graphs, C.-Q. ZHANG
400 Dense sphere packings: a blueprint for formal proofs, T. HALES
401 A double Hall algebra approach to affine quantum Schur–Weyl theory, B. DENG, J. DU & Q. FU
402 Mathematical aspects of fluid mechanics, J.C. ROBINSON, J.L. RODRIGO & W. SADOWSKI (eds)
403 Foundations of computational mathematics, Budapest 2011, F. CUCKER, T. KRICK, A. PINKUS & A. SZANTO (eds)
404 Operator methods for boundary value problems, S. HASSI, H.S.V. DE SNOO & F.H. SZAFRANIEC (eds)
405 Torsors, étale homotopy and applications to rational points, A.N. SKOROBOGATOV (ed)
406 Appalachian set theory, J. CUMMINGS & E. SCHIMMERLING (eds)
407 The maximal subgroups of the low-dimensional finite classical groups, J.N. BRAY, D.F. HOLT & C.M. RONEY-DOUGAL
408 Complexity science: the Warwick master's course, R. BALL, V. KOLOKOLTSOV & R.S. MACKAY (eds)
409 Surveys in combinatorics 2013, S.R. BLACKBURN, S. GERKE & M. WILDON (eds)
410 Representation theory and harmonic analysis of wreath products of finite groups, T. CECCHERINI-SILBERSTEIN, F. SCARABOTTI & F. TOLLI
411 Moduli spaces, L. BRAMBILA-PAZ, O. GARCÍA-PRADA, P. NEWSTEAD & R.P. THOMAS (eds)
412 Automorphisms and equivalence relations in topological dynamics, D.B. ELLIS & R. ELLIS
413 Optimal transportation, Y. OLLIVIER, H. PAJOT & C. VILLANI (eds)
414 Automorphic forms and Galois representations I, F. DIAMOND, P.L. KASSAEI & M. KIM (eds)
415 Automorphic forms and Galois representations II, F. DIAMOND, P.L. KASSAEI & M. KIM (eds)
416 Reversibility in dynamics and group theory, A.G. O'FARRELL & I. SHORT
417 Recent advances in algebraic geometry, C.D. HACON, M. MUSTAŢĂ & M. POPA (eds)
418 The Bloch–Kato conjecture for the Riemann zeta function, J. COATES, A. RAGHURAM, A. SAIKIA & R. SUJATHA (eds)
419 The Cauchy problem for non-Lipschitz semi-linear parabolic partial differential equations, J.C. MEYER & D.J. NEEDHAM
420 Arithmetic and geometry, L. DIEULEFAIT *et al* (eds)
421 O-minimality and Diophantine geometry, G.O. JONES & A.J. WILKIE (eds)
422 Groups St Andrews 2013, C.M. CAMPBELL *et al* (eds)
423 Inequalities for graph eigenvalues, Z. STANIĆ
424 Surveys in combinatorics 2015, A. CZUMAJ *et al* (eds)
425 Geometry, topology and dynamics in negative curvature, C.S. ARAVINDA, F.T. FARRELL & J.-F. LAFONT (eds)
426 Lectures on the theory of water waves, T. BRIDGES, M. GROVES & D. NICHOLLS (eds)
427 Recent advances in Hodge theory, M. KERR & G. PEARLSTEIN (eds)
428 Geometry in a Fréchet context, C. T. J. DODSON, G. GALANIS & E. VASSILIOU
429 Sheaves and functions modulo p, L. TAELMAN
430 Recent progress in the theory of the Euler and Navier-Stokes equations, J.C. ROBINSON, J.L. RODRIGO, W. SADOWSKI & A. VIDAL-LÓPEZ (eds)
431 Harmonic and subharmonic function theory on the real hyperbolic ball, M. STOLL
432 Topics in graph automorphisms and reconstruction (2nd Edition), J. LAURI & R. SCAPELLATO
433 Regular and irregular holonomic D-modules, M. KASHIWARA & P. SCHAPIRA
434 Analytic semigroups and semilinear initial boundary value problems (2nd Edition), K. TAIRA
435 Graded rings and graded Grothendieck groups, R. HAZRAT
436 Groups, graphs and random walks, T. CECCHERINI-SILBERSTEIN, M. SALVATORI & E. SAVA-HUSS (eds)
437 Dynamics and analytic number theory, D. BADZIAHIN, A. GORODNIK & N. PEYERIMHOFF (eds)

London Mathematical Society Lecture Note Series: 437

Dynamics and Analytic Number Theory

Proceedings of the Durham Easter School 2014

Edited by

DZMITRY BADZIAHIN
University of Durham

ALEXANDER GORODNIK
University of Bristol

NORBERT PEYERIMHOFF
University of Durham

CAMBRIDGE
UNIVERSITY PRESS

University Printing House, Cambridge CB2 8BS, United Kingdom

One Liberty Plaza, 20th Floor, New York, NY 10006, USA

477 Williamstown Road, Port Melbourne, VIC 3207, Australia

314-321, 3rd Floor, Plot 3, Splendor Forum, Jasola District Centre, New Delhi - 110025, India

79 Anson Road, #06-04/06, Singapore 079906

Cambridge University Press is part of the University of Cambridge.

It furthers the University's mission by disseminating knowledge in the pursuit of education, learning and research at the highest international levels of excellence.

www.cambridge.org
Information on this title: www.cambridge.org/9781107552371
© Cambridge University Press 2016

This publication is in copyright. Subject to statutory exception and to the provisions of relevant collective licensing agreements, no reproduction of any part may take place without the written permission of Cambridge University Press.
First published 2016

A catalogue record for this publication is available from the British Library

Library of Congress Cataloging in Publication data
Names: Durham Easter School (2014 : University of Durham) | Badziahin, Dmitry (Dmitry A.), editor. | Gorodnik, Alexander, 1975– editor. | Peyerimhoff, Norbert, 1964– editor.
Title: Dynamics and analytic number theory : proceedings of the Durham Easter School 2014 / edited by Dzmitry Badziahin, University of Durham, Alexander Gorodnik, University of Bristol, Norbert Peyerimhoff, University of Durham.
Description: Cambridge : Cambridge University Press, [2016] |
Series: London Mathematical Society lecture note series ; 437 |
Includes bibliographical references and index.
Identifiers: LCCN 2016044609 | ISBN 9781107552371 (alk. paper)
Subjects: LCSH: Number theory – Congresses. | Dynamics – Congresses.
Classification: LCC QA241 .D87 2014 | DDC 512.7/3–dc23
LC record available at https://lccn.loc.gov/2016044609

ISBN 978-1-107-55237-1 Paperback

Cambridge University Press has no responsibility for the persistence or accuracy of URLs for external or third-party internet websites referred to in this publication, and does not guarantee that any content on such websites is, or will remain, accurate or appropriate.

Contents

Contributors

Tim Austin
Courant Institute, NYU
New York, NY 10012, USA
Email: tim@cims.nyu.edu

Victor Beresnevich
Department of Mathematics, University of York
Heslington, York, Y010 5DD, United Kingdom
Email: victor.beresnevich@york.ac.uk

Yann Bugeaud
Département de mathématiques, Université de Strasbourg
F-67084 Strasbourg, France
Email: bugeaud@math.unistra.fr

Manfred Einsiedler
Departement Mathematik, ETH Zürich
8092 Zürich, Switzerland
Email: manfred.einsiedler@math.ethz.ch

Giovanni Forni
Department of Mathematics, University of Maryland
College Park, MD 20742-4015, USA
Email: gforni@math.umd.edu

Alex Kontorovich
Department of Mathematics, Rutgers University
Piscataway, NJ 08854, USA
Email: alex.kontorovich@rutgers.edu

Felipe Ramírez
Department of Mathematics and Computer Science
Wesleyan University, Middletown, CT 06459, USA
Email: framirez@wesleyan.edu

Sanju Velani
Department of Mathematics, University of York
Heslington, York, YO10 5DD, United Kingdom
Email: slv3@york.ac.uk

Tom Ward
Executive Office, Palatine Centre, Durham University
Durham DH1 3LE, United Kingdom
Email: t.b.ward@durham.ac.uk

Preface

This book is devoted to some of the interesting recently discovered interactions between Analytic Number Theory and the Theory of Dynamical Systems. Analytical Number Theory has a very long history. Many people associate its starting point with the work of Dirichet on L-functions in 1837, where he proved his famous result about infinitely many primes in arithmetic progressions. Since then, analytical methods have played a crucial role in proving many important results in Number Theory. For example, the study of the Riemann zeta function allowed to uncover deep information about the distribution of prime numbers. Hardy and Littlewood developed their circle method to establish first explicit general estimates for the Waring problem. Later, Vinogradov used the idea of the circle method to create his own method of exponential sums which allowed him to solve, unconditionally of the Riemann hypothesis, the ternary Goldbach conjecture for all but finitely many natural numbers. Roth also used exponential sums to prove the existence of three-term arithmetic progressions in subsets of positive density. One of the fundamental questions which arise in the investigation of exponential sums, as well as many other problems in Number Theory, is how rational numbers/vectors are distributed and how well real numbers/vectors can be approximated by rationals. Understanding various properties of sets of numbers/vectors that have prescribed approximational properties, such as their size, is the subject of the metric theory of Diophantine approximation, which involves an interesting interplay between Arithmetic and Measure Theory. While these topics are now considered as classical, the behaviour of exponential sums is still not well understood today, and there are still many challenging open problems in Diophantine approximation. On the other hand, in the last decades there have been several important breakthroughs in these areas of Number Theory where progress on long-standing open problems has been achieved by utilising techniques which originated from the Theory of Dynamical Systems. These developments have uncovered many profound and very promising connections between number-theoretic and dynamical objects that are at the forefront of current research. For instance, it turned out that properties of exponential sums are intimately related to the behaviour of orbits of flows on nilmanifolds; the existence of given combinatorial configurations (e.g. arithmetic progressions) in subsets of integers can be established through the study of multiple recurrence properties for dynamical systems; and Diophantine properties of vectors in the Euclidean spaces can be characterised in terms of excursions of orbits of suitable flows in the space of lattices.

The material of this book is based on the Durham Easter School, 'Dynamics and Analytic Number Theory', that was held at the University of Durham in Spring 2014. The intention of this school was to communicate some of these remarkable developments at the interface between Number Theory and Dynamical Systems to young researchers. The Easter School consisted of a series of mini-courses (with two to three lectures each) given by Tim Austin, Manfred Einsiedler, Giovanni Forni, Alex Kontorovich, Sanju Velani and Trevor Wooley, and a talk by Yann Bugeaud presenting a collection of recent results and open problems in Diophantine approximation. The event was very well received by more than 60 participants, many of them PhD students from all around the world. Because of the great interest of young researchers in this topic, we decided to encourage the speakers to write contributions to this Proceedings volume.

One of the typical examples where both classical and dynamical approaches are now actively developing and producing deep results is the theory of Diophantine approximation. One of the classical problems in this area asks how well a given n-dimensional vector $\mathbf{x} \in \mathbb{R}^n$ can be approximated by vectors with rational coefficients. More specifically, one can ask: what is the supremum $\lambda(\mathbf{x})$ of the values λ such that the inequality

$$||q\mathbf{x} - \mathbf{p}||_\infty < Q^{-\lambda} \tag{1}$$

has infinitely many integer solutions $Q \in \mathbb{N}, q \in \mathbb{N}, \mathbf{p} \in \mathbb{Z}^n$ satisfying $q \leq Q$? This type of problem is referred to as a simultaneous Diophantine approximation. There is also a dual Diophantine approximation problem which asks for the supremum $\omega(\mathbf{x})$ of the values ω such that the inequality

$$|(\mathbf{x}, \mathbf{q}) - p| < Q^{-\omega} \tag{2}$$

has infinitely many solutions $Q \in \mathbb{N}$, $\mathbf{q} \in \mathbb{Z}^n$, $p \in \mathbb{Z}$ with $\mathbf{q} \neq \mathbf{0}$ and $||\mathbf{q}||_\infty \leq Q$. It turns out that there are various relations between the exponents $\lambda(\mathbf{x})$ and $\omega(\mathbf{x})$. Chapter 2 provides an overview of known relations between these and some other similar exponents. It mostly concentrates on the case where $\mathbf{x}$ lies on the so-called Veronese curve which is defined by $\mathbf{x}(t) := (t, t^2, \ldots, t^n)$ with real t. This case is of particular importance for number theorists since it has implications for the question about the distribution of algebraic numbers of bounded degree. For example, condition (2) in this case transforms to $|P(t)| < Q^{-\omega}$ where $P(t)$ is a polynomial with integer coefficients. For large Q this implies that x is very close to the root of P, which is an algebraic number.

Metric theory of Diophantine approximation does not work with particular vectors $\mathbf{x}$. Instead it deals with the sets of all vectors $\mathbf{x}$ satisfying inequalities like (1) or (2) for infinitely many $Q \in \mathbb{N}, q \in \mathbb{N}, \mathbf{p} \in \mathbb{Z}^n$ (respectively, $Q \in \mathbb{N}$, $\mathbf{q} \in \mathbb{Z}^n$, $p \in \mathbb{Z}$, $\mathbf{q} \neq \mathbf{0}$). The central problem is to estimate the measure and the Hausdorff dimension of such sets. This area of Number Theory was founded at the beginning of the twentieth century with Khintchine's work which was later generalised by Groshev. In the most general way they showed that, given a function $\psi : \mathbb{R}_{\geq 0} \to \mathbb{R}_{\geq 0}$, the set of $m \times n$ matrices A which satisfy the inequality

$$||A\mathbf{q} - \mathbf{p}||_\infty < \psi(||\mathbf{q}||_\infty)$$

with $\mathbf{p} \in \mathbb{Z}^n$ and $\mathbf{q} \in \mathbb{Z}^m$, has either zero or full Lebesgue measure. The matrices A satisfying this property are usually called *ψ-well approximable*. Furthermore, with some mild conditions on ψ, the Lebesgue measure of the set of ψ-well approximable matrices is determined by the convergence of a certain series which involves ψ. Later, many other results of this type were established, some of them with help of the classical methods and others by using the ideas from homogeneous dynamics.

Chapter 1 describes several powerful 'classical' techniques used in metric theory of Diophantine approximation, such as the Mass Transference Principle, ubiquitous systems, Cantor sets constructions and winning sets. The Mass Transference Principle allows us to get results about the more sensitive Hausdorff measure and Hausdorff dimension of sets of well approximable matrices or similar objects as soon as results about their Lebesgue measure are known. Ubiquitous systems provide another powerful method originating from works of A. Baker and W. Schmidt. It enables us to obtain the 'full Lebesgue measure'-type results in various analogues of the Khintchine–Groshev theorem. Finally, Chapter 1 introduces the generalised Cantor set construction technique, which helps in investigating badly approximable numbers or vectors. It also relates such sets with so-called winning sets developed by W. Schmidt. The winning sets have several surprising properties. For example, they have the maximal possible Hausdorff dimension and, even though such sets may be null in terms of Lebesgue measure, their countable intersection must also be winning.

Chapter 3 is devoted to the study of exponential sums. Given a real polynomial $P(x) = a_k x^k + \cdots + a_1 x + a_0$, the Weyl sums are defined as

$$W_N := \sum_{n=0}^{N-1} e^{2\pi i P(n)}.$$

The study of Weyl sums has a long history that goes back to foundational works of Hardy, Littlewood, and Weyl. When the coefficients of the polynomial $P(X)$ satisfy a suitable irrationality condition, then it is known that for some $w \in (0, 1)$,

$$W_N = O(N^{1-w}) \quad \text{as } N \to \infty,$$

and improving the value of the exponent in this estimate is a topic of current research. This problem has been approached recently by several very different methods. The method of Wooley is based on refinements of the Vinogradov mean value theorem and a new idea of efficient congruencing, and the method of Flaminio and Forni involves the investigation of asymptotic properties of flows on nilmanifolds using renormalisation techniques. It is quite remarkable that the exponents w obtained by the Flaminio–Forni approach, which is determined by optimal scaling of invariant distributions, essentially coincide with the exponents derived by Wooley using his method of efficient congruencing.

As discussed in Chapter 3, flows on nilmanifolds provide a very convenient tool for investigating the distribution of polynomial sequences modulo one and modelling Weyl sums. We illustrate this by a simple example. Let

$$N := \left\{ [p, q, r] := \begin{pmatrix} 1 & p & r \\ 0 & 1 & q \\ 0 & 0 & 1 \end{pmatrix} : p, q, r \in \mathbb{R} \right\}$$

denote the three-dimensional Heisenberg group, and Γ be the subgroup consisting of matrices with integral entries. Then the factor space $M := \Gamma \backslash N$ provides the simplest example of a nilmanifold. Given an upper triangular nilpotent matrix $X = (x_{ij})$, the flow generated by X is defined by

$$\phi_t^X(m) = m \exp(tX) \quad \text{with } m \in M.$$

More explicitly, $\exp(tX) = [x_{12}t, x_{23}t, x_{13}t + x_{12}x_{23}t^2/2]$. The space M contains a two-dimensional subtorus T defined by the condition $q = 0$. If we take $x_{23} = 1$, then the intersection of the orbit $\phi_t^X(\Gamma e)$ with this torus gives the sequence of points $[x_{12}n, 0, x_{13}n + x_{12}n^2/2]$ with $n \in \mathbb{N}$. Hence, choosing suitable matrices X, the flows ϕ_t^X can be used to model values of general quadratic polynomials P modulo one. Moreover, this relation can be made much more precise. In particular, with a suitable choice of a test function F on M and $m \in M$,

$$\sum_{n=0}^{N-1} e^{2\pi i P(n)} = \int_0^N F(\phi_t^X(m))\, dt + O(1).$$

This demonstrates that quadratic Weyl sums are intimately related to averages of one-parameter flows on the Heisenberg manifold. A more elaborate construction discussed in detail in Chapter 3 shows that general Weyl sums can be approximated by integrals along orbits on higher-dimensional nilmanifolds. Chapter 3 discusses asymptotic behaviour of orbits averages on nilmanifolds and related estimates for Weyl sums.

Dynamical systems techniques also provide powerful tools to analyse combinatorial structures of large subsets of integers and of more general groups. This active research field fusing ideas from Ramsey Theory, Additive Combinatorics, and Ergodic Theory is surveyed in Chapter 4. We say that a subset $E \subset \mathbb{Z}$ has positive upper density if

$$\bar{\mathrm{d}}(E) := \limsup_{N-M\to\infty} \frac{|E \cap [M, N]|}{N - M} > 0.$$

Surprisingly, this soft analytic condition on the set E has profound combinatorial consequences, one of the most remarkable of which is the Szemerédi theorem. It states that every subset of positive density contains arbitrarily long arithmetic progressions: namely, configurations of the form $a, a+n, \ldots, a+(k-1)n$ with arbitrary large k. It should be noted that the existence of three-term arithmetic progressions had previously been established by Roth using a variant of the circle method, but the case of general progressions required substantial new ideas. Shortly after Szemerédi's work appeared, Furstenberg discovered a very different ingenious approach to this problem that used ergodic-theoretic techniques. He realised that the Szemerédi theorem is equivalent to a new ergodic-theoretic phenomenon called *multiple recurrence*. This unexpected connection is summarised by the Furstenberg correspondence principle which shows that, given a subset $E \subset \mathbb{Z}$, one can construct a probability space (X, μ), a measure-preserving transformation $T : X \to X$, and a measurable subset $A \subset X$ such that $\mu(A) = \bar{\mathrm{d}}(E)$ and

$$\bar{\mathrm{d}}(E \cap (E-n) \cap \cdots \cap (E-(k-1)n)) \geq \mu(A \cap T^{-n}(A) \cap \cdots \cap T^{-(k-1)n}(A)).$$

This allows the proof of Szemerédi's theorem to be reduced to establishing the multiple recurrence property, which shows that if $\mu(A) > 0$ and $k \geq 1$, then there exists $n \geq 1$ such that

$$\mu(A \cap T^{-n}A \cap \cdots \cap T^{-(k-1)n}A) > 0.$$

This result is the crux of Furstenberg's approach, and in order to prove it, a deep structure theorem for general dynamical systems is needed. Furstenberg's work has opened a number of promising vistas for future research and started a new field of Ergodic Theory – ergodic Ramsey theory, which explores the existence of combinatorial structures in large subsets of groups. This is the subject of Chapter 4. In view of the above connection it is of fundamental importance to explore asymptotics of the averages

$$\frac{1}{N}\sum_{n=0}^{N-1}\mu(A\cap T^{-n}(A)\cap\cdots\cap T^{-(k-1)n}(A)),$$

and more generally, the averages

$$\frac{1}{N}\sum_{n=0}^{N-1}(f_1\circ T^n)\cdots(f_{k-1}\circ T^{(k-1)n}) \tag{3}$$

for test functions $f_1,\ldots,f_{k-1}\in L^\infty(\mu)$. The existence of limits for these averages was established in the groundbreaking works of Host, Kra, and Ziegler. Chapter 4 explains an elegant argument of Austin which permits the proof of the existence of limits for these multiple averages as well as multiple averages for actions of the group $\mathbb{Z}^d$.

A number of important applications of the Theory of Dynamical Systems to Number Theory involve analysing the distribution of orbits on the space of unimodular lattices in $\mathbb{R}^{d+1}$. This space, which will be denoted by X_{d+1}, consists of discrete cocompact subgroups of $\mathbb{R}^{d+1}$ with covolume one. It can be realised as a homogeneous space

$$\mathsf{X}_{d+1}\simeq \mathrm{SL}_{d+1}(\mathbb{R})/\mathrm{SL}_{d+1}(\mathbb{Z}).$$

which allows us to equip X_{d+1} with coordinate charts and an invariant finite measure. Some of the striking applications of dynamics on the space X_{d+1} to problems of Diophantine approximation are explored in Chapter 5. It was realised by Dani that information about the distribution of suitable orbits on X_{d+1} can be used to investigate the existence of solutions of Diophantine inequalities. In particular, this allows a convenient dynamical characterisation of many Diophantine classes of vectors in $\mathbb{R}^d$ discussed in Chapters 1 and 2 to be obtained, such as, for instance, badly approximable vectors, very well approximable vectors, singular vectors. This connection is explained by the following construction. Given a vector $\mathbf{v}\in\mathbb{R}^d$, we consider the lattice

$$\Lambda_{\mathbf{v}} := \{(q, q\mathbf{v}+\mathbf{p}) : (q,\mathbf{p})\in\mathbb{Z}\times\mathbb{Z}^d\},$$

and a subset of the space X_{d+1} defined as

$$\mathsf{X}_{d+1}(\varepsilon) := \{\Lambda : \Lambda \cap [-\varepsilon, \varepsilon]^{d+1} \neq \{0\}\}.$$

Let $g_Q := \mathrm{diag}(Q^{-d}, Q, \ldots, Q)$. If we establish that the orbit $g_Q\Lambda$ visits the subset $\mathsf{X}_{d+1}(\varepsilon)$, then this will imply that the systems of inequalities

$$|q| \leq \varepsilon Q^d \quad \text{and} \quad \|q\mathbf{v} - \mathbf{p}\|_\infty \leq \frac{\varepsilon}{Q}$$

have a non-trivial integral solution $(q, \mathbf{p}) \in \mathbb{Z} \times \mathbb{Z}^d$. When $\varepsilon \geq 1$, the existence of solutions is a consequence of the classical Dirichlet theorem, but for $\varepsilon < 1$, this is a delicate property which was studied by Davenport and Schmidt. Vectors for which the above system of inequalities has a non-trivial solution for some $\varepsilon \in (0, 1)$ and all sufficiently large Q are called *Dirichlet-improvable*. Chapter 5 explains how to study this property using dynamical systems tools such as the theory of unipotent flow. This approach proved to be very successful. In particular, it was used by Shah to solve the problem posed by Davenport and Schmidt in the 60s. He proved that if $\phi : (0, 1) \to \mathbb{R}^d$ is an analytic curve whose image is not contained in a proper affine subspace, then the vector $\phi(t)$ is not Dirichlet-improvable for almost all t. Chapter 5 explains Shah's proof of this result.

Chapter 5 also discusses how dynamical systems techniques can be used to derive asymptotic counting results. Although this approach is applicable in great generality, its essence can be illustrated by a simple example: counting points in lattice orbits on the hyperbolic upper half-plane $\mathbb{H}$. We recall that the group $G = \mathrm{PSL}_2(\mathbb{R})$ acts on $\mathbb{H}$ by isometries. Given $\Gamma = \mathrm{PSL}_2(\mathbb{Z})$ (or, more generally, a discrete subgroup Γ of G with finite covolume), we consider the orbit $\Gamma \cdot i$ in $\mathbb{H}$. We will be interested in asymptotics of the counting function

$$N(R) := |\{\gamma \cdot i : \mathsf{d}_{\mathbb{H}}(\gamma \cdot i, i) < R, \gamma \in \Gamma\}|,$$

where $d_{\mathbb{H}}$ denotes the hyperbolic distance in $\mathbb{H}$. Since $\mathbb{H} \simeq G/K$ with $K = \mathrm{PSO}(2)$, the following diagram:

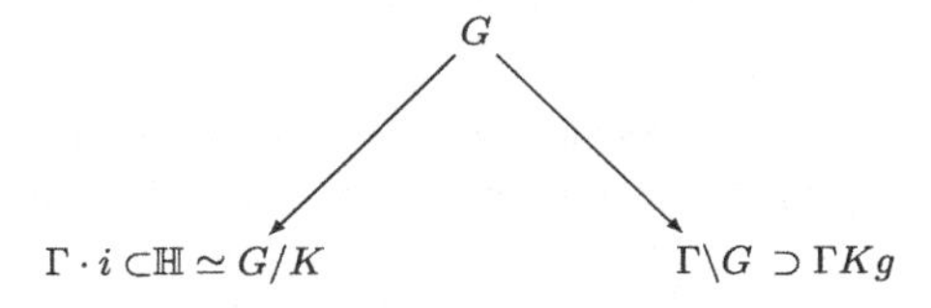

suggests that the counting function $N(R)$ can be expressed in terms of the space

$$X := \Gamma \backslash G.$$

This idea, which goes back to the work of Duke, Rudnick, and Sarnak, is explained in detail in Chapter 5. Ultimately, one shows that $N(R)$ can be approximated by combinations of averages along orbits $\Gamma K g$ as g varies over some subset of G. This argument reduces the original problem to analysing the distribution of the sets $\Gamma K g$ inside the space X which can be carried out using dynamical systems techniques.

The space X introduced above is of fundamental importance in the Theory of Dynamical Systems and Geometry because it can be identified with the unit tangent bundle of the modular surface $\Gamma\backslash\mathbb{H}$. Of particular interest is the geodesic flow defined on this space, which plays a central role in Chapter 6. This chapter discusses recent striking applications of the sieving theory for thin groups, developed by Bourgain and Kontorovich, to the arithmetics of continued fractions and the distribution of periodic geodesic orbits. It is well known in the theory of hyperbolic dynamical systems that one can construct periodic geodesic orbits with prescribed properties. In particular, a single periodic geodesic orbit may exhibit a very peculiar behaviour. Surprisingly, it turns out that the finite packets of periodic geodesic orbits corresponding to a given fundamental discriminant D become equidistributed as $D \to \infty$. This remarkable result was proved in full generality by Duke, generalising previous works of Linnik and Skubenko. While Duke's proof uses elaborate tools from analytic number theory (in particular, the theory of half-integral modular forms), now there is also a dynamical approach developed by Einsiedler, Lindenstrauss, Michel, and Venkatesh. They raised a question whether there exist infinitely many periodic geodesic orbits corresponding to fundamental discriminants which are contained in a fixed bounded subset of X. Chapter 6 outlines an approach to this problem, which uses that the geodesic flow dynamics is closely related to the symbolic dynamics of the continued fractions expansions. In particular, a quadratic irrational with a periodic continued fraction expansion

$$\alpha = [\overline{a_0, a_1, \ldots, a_\ell}]$$

corresponds to a periodic geodesic orbit. Moreover, the property of having a fundamental discriminant can be characterised in terms of the trace of the matrix

$$M_\alpha := \begin{pmatrix} a_0 & 1 \\ 1 & 0 \end{pmatrix}\begin{pmatrix} a_1 & 1 \\ 1 & 0 \end{pmatrix}\cdots\begin{pmatrix} a_\ell & 1 \\ 1 & 0 \end{pmatrix},$$

and the corresponding geodesic orbit lies in a fixed bounded set of X if $a_i \leq A$ for all i for a fixed $A > 0$. Hence, the original question reduces to the investigation of the semigroup

$$\Gamma_A := \left\langle \begin{pmatrix} a & 1 \\ 1 & 0 \end{pmatrix} : a \leq A \right\rangle^+ \cap \mathrm{SL}_2(\mathbb{R}),$$

and the trace map $\mathrm{tr} : \Gamma_A \to \mathbb{N}$. The semigroup Γ_A arises naturally in connection with several other deep problems involving periodic geodesic orbits and continued fractions. Chapter 6 outlines a promising approach to the Arithmetic Chaos Conjecture formulated by McMullen, which predicts that there exists a fixed bounded subset of the space X such that, for all real quadratic fields K, the closure of the set of periodic geodesic orbits defined over K and contained in this set has positive entropy. Equivalently, in the language of continued fractions, McMullen's conjecture predicts that for some $A < \infty$, the set

$$\{\alpha = [\overline{a_0, a_1, \ldots, a_\ell}] \in K : \text{all } a_j \leq A\}$$

has exponential growth as $\ell \to \infty$. Since

$$\alpha \in \mathbb{Q}(\sqrt{\mathrm{tr}(M_\alpha)^2 - 4}),$$

this problem also reduces to the analysis of the map $\mathrm{tr} : \Gamma_A \to \mathbb{N}$. Chapter 6 also discusses progress on the Zaremba conjecture regarding continued fraction expansions of rational fractions. As is explained in Chapter 6, all these problems can be unified by the far-reaching Local-Global Conjectures describing the distribution of solutions of $F(\gamma) = n$, $\gamma \in \Gamma_A$, where F is a suitable polynomial map.

We hope that this book will help to communicate the exciting material written by experts in the field and covering a wide range of different topics which are, nevertheless, in many ways connected to a broad circle of young researchers as well as to other experts working in Number Theory or Dynamical Systems.

Dmitry Badziahin
Department of Mathematical Sciences, Durham University
Durham, DH1 3LE, UK.
email: dzmitry.badziahin@durham.ac.uk
www.maths.dur.ac.uk/users/dzmitry.badziahin/
DB_webpage.htm

Alex Gorodnik
School of Mathematics, University of Bristol
Bristol BS8 1SD, UK.
email: a.gorodnik@bristol.ac.uk
www.maths.bris.ac.uk/~mazag/

Norbert Peyerimhoff
Department of Mathematical Sciences, Durham University
Durham, DH1 3LE, UK.
email: norbert.peyerimhoff@durham.ac.uk
www.maths.dur.ac.uk/~dma0np/

Acknowledgements

We, as part of the organising committee of the Durham Easter School 2014, which followed a one-day LMS Northern Regional Meeting dedicated to the same topic, would like to thank all the people who helped to make this event highly enjoyable and successful. We are also grateful to everybody who made this Proceedings volume possible. Specifically, we would like to acknowledge the help of:

- our Easter School co-organisers, T. Ward, A. Ghosh, and B. Weiss, for their support in this event and the selection of a very impressive list of speakers;
- Sam Harrison and his colleagues from Cambridge University Press for all their encouragement and support in producing this book;
- our distinguished speakers for their interesting talks;
- all participants of the Easter School for their interest and help to create a positive and inspiring atmosphere;
- all contributors to this volume, for their work in producing their chapters within the given timeframe.

Particular thanks are due to Durham University for its great hospitality in hosting the Easter School.

Last, but not least, we would like to stress the fact that the Easter School would not have been possible without the generous funding received, via the University of Bristol, from the ERC grant 239606 and the financial support of the London Mathematical Society.

1

Metric Diophantine Approximation: Aspects of Recent Work

Victor Beresnevich[1], *Felipe Ramírez*[2] *and Sanju Velani*[1]

Abstract

In these notes, we begin by recalling aspects of the classical theory of metric Diophantine approximation, such as theorems of Khintchine, Jarník, Duffin–Schaeffer and Gallagher. We then describe recent strengthening of various classical statements as well as recent developments in the area of Diophantine approximation on manifolds. The latter includes the well approximable, the badly approximable and the inhomogeneous aspects.

1.1 Background: Dirichlet and Bad

1.1.1 Dirichlet's Theorem and Two Important Consequences

Diophantine approximation is a branch of number theory that can loosely be described as a quantitative analysis of the density of the rationals $\mathbb{Q}$ in the reals $\mathbb{R}$. Recall that to say that $\mathbb{Q}$ is dense in $\mathbb{R}$ is to say that:

> for any real number x and $\epsilon > 0$ there exists a rational number p/q $(q > 0)$ such that $|x - p/q| < \epsilon$.

In other words, any real number can be approximated by a rational number with any assigned degree of accuracy. But how 'rapidly' can we approximate a given $x \in \mathbb{R}$?

> Given $x \in \mathbb{R}$ and $q \in \mathbb{N}$, how small can we make ϵ? Trivially, we can take any $\epsilon > 1/2q$. Can we do better than $1/2q$?

[1] VB and SV are supported in part by EPSRC Programme Grant: EP/J018260/1.
[2] FR is supported by EPSRC Programme Grant: EP/J018260/1.

The following rational numbers all lie within $1/(\text{denominator})^2$ of the circle constant $\pi = 3.141\ldots$:

$$\frac{3}{1}, \frac{22}{7}, \frac{333}{106}, \frac{355}{113}, \frac{103993}{33102}. \tag{1.1}$$

This shows that, at least sometimes, the answer to the last question is 'yes'. A more complete answer is given by *Dirichlet's theorem*, which is itself a simple consequence of the following powerful fact.

Pigeonhole Principle *If n objects are placed in m boxes and $n > m$, then some box will contain at least two objects.*

Theorem 1.1.1 (Dirichlet, 1842) *For any $x \in \mathbb{R}$ and $N \in \mathbb{N}$, there exist $p, q \in \mathbb{Z}$ such that*

$$\left|x - \frac{p}{q}\right| < \frac{1}{qN} \quad \textit{and} \quad 1 \le q \le N. \tag{1.2}$$

The proof can be found in most elementary number theory books. However, given the important consequences of the theorem and its various hybrids, we have decided to include the proof.

Proof As usual, let $[x] := \max\{n \in \mathbb{Z} : n \le x\}$ denote the integer part of the real number x and let $\{x\} = x - [x]$ denote the fractional part of x. Note that for any $x \in \mathbb{R}$ we have that $0 \le \{x\} < 1$.

Consider the $N + 1$ numbers

$$\{0x\}, \{x\}, \{2x\}, \ldots, \{Nx\} \tag{1.3}$$

in the unit interval $[0, 1)$. Divide $[0, 1)$ into N equal semi-open subintervals as follows:

$$[0, 1) = \bigcup_{u=0}^{N-1} I_u \quad \text{where} \quad I_u := \left[\frac{u}{N}, \frac{u+1}{N}\right), \quad u = 0, 1, \ldots, N-1. \tag{1.4}$$

Since the $N + 1$ points (1.3) are situated in the N subintervals (1.4), the Pigeonhole principle guarantees that some subinterval contains at least two points, say $\{q_2 x\}, \{q_1 x\} \in I_u$, where $0 \le u \le N - 1$ and $q_1, q_2 \in \mathbb{Z}$ with $0 \le q_1 < q_2 \le N$. Since the length of I_u is N^{-1} and I_u is semi-open we have that

$$|\{q_2 x\} - \{q_1 x\}| < \frac{1}{N}. \tag{1.5}$$

We have that $q_i x = p_i + \{q_i x\}$ where $p_i = [q_i x] \in \mathbb{Z}$ for $i = 1, 2$. Returning to (1.5) we get

$$|\{q_2 x\} - \{q_1 x\}| = |q_2 x - p_2 - (q_1 x - p_1)| = |(q_2 - q_1)x - (p_2 - p_1)|. \quad (1.6)$$

Now define $q = q_2 - q_1 \in \mathbb{Z}$ and $p = p_2 - p_1 \in \mathbb{Z}$. Since $0 \le q_1, q_2 \le N$ and $q_1 < q_2$ we have that $1 \le q \le N$. By (1.5) and (1.6), we get

$$|qx - p| < \frac{1}{N}$$

whence (1.2) readily follows. □

The following statement is an important consequence of Dirichlet's theorem.

Theorem 1.1.2 (Dirichlet, 1842) *Let $x \in \mathbb{R} \setminus \mathbb{Q}$. Then there exist infinitely many integers q, p such that* $\gcd(p, q) = 1$, $q > 0$ *and*

$$\left| x - \frac{p}{q} \right| < \frac{1}{q^2}. \quad (1.7)$$

Remark 1.1.3 Theorem 1.1.2 is true for all $x \in \mathbb{R}$ if we remove the condition that p and q are coprime; that is, if we allow approximations by non-reduced rational fractions.

Proof Observe that Theorem 1.1.1 is valid with $\gcd(p, q) = 1$. Otherwise $p/q = p'/q'$ with $\gcd(p', q') = 1$ and $0 < q' < q \le N$ and $|x - p/q| = |x - p'/q'| < 1/(qN) < 1/(q'N)$.

Suppose x is irrational and that there are only finitely many rationals

$$\frac{p_1}{q_1}, \frac{p_2}{q_2}, \ldots, \frac{p_n}{q_n},$$

where $\gcd(p_i, q_i) = 1$, $q_i > 0$ and

$$\left| x - \frac{p_i}{q_i} \right| < \frac{1}{q_i^2}$$

for all $i = 1, 2, \ldots, n$. Since x is irrational, $x - \frac{p_i}{q_i} \neq 0$ for $i = 1, \ldots, n$. Then there exists $N \in \mathbb{N}$ such that

$$\left| x - \frac{p_i}{q_i} \right| > \frac{1}{N} \qquad \text{for all } 1 \le i \le n.$$

By Theorem 1.1.1, there exists a reduced fraction $\frac{p}{q}$ such that

$$\left| x - \frac{p}{q} \right| < \frac{1}{qN} \le \frac{1}{N} \qquad (1 \le q \le N).$$

Therefore, $\frac{p}{q} \neq \frac{p_i}{q_i}$ for any i but satisfies (1.7). A contradiction. □

Theorem 1.1.2 tells us in particular that the list (1.1) of good rational approximations to π is not just a fluke. This list can be extended to an infinite sequence, and furthermore, such a sequence of good approximations exists for *every* irrational number. (See §1.1.2.)

Another important consequence of Theorem 1.1.1 is Theorem 1.1.4. Unlike Theorem 1.1.2, its significance is not so immediately clear. However, it will become apparent during the course of these notes that it is the key to the two fundamental theorems of classical metric Diophantine approximation: namely, the theorems of Khintchine and Jarník.

First, some notational matters. Unless stated otherwise, given a set $X \subset \mathbb{R}$, we will denote by $m(X)$ the one-dimensional Lebesgue measure of X. And we will use $B(x, r)$ to denote $(x - r, x + r) \subset \mathbb{R}$, the ball around $x \in \mathbb{R}$ of radius $r > 0$.

Theorem 1.1.4 *Let $[a, b] \subset \mathbb{R}$ be an interval and $k \geq 6$ be an integer. Then*

$$m\left([a,b] \cap \bigcup_{k^{n-1}<q\leq k^n} \bigcup_{p\in\mathbb{Z}} B\left(\tfrac{p}{q}, \tfrac{k}{k^{2n}}\right)\right) \geq \tfrac{1}{2}(b-a)$$

for all sufficiently large $n \in \mathbb{N}$.

Proof By Dirichlet's theorem, for any $x \in I := [a, b]$ there are coprime integers p, q with $1 \leq q \leq k^n$ satisfying $|x - p/q| < (qk^n)^{-1}$. We therefore have that

$$\begin{aligned} m(I) &= m\left(I \cap \bigcup_{q\leq k^n} \bigcup_{p\in\mathbb{Z}} B\left(\frac{p}{q}, \frac{1}{qk^n}\right)\right) \\ &\leq m\left(I \cap \bigcup_{q\leq k^{n-1}} \bigcup_{p\in\mathbb{Z}} B\left(\frac{p}{q}, \frac{1}{qk^n}\right)\right) + m\left(I \cap \bigcup_{k^{n-1}<q\leq k^n} \bigcup_{p\in\mathbb{Z}} B\left(\frac{p}{q}, \frac{k}{k^{2n}}\right)\right). \end{aligned}$$

Also, notice that

$$\begin{aligned} m\left(I \cap \bigcup_{q\leq k^{n-1}} \bigcup_{p\in\mathbb{Z}} B\left(\frac{p}{q}, \frac{1}{qk^n}\right)\right) &= m\left(I \cap \bigcup_{q\leq k^{n-1}} \bigcup_{p=aq-1}^{bq+1} B\left(\frac{p}{q}, \frac{1}{qk^n}\right)\right) \\ &\leq 2 \sum_{q\leq k^{n-1}} \frac{1}{qk^n}\left(m(I)q + 3\right) \leq \frac{3}{k} m(I) \end{aligned}$$

for large n. It follows that for $k \geq 6$,

$$m\left(I \cap \bigcup_{k^{n-1}<q\leq k^n} \bigcup_{p\in\mathbb{Z}} B\left(\tfrac{p}{q}, \tfrac{k}{k^{2n}}\right)\right) \geq m(I) - \tfrac{3}{k}\, m(I) \geq \tfrac{1}{2}\, m(I)$$

for large n. □

1.1.2 Basics of Continued Fractions

From Dirichlet's theorem we know that for any real number x there are infinitely many 'good' rational approximates p/q; but how can we find them? The theory of continued fractions provides a simple mechanism for generating them. We collect some basic facts about continued fractions in this section. For proofs and a more comprehensive account, see, for example, [57, 66, 80].

Let x be an irrational number and let $[a_0; a_1, a_2, a_3, \ldots]$ denote its continued fraction expansion. Denote its nth convergent by

$$\frac{p_n}{q_n} := [a_0; a_1, a_2, a_3, \ldots, a_n].$$

Recall that the convergents can be obtained by the following recursion

$$\begin{aligned} p_0 &= a_0, & q_0 &= 1, & \\ p_1 &= a_1 a_0 + 1, & q_1 &= a_1, & \\ p_k &= a_k p_{k-1} + p_{k-2}, & q_k &= a_k q_{k-1} + q_{k-2} & \text{for } k \geq 2, \end{aligned}$$

and that they satisfy the inequalities

$$\frac{1}{q_n(q_{n+1} + q_n)} \leq \left| x - \frac{p_n}{q_n} \right| < \frac{1}{q_n q_{n+1}}. \tag{1.8}$$

From this it is clear that the convergents provide explicit solutions to the inequality in Theorem 1.1.2 (Dirichlet); that is,

$$\left| x - \frac{p_n}{q_n} \right| \leq \frac{1}{q_n^2} \qquad \forall n \in \mathbb{N}.$$

In fact, it turns out that for irrational x the convergents are *best approximates* in the sense that if $1 \leq q < q_n$, then any rational $\frac{p}{q}$ satisfies

$$\left| x - \frac{p_n}{q_n} \right| < \left| x - \frac{p}{q} \right|.$$

Regarding $\pi = 3.141\ldots$, the rationals (1.1) are the first five convergents.

1.1.3 Competing with Dirichlet and Losing Badly

We have presented Dirichlet's theorem as an answer to whether the trivial inequality $|x - p/q| \leq 1/2q$ can be beaten. Naturally, one may also ask if we can do any better than Dirichlet's theorem. Let us formulate this a little more precisely. For $x \in \mathbb{R}$, let

$$\|x\| := \min\{|x - m| : m \in \mathbb{Z}\}$$

denote the distance from x to the nearest integer. Dirichlet's theorem (Theorem 1.1.2) can be restated as follows: *for any $x \in \mathbb{R}$, there exist infinitely many integers $q > 0$ such that*

$$q\,\|qx\| \leq 1. \tag{1.9}$$

Can we replace right-hand side of (1.9) by arbitrary $\epsilon > 0$? In other words, is it true that $\liminf_{q\to\infty} q\,\|qx\| = 0$ for every x? One might notice that (1.8) implies that there certainly do exist x for which this is true. (One can write down a continued fraction whose partial quotients grow as fast as one pleases.) Still, the answer to the question is 'no'. It was proved by Hurwitz (1891) that, for every $x \in \mathbb{R}$, we have $q\,\|qx\| < \epsilon = 1/\sqrt{5}$ for infinitely many $q > 0$, and that this is the best possible answer in the sense that the statement becomes false if $\epsilon < 1/\sqrt{5}$.

The fact that $1/\sqrt{5}$ is the best possible answer is relatively easy to see. Assume that it can be replaced by

$$\frac{1}{\sqrt{5}+\epsilon} \qquad (\epsilon > 0,\ \text{arbitrary}).$$

Consider the Golden Ratio $x_1 = \frac{\sqrt{5}+1}{2}$, root of the polynomial

$$f(t) = t^2 - t - 1 = (t - x_1)(t - x_2),$$

where $x_2 = \frac{1-\sqrt{5}}{2}$. Assume there exists a sequence of rationals $\frac{p_i}{q_i}$ satisfying

$$\left|x_1 - \frac{p_i}{q_i}\right| < \frac{1}{(\sqrt{5}+\epsilon)q_i^2}.$$

Then, for sufficiently large values of i, the right-hand side of the above inequality is less than ϵ and so

$$\left|x_2 - \frac{p_i}{q_i}\right| \leq |x_2 - x_1| + \left|x_1 - \frac{p_i}{q_i}\right| < \sqrt{5}+\epsilon.$$

It follows that

$$0 \neq \left|f\left(\frac{p_i}{q_i}\right)\right| < \frac{1}{(\sqrt{5}+\epsilon)q_i^2}\cdot(\sqrt{5}+\epsilon)$$

$$\implies \left|q_i^2 f\left(\frac{p_i}{q_i}\right)\right| < 1.$$

However, the left-hand side is a strictly positive integer. This is a contradiction, for there are no integers in $(0, 1)$ – an extremely useful fact.

The above argument shows that if $x = \frac{\sqrt{5}+1}{2}$ then there are at most finitely many rationals p/q such that

$$\left| x - \frac{p}{q} \right| < \frac{1}{(\sqrt{5}+\epsilon)q^2}.$$

Therefore, there exists a constant $c(x) > 0$ such that

$$\left| x - \frac{p}{q} \right| > \frac{c(x)}{q^2} \qquad \forall\ p/q \in \mathbb{Q}.$$

All of this shows that there exist numbers for which we cannot improve Dirichlet's theorem arbitrarily. These are called *badly approximable numbers* and are defined by

$$\begin{aligned} \mathbf{Bad} &:= \{x \in \mathbb{R} : \inf_{q \in \mathbb{N}} q\|qx\| > 0\} \\ &= \{x \in \mathbb{R} : c(x) := \liminf_{q \to \infty} q\|qx\| > 0\}. \end{aligned}$$

Note that if x is badly approximable then for the associated badly approximable constant $c(x)$ we have that

$$0 < c(x) \leq \frac{1}{\sqrt{5}}.$$

Clearly, **Bad** $\neq \varnothing$ since the Golden Ratio is badly approximable. Indeed, if $x \in$ **Bad** then $tx \in$ **Bad** for any $t \in \mathbb{Z} \setminus \{0\}$ and so **Bad** is at least countable.

Bad has a beautiful characterisation via continued fractions.

Theorem 1.1.5 *Let $x = [a_0; a_1, a_2, a_3, \ldots]$ be irrational. Then*

$$x \in \mathbf{Bad} \iff \exists\ M = M(x) \geq 1 \text{ such that } a_i \leq M \ \ \forall i.$$

That is, **Bad** *consists exactly of the real numbers whose continued fractions have bounded partial quotients.*

Proof It follows from (1.8) that

$$\frac{1}{q_n^2(a_{n+1}+2)} \leq \left| x - \frac{p_n}{q_n} \right| < \frac{1}{a_{n+1}q_n^2}, \tag{1.10}$$

and from this it immediately follows that if $x \in$ **Bad**, then

$$a_n \leq \max\{|a_o|, 1/c(x)\}.$$

Conversely, suppose the partial quotients of x are bounded, and take any $q \in \mathbb{N}$. Then there is $n \geq 1$ such that $q_{n-1} \leq q < q_n$. On using the fact that convergents are best approximates, it follows that

$$\left| x - \frac{p}{q} \right| \geq \left| x - \frac{p_n}{q_n} \right| \geq \frac{1}{q_n^2(M+2)} = \frac{1}{q^2(M+2)} \frac{q^2}{q_n^2}.$$

It is easily seen that

$$\frac{q}{q_n} \geq \frac{q_{n-1}}{q_n} \geq \frac{1}{M+1},$$

which proves that

$$c(x) \geq \frac{1}{(M+2)(M+1)^2} > 0,$$

hence $x \in \mathbf{Bad}$. □

Recall that a continued fraction is said to be *periodic* if it is of the form $x = [a_0; \ldots, a_n, \overline{a_{n+1}, \ldots, a_{n+m}}]$. Also, recall that an irrational number α is called a *quadratic irrational* if α is a solution to a quadratic equation with integer coefficients:

$$ax^2 + bx + c = 0 \qquad (a, b, c \in \mathbb{Z},\ a \neq 0).$$

It is a well-known fact that an irrational number x has periodic continued fraction expansion if and only if x is a quadratic irrational. This and Theorem 1.1.5 imply the following corollary.

Corollary 1.1.6 *Every quadratic irrational is badly approximable.*

The simplest instance of this is the Golden Ratio, a root of $x^2 - x - 1$, whose continued fraction is

$$\frac{\sqrt{5}+1}{2} = [1; 1, 1, 1, \ldots] := [\overline{1}],$$

with partial quotients clearly bounded.

Indeed, much is known about the badly approximable numbers, yet several simple questions remain unanswered. For example:

Folklore Conjecture *The only algebraic irrationals that are in* **Bad** *are the quadratic irrationals.*

Remark 1.1.7 Though this conjecture is widely believed to be true, there is no direct evidence for it. That is, there is no single algebraic irrational of degree greater than two whose membership (or non-membership) in **Bad** has been verified.

A particular goal of these notes is to investigate the 'size' of **Bad**. We will show:

(a) $m(\mathbf{Bad}) = 0$,
(b) $\dim \mathbf{Bad} = 1$,

where dim refers to the Hausdorff dimension (see §1.3.1). In other words, we will see that **Bad** is a small set in that it has measure zero in $\mathbb{R}$, but it is a large set in that it has the same (Hausdorff) dimension as $\mathbb{R}$.

Let us now return to Dirichlet's theorem (Theorem 1.1.2). Every $x \in \mathbb{R}$ can be approximated by rationals p/q with 'rate of approximation' given by q^{-2} – the right-hand side of inequality (1.7) determines the 'rate' or 'error' of approximation by rationals. The above discussion shows that this rate of approximation cannot be improved by an arbitrary constant for every real number – **Bad** is non-empty. On the other hand, we have stated above that **Bad** is a zero-measure set, meaning that the set of points for which we *can* improve Dirichlet's theorem by an arbitrary constant is full. In fact, we will see that if we exclude a set of real numbers of measure zero, then from a measure theoretic point of view the rate of approximation can be improved not just by an arbitrary constant but by a logarithm (see Remark 1.2.8).

1.2 Metric Diophantine Approximation: The Classical Lebesgue Theory

In the previous section, we have been dealing with variations of Dirichlet's theorem in which the right-hand side or rate of approximation is of the form ϵq^{-2}. It is natural to broaden the discussion to general approximating functions. More precisely, for a function $\psi : \mathbb{N} \to \mathbb{R}^+ = [0, \infty)$, a real number x is said to be *ψ-approximable* if there are infinitely many $q \in \mathbb{N}$ such that

$$\|qx\| < \psi(q). \tag{1.11}$$

The function ψ governs the 'rate' at which the rationals approximate the reals and will be referred to as an *approximating function*.

One can readily verify that the set of ψ-approximable numbers is invariant under translations by integer vectors. Therefore, without any loss of generality, and to ease the 'metrical' discussion which follows, we shall restrict our attention to ψ-approximable numbers in the unit interval $\mathrm{I} := [0, 1)$. The set of such numbers is clearly a subset of I and will be denoted by $W(\psi)$; i.e.

$$W(\psi) := \{x \in \mathrm{I}\colon \|qx\| < \psi(q) \text{ for infinitely many } q \in \mathbb{N}\}.$$

Notice that in this notation we have that

$$\text{Dirichlet's theorem (Theorem 1.1.2)} \quad \Longrightarrow \quad W(\psi) = \mathrm{I} \text{ if } \psi(q) = q^{-1}.$$

Yet, the existence of badly approximable numbers implies that there exist approximating functions ψ for which $W(\psi) \neq I$. Furthermore, the fact that $m(\mathbf{Bad}) = 0$ implies that we can have $W(\psi) \neq I$ while $m(W(\psi)) = 1$.

A key aspect of the classical theory of Diophantine approximation is to determine the 'size' of $W(\psi)$ in terms of:

(a) Lebesgue measure;
(b) Hausdorff dimension; and
(c) Hausdorff measure.

From a measure theoretic point of view, as we move from (a) to (c) in the above list, the notion of size becomes subtler. In this section we investigate the 'size' of $W(\psi)$ in terms of one-dimensional Lebesgue measure m.

We start with the important observation that $W(\psi)$ is a lim sup set of balls. For a fixed $q \in \mathbb{N}$, let

$$A_q(\psi) := \{x \in \mathrm{I} : \|qx\| < \psi(q)\}$$

$$:= \bigcup_{p=0}^{q} B\Big(\frac{p}{q}, \frac{\psi(q)}{q}\Big) \cap \mathrm{I}. \tag{1.12}$$

Note that

$$m\big(A_q(\psi)\big) \leqslant 2\psi(q) \tag{1.13}$$

with equality when $\psi(q) < 1/2$ since then the intervals in (1.12) are disjoint.

The set $W(\psi)$ is simply the set of real numbers in I which lie in infinitely many sets $A_q(\psi)$ with $q = 1, 2, \ldots$ i.e.

$$W(\psi) = \limsup_{q\to\infty} A_q(\psi) := \bigcap_{t=1}^{\infty} \bigcup_{q=t}^{\infty} A_q(\psi)$$

is a lim sup set. Now notice that for each $t \in \mathbb{N}$

$$W(\psi) \subset \bigcup_{q=t}^{\infty} A_q(\psi),$$

i.e. for each t, the collection of balls $B(p/q, \psi(q)/q)$ associated with the sets $A_q(\psi) : q = t, t+1, \ldots$ form a cover for $W(\psi)$. Thus, it follows via (1.13) that

$$\begin{aligned} m\big(W(\psi)\big) &\leq m\left(\bigcup_{q=t}^{\infty} A_q(\psi)\right) \\ &\leq \sum_{q=t}^{\infty} m\big(A_q(\psi)\big) \\ &\leq 2\sum_{q=t}^{\infty} \psi(q). \end{aligned} \tag{1.14}$$

Now suppose

$$\sum_{q=1}^{\infty} \psi(q) < \infty.$$

Then given any $\epsilon > 0$, there exists t_0 such that for all $t \geq t_0$

$$\sum_{q=t}^{\infty} \psi(q) < \frac{\epsilon}{2}.$$

It follows from (1.14), that

$$m\big(W(\psi)\big) < \epsilon.$$

But $\epsilon > 0$ is arbitrary, whence

$$m\big(W(\psi)\big) = 0$$

and we have established the following statement.

Theorem 1.2.1 *Let $\psi : \mathbb{N} \to \mathbb{R}^+$ be a function such that*

$$\sum_{q=1}^{\infty} \psi(q) < \infty.$$

Then

$$m(W(\psi)) = 0.$$

This theorem is in fact a simple consequence of a general result in probability theory.

1.2.1 The Borel–Cantelli Lemma

Let $(\Omega, \mathcal{A}, \mu)$ be a measure space with $\mu(\Omega) < \infty$ and let E_q $(q \in \mathbb{N})$ be a family of measurable sets in Ω. Also, let

$$E_\infty := \limsup_{q\to\infty} E_q := \bigcap_{t=1}^{\infty}\bigcup_{q=t}^{\infty} E_q\ ;$$

i.e. E_∞ is the set of $x \in \Omega$ such that $x \in E_i$ for infinitely many $i \in \mathbb{N}$.

The proof of the Theorem 1.2.1 mimics the proof of the following fundamental statement from probability theory.

Lemma 1.2.2 (Convergence Borel–Cantelli) *Suppose that* $\sum_{q=1}^{\infty} \mu(E_q) < \infty$. *Then*

$$\mu(E_\infty) = 0.$$

Proof The proof is left as an exercise for the reader. □

To see that Theorem 1.2.1 is a trivial consequence of the above lemma, simply put $\Omega = \mathrm{I} = [0, 1]$, $\mu = m$ and $E_q = A_q(\psi)$ and use (1.13).

Now suppose we are in a situation where the sum of the measures diverges. Unfortunately, as the following example demonstrates, it is not the case that if $\sum \mu(E_q) = \infty$ then $\mu(E_\infty) = \mu(\Omega)$ or indeed that $\mu(E_\infty) > 0$.

Example Let $E_q = (0, \frac{1}{q})$. Then $\sum_{q=1}^{\infty} m(E_q) = \sum_{q=1}^{\infty} \frac{1}{q} = \infty$. However, for any $t \in \mathbb{N}$ we have that

$$\bigcup_{q=t}^{\infty} E_q = E_t\,,$$

and thus

$$E_\infty = \bigcap_{t=1}^{\infty} E_t = \bigcap_{t=1}^{\infty} (0, \tfrac{1}{t}) = \varnothing,$$

implying that $m(E_\infty) = 0$.

The problem in the above example is that the sets E_q overlap 'too much' – in fact, they are nested. The upshot is that in order to have $\mu(E_\infty) > 0$, we not only need the sum of the measures to diverge but also that the sets E_q $(q \in \mathbb{N})$ are in some sense independent. Indeed, it is well known that if we had pairwise independence in the standard sense; *i.e.* if

$$\mu(E_s \cap E_t) = \mu(E_s)\mu(E_t) \qquad \forall s \neq t,$$

then we would have $\mu(E_\infty) = \mu(\Omega)$. However, we very rarely have this strong form of independence in our applications. What is much more useful to us is the following statement, whose proof can be found in [58, 90].

Lemma 1.2.3 (Divergence Borel–Cantelli) *Suppose that* $\sum_{q=1}^{\infty} \mu(E_q) = \infty$ *and that there exists a constant* $C > 0$ *such that*

$$\sum_{s,t=1}^{Q} \mu(E_s \cap E_t) \leq C \left(\sum_{s=1}^{Q} \mu(E_s) \right)^2 \tag{1.15}$$

holds for infinitely many $Q \in \mathbb{N}$. Then

$$\mu(E_\infty) \geq 1/C\,.$$

The independence condition (1.15) is often referred to as *quasi-independence on average*, and, together with the divergent sum condition, it guarantees that the associated lim sup set has positive measure. It does not guarantee *full* measure (i.e. that $\mu(E_\infty) = \mu(\Omega)$), which is what we are trying to prove, for example, in Khintchine's theorem. But this is not an issue if we already know (by some other means) that E_∞ satisfies a zero-full law (which is also often called a zero-one law) with respect to the measure μ, meaning a statement guaranteeing that

$$\mu(E_\infty) = 0 \quad \text{or} \quad \mu(\Omega).$$

Happily, this is the case with the lim sup set $W(\psi)$ of ψ-well approximable numbers [38, 37, 58].

Alternatively, assuming Ω is equipped with a metric such that μ becomes a doubling Borel measure, we can guarantee that $\mu(E_\infty) = \mu(\Omega)$ if we can establish *local quasi-independence on average* [14, §8]; *i.e.* we replace (1.15) in the above lemma by the condition that

$$\sum_{s,t=1}^{Q} \mu\big((B \cap E_s) \cap (B \cap E_t)\big) \leq \frac{C}{\mu(B)} \left(\sum_{s=1}^{Q} \mu(B \cap E_s) \right)^2 \tag{1.16}$$

for any sufficiently small ball B with centre in Ω and $\mu(B) > 0$. The constant C is independent of the ball B. Recall that μ is doubling if $\mu(2B) \ll \mu(B)$ for balls B centred in Ω. In some literature such measures are also referred to as Federer measures.

The Divergence Borel–Cantelli Lemma is key to determining $m(W(\psi))$ in the case where $\sum_{q=1}^{\infty} \psi(q)$ diverges – the subject of the next section and the main substance of Khintchine's theorem. Before turning to this, let us ask ourselves one final question regarding quasi-independence on average and positive measure of lim sup sets.

Question. Is the converse to Divergence Borel–Cantelli true? More precisely, if $\mu(E_\infty) > 0$ then is it true that the sets E_t are quasi-independent on average?

The following theorem is a consequence of a more general result established in [29].

Theorem 1.2.4 *Let (Ω, d) be a compact metric space equipped with a Borel probability measure μ. Let E_q $(q \in \mathbb{N})$ be a sequence of balls in Ω such that $\mu(E_\infty) > 0$. Then, there exists a strictly increasing sequence of integers $(q_k)_{k\in\mathbb{N}}$ such that $\sum_{k=1}^{\infty} \mu(E_{q_k}) = \infty$ and the balls E_{q_k} $(k \in \mathbb{N})$ are quasi-independent on average.*

1.2.2 Khintchine's Theorem

The following fundamental statement in metric Diophantine approximation (of which Theorem 1.2.1 is the 'easy case') provides an elegant criterion for the 'size' of the set $W(\psi)$ expressed in terms of Lebesgue measure.

Theorem 1.2.5 (Khintchine, 1924) *Let $\psi : \mathbb{N} \to \mathbb{R}^+$ be a monotonic function. Then*

$$m(W(\psi)) = \begin{cases} 0 & \text{if } \sum_{q=1}^{\infty} \psi(q) < \infty\,, \\ 1 & \text{if } \sum_{q=1}^{\infty} \psi(q) = \infty\,. \end{cases}$$

Remark 1.2.6 It is worth mentioning that Khintchine's original statement [64] made the stronger assumption that $q\psi(q)$ is monotonic.

Remark 1.2.7 The assumption that ψ is monotonic is only required in the divergent case. It cannot in general be removed – see §1.2.2.1.

Remark 1.2.8 Khintchine's theorem implies that

$$m(W(\psi)) = 1 \quad \text{if} \quad \psi(q) = \frac{1}{q \log q}.$$

Thus, from a measure theoretic point of view the 'rate' of approximation given by Dirichlet's theorem can be improved by a logarithm.

Remark 1.2.9 As mentioned in the previous section, in view of Cassels' zero-full law [38] (also known as zero-one) we know that $m(W(\psi)) = 0$ or 1 regardless of whether or not ψ is monotonic.

Remark 1.2.10 A key ingredient to directly establishing the divergent part is to show that the sets

$$A_s^* = A_s^*(\psi) := \bigcup_{2^{s-1} \le q < 2^s} \bigcup_{p=0}^{q} B\left(\frac{p}{q}, \frac{\psi(2^s)}{2^s}\right) \cap \mathrm{I}$$

are quasi-independent on average. Notice that:

- for ψ monotonic, $W(\psi) \supset W^*(\psi) := \limsup_{s\to\infty} A_s^*(\psi)$;
- if $\psi(q) < q^{-1}$, the balls in $A_s^*(\psi)$ are disjoint and so

$$m(A_s^*(\psi)) \asymp 2^s \psi(2^s);$$

- for ψ monotonic, $\sum \psi(q) \asymp \sum 2^s \psi(2^s)$.

Notation. Throughout, the Vinogradov symbols $\ll$ and $\gg$ will be used to indicate an inequality with an unspecified positive multiplicative constant. If $a \ll b$ and $a \gg b$, we write $a \asymp b$ and say that the two quantities a and b are comparable.

The following is a simple consequence of Khintchine's theorem.

Corollary 1.2.11 *Let* **Bad** *be the set of badly approximable numbers. Then*

$$m(\mathbf{Bad}) = 0.$$

Proof Consider the function $\psi(q) = 1/(q \log q)$ and observe that

$$\mathbf{Bad} \cap \mathrm{I} \subset \mathbf{Bad}(\psi) := \mathrm{I} \setminus W(\psi).$$

By Khintchine's theorem, $m(W(\psi)) = 1$. Thus $m(\mathbf{Bad}(\psi)) = 0$ and so $m(\mathbf{Bad} \cap \mathrm{I}) = 0$. □

1.2.2.1 The Duffin–Schaeffer Conjecture

The main substance of Khintchine's theorem is the divergent case and it is where the assumption that ψ is monotonic is necessary. In 1941, Duffin and Schaeffer [48] constructed a non-monotonic approximating function ϑ for which the sum $\sum_q \vartheta(q)$ diverges but $m(W(\vartheta)) = 0$. We now discuss the construction. We start by recalling two well-known facts: for any $N \in \mathbb{N}$, p prime, and $s > 0$,

Fact 1. $\sum_{q|N} q = \prod_{p|N}(1+p)$;
Fact 2. $\prod_p (1+p^{-s}) = \zeta(s)/\zeta(2s)$.

In view of Fact 2, we have that

$$\prod_p (1+p^{-1}) = \infty.$$

Thus, we can find a sequence of square-free positive integers N_i $(i = 1, 2, \ldots)$ such that $(N_i, N_j) = 1$ $(i \neq j)$ and

$$\prod_{p|N_i} (1+p^{-1}) > 2^i + 1. \tag{1.17}$$

Now let

$$\vartheta(q) = \begin{cases} 2^{-i-1} q/N_i & \text{if } q > 1 \text{ and } q|N_i \text{ for some } i\,, \\ 0 & \text{otherwise.} \end{cases} \tag{1.18}$$

As usual, let

$$A_q := A_q(\vartheta) = \bigcup_{p=0}^{q} B\Big(\frac{p}{q}, \frac{\vartheta(q)}{q}\Big) \cap \mathrm{I}$$

and observe that if $q|N_i$ $(q > 1)$ then $A_q \subseteq A_{N_i}$ and so

$$\bigcup_{q|N_i} A_q = A_{N_i}.$$

In particular,

$$m\Big(\bigcup_{q|N_i} A_q\Big) = m(A_{N_i}) = 2\vartheta(N_i) = 2^{-i}.$$

By definition,

$$W(\vartheta) = \limsup_{q\to\infty} A_q = \limsup_{i\to\infty} A_{N_i}.$$

Now

$$\sum_{i=1}^{\infty} m(A_{N_i}) = 1$$

and so the convergence Borel–Cantelli lemma implies that

$$m(W(\vartheta)) = 0.$$

However, it can be verified (*exercise*) by using Fact 1 together with (1.17) that

$$\sum_{q=1}^{\infty} \vartheta(q) = \sum_{i=1}^{\infty} 2^{-i-1} \frac{1}{N_i} \sum_{q>1:\,q|N_i} q = \infty.$$

In the same paper [48], Duffin and Schaeffer provided an appropriate statement for arbitrary ψ that we now discuss. The now famous Duffin–Schaeffer conjecture represents a key open problem in number theory. The integer p implicit in the inequality (1.11) satisfies

$$\left|x - \frac{p}{q}\right| < \frac{\psi(q)}{q}. \tag{1.19}$$

To relate the rational p/q with the error of approximation $\psi(q)/q$ uniquely, we impose the coprimeness condition $(p, q) = 1$. In this case, let $W'(\psi)$ denote the set of x in I for which the inequality (1.19) holds for infinitely many $(p, q) \in \mathbb{Z} \times \mathbb{N}$ with $(p, q) = 1$. Clearly, $W'(\psi) \subset W(\psi)$. For any approximating function $\psi : \mathbb{N} \to \mathbb{R}^+$ one easily deduces that

$$m(W'(\psi)) = 0 \quad \text{if} \quad \sum_{q=1}^{\infty} \varphi(q) \frac{\psi(q)}{q} < \infty.$$

Here, and throughout, φ is the Euler function.

Conjecture 1.2.12 (Duffin–Schaeffer, 1941) *For any function* $\psi : \mathbb{N} \to \mathbb{R}^+$

$$m(W'(\psi)) = 1 \quad if \quad \sum_{q=1}^{\infty} \varphi(q) \frac{\psi(q)}{q} = \infty.$$

Remark 1.2.13 Let ϑ be given by (1.18). On using the fact that $\sum_{d|n} \varphi(d) = n$, it is relatively easy to show (*exercise*) that

$$\sum_{q=1}^{\infty} \varphi(q) \frac{\vartheta(q)}{q} < \infty.$$

Thus, although ϑ provides a counterexample to Khintchine's theorem without monotonicity, it is not a counterexample to the Duffin–Schaeffer conjecture.

Remark 1.2.14 It is known that $m(W'(\psi)) = 0$ or 1. This is Gallagher's zero-full law [52] and is the natural analogue of Cassels' zero-full law for $W(\psi)$.

Although various partial results have been established (see [58, 90]), the full conjecture is one of the most difficult and profound unsolved problems in metric number theory. In the case where ψ is monotonic it is relatively straightforward to show that Khintchine's theorem and the Duffin–Schaeffer conjecture are equivalent statements (*exercise*).

1.2.3 A Limitation of the Lebesgue Theory

Let $\tau > 0$ and write $W(\tau)$ for $W(\psi : q \to q^{-\tau})$. The set $W(\tau)$ is usually referred to as the set of *τ-well approximable numbers*. Note that in view of Dirichlet (Theorem 1.1.2) we have that $W(\tau) = \mathrm{I}$ if $\tau \leq 1$ and so trivially $m(W(\tau)) = 1$ if $\tau \leq 1$. On the other hand, if $\tau > 1$

$$\sum_{q=1}^{\infty} q^{-\tau} < \infty$$

and Khintchine's theorem implies that $m(W(\tau)) = 0$. So for any $\tau > 1$, the set of τ-well approximable numbers is of measure zero. We cannot obtain any further information regarding the 'size' of $W(\tau)$ in terms of Lebesgue measure – it is always zero. Intuitively, the 'size' of $W(\tau)$ should decrease as rate of approximation governed by τ increases. For example, we would expect that $W(2015)$ is 'smaller' than $W(2)$ – clearly $W(2015) \subset W(2)$, but Lebesgue measure is unable to distinguish between them. In short, we require a more delicate notion of 'size' than simply Lebesgue measure. The appropriate notion of 'size' best suited for describing the finer measure theoretic structures of $W(\tau)$ and indeed $W(\psi)$ is that of Hausdorff measures.

1.3 Metric Diophantine Approximation: The Classical Hausdorff Theory

1.3.1 Hausdorff Measure and Dimension

In what follows, a *dimension function* $f : \mathbb{R}^+ \to \mathbb{R}^+$ is a left continuous, monotonic function such that $f(0) = 0$. Suppose F is a subset of $\mathbb{R}^n$. Given a ball B in $\mathbb{R}^n$, let $r(B)$ denote the radius of B. For $\rho > 0$, a countable collection $\{B_i\}$ of balls in $\mathbb{R}^n$ with $r(B_i) \leq \rho$ for each i such that $F \subset \bigcup_i B_i$ is called a *ρ-cover for F*. Define

$$\mathcal{H}^f_\rho(F) := \inf \sum_i f(r(B_i)),$$

where the infimum is taken over all ρ-covers of F. Observe that as ρ decreases the class of allowed ρ-covers of F is reduced and so $\mathcal{H}^f_\rho(F)$ increases. Therefore, the following (finite or infinite) limit exists

$$\mathcal{H}^f(F) := \lim_{\rho \to 0+} \mathcal{H}^f_\rho(F) = \sup_{\rho > 0} \mathcal{H}^f_\rho(F),$$

and is referred to as the *Hausdorff f-measure of F*. In the case that

$$f(r) = r^s \ (s \geq 0),$$

the measure $\mathcal{H}^f$ is the more common *s-dimensional Hausdorff measure* $\mathcal{H}^s$, the measure $\mathcal{H}^0$ being the cardinality of F. Note that when s is a positive integer, $\mathcal{H}^s$ is a constant multiple of Lebesgue measure in $\mathbb{R}^s$. (The constant is explicitly known!) Thus if the s-dimensional Hausdorff measure of a set is known for each $s > 0$, then so is its n-dimensional Lebesgue measure for each $n \geq 1$. The following easy property

$$\mathcal{H}^s(F) < \infty \quad \Longrightarrow \quad \mathcal{H}^{s'}(F) = 0 \qquad \text{if } s' > s$$

implies that there is a unique real point s at which the Hausdorff s-measure drops from infinity to zero (unless the set F is finite so that $\mathcal{H}^s(F)$ is never infinite). This point is called the *Hausdorff dimension* of F and is formally defined as

$$\dim F := \inf\left\{s > 0 : \mathcal{H}^s(F) = 0\right\}.$$

- By the definition of $\dim F$ we have that

$$\mathcal{H}^s(F) = \begin{cases} 0 & \text{if } s > \dim F \\ \infty & \text{if } s < \dim F. \end{cases}$$

- If $s = \dim F$, then $\mathcal{H}^s(F)$ may be zero or infinite or may satisfy

$$0 < \mathcal{H}^s(F) < \infty;$$

  in this case F is said to be an s-set.
- Let $\mathrm{I} = [0, 1]$. Then $\dim \mathrm{I} = 1$ and

$$2\mathcal{H}^s(\mathrm{I}) = \begin{cases} 0 & \text{if } s > 1 \\ 1 & \text{if } s = 1 \\ \infty & \text{if } s < 1. \end{cases}$$

  Thus, $2\mathcal{H}^1(\mathrm{I}) = m(\mathrm{I})$ and I is an example of an s-set with $s = 1$. Note that the presence of the factor '2' here is because, in the definition of the Hausdorff measure, we have used the radii of balls rather than their diameters.

The Hausdorff dimension has been established for many number theoretic sets, e.g. $W(\tau)$ (this is the Jarník–Besicovitch theorem discussed below), and is easier than determining the Hausdorff measure. Further details regarding Hausdorff measure and dimension can be found in [50, 72].

To calculate $\dim F$ (say $\dim F = \alpha$), it is usually the case that we establish the upper bound $\dim F \leq \alpha$ and lower bound $\dim F \geq \alpha$ separately. If we can exploit a 'natural' cover of F, then upper bounds are usually easier.

Example 1.3.1 Consider the middle third Cantor set K defined as follows: starting with $I_0 = [0,1]$ remove the open middle thirds part of the interval. This gives the union of two intervals $[0, \frac{1}{3}]$ and $[\frac{2}{3}, 1]$. Then repeat the procedure of removing the middle third part from each of the intervals in your given collection. Thus, at 'level' n of the construction we will have the union E_n of 2^n closed intervals, each of length 3^{-n}. The middle third Cantor set is defined by

$$K = \bigcap_{n=0}^{\infty} E_n .$$

This set consists exactly of all real numbers such that their expansion to the base 3 does not contain the 'digit' 1.

Let $\{I_{n,j}\}$ be the collection of intervals in E_n. This is a collection of 2^n intervals, each of length 3^{-n}. Naturally, $\{I_{n,j}\}$ is a cover of K. Furthermore, for any $\rho > 0$ there is a sufficiently large n such that $\{I_{n,j}\}$ is a ρ-cover of K. It follows that

$$\mathcal{H}^s_\rho(K) \leq \sum_j r(I_{n,j})^s \asymp 2^n 2^{-s} 3^{-ns} \ll \left(\frac{2}{3^s}\right)^n \to 0$$

as $n \to \infty$ (i.e. $\rho \to 0$) if

$$\frac{2}{3^s} < 1 \Rightarrow s > \frac{\log 2}{\log 3}.$$

In other words,

$$\mathcal{H}^s(K) = 0 \text{ if } s > \frac{\log 2}{\log 3}.$$

It follows from the definition of the Hausdorff dimension

$$\dim K = \inf\{s : \mathcal{H}^s(K) = 0\}$$

that $\dim K \leqslant \frac{\log 2}{\log 3}$.

In fact, $\dim K = \frac{\log 2}{\log 3}$. To prove that

$$\dim K \geqslant \frac{\log 2}{\log 3}$$

we need to work with arbitrary covers of K and this is much harder. Let $\{B_i\}$ be an arbitrary ρ-cover with $\rho < 1$. K is bounded and closed (intersection of closed intervals), i.e. K is compact. Hence, without loss of generality, we can assume that $\{B_i\}$ is finite. For each B_i, let r_i and d_i denote its radius and diameter, respectively, and let k be the unique integer such that

$$3^{-(k+1)} \leqslant d_i < 3^{-k}. \tag{1.20}$$

Then B_i intersects at most one interval of E_k as the intervals in E_k are separated by at least 3^{-k}.

If $j \geqslant k$, then B_i intersects at most

$$2^{j-k} = 2^j 3^{-sk} \leqslant 2^j 3^s d_i^s \tag{1.21}$$

intervals of E_j, where $s := \frac{\log 2}{\log 3}$ and the final inequality makes use of (1.20). These are the intervals that are contained in the unique interval of E_k that intersects B_i.

Now choose j large enough so that

$$3^{-(j+1)} \leqslant d_i \quad \forall B_i \in \{B_i\}.$$

This is possible because the collection $\{B_i\}$ is finite. Then $j \geqslant k$ for each B_i and (1.21) is valid. Furthermore, since $\{B_i\}$ is a cover of K, it must intersect every interval of E_j. There are 2^j intervals in E_j. Thus

$$\begin{aligned} 2^j &= \#\{I \in E_j : \cup B_i \cap I \neq \varnothing\} \\ &\leq \sum_i \#\{I \in E_j : B_i \cap I \neq \varnothing\} \\ &\leq \sum_i 2^j 3^s d_i^s. \end{aligned}$$

The upshot is that for any arbitrary cover $\{B_i\}$, we have that

$$2^s \sum r_i^s \asymp \sum d_i^s \geq 3^{-s} = \frac{1}{2}.$$

By definition, this implies that $\mathcal{H}^s(K) \geq 2^{-(1+s)}$ and so $\dim K \geq \frac{\log 2}{\log 3}$.

Even for this simple Cantor set example, the lower bound for $\dim K$ is much more involved than the upper bound. This is usually the case and the number theoretic sets $W(\psi)$ and $W(\tau)$ are no exception.

1.3.2 The Jarník–Besicovitch Theorem

Recall, the lim sup nature of $W(\psi)$; namely that

$$W(\psi) = \limsup_{q\to\infty} A_q(\psi) := \bigcap_{t=1}^{\infty} \bigcup_{q=t}^{\infty} A_q(\psi)$$

where

$$A_q(\psi) = \bigcup_{p=0}^{q} B\Big(\frac{p}{q}, \frac{\psi(q)}{q}\Big) \cap \mathrm{I}.$$

By definition, for each t, the collection of balls $B(p/q, \psi(q)/q)$ associated with the sets $A_q(\psi) : q = t, t+1, \ldots$ form a cover for $W(\psi)$. Suppose for the moment that ψ *is monotonic* and $\psi(q) < 1$ for q large. Now, for any $\rho > 0$, choose t large enough so that $\rho > \psi(t)/t$. Then the balls in $\{A_q(\psi)\}_{q \geqslant t}$ form a ρ cover of $W(\psi)$. Thus,

$$\mathcal{H}^s_\rho\big(W(\psi)\big) \leq \sum_{q=t}^{\infty} q\big(\psi(q)/q\big)^s \to 0$$

as $t \to \infty$ (i.e. $\rho \to 0$) if

$$\sum_{q=1}^{\infty} q^{1-s}\psi^s(q) < \infty;$$

i.e. $\mathcal{H}^s\big(W(\psi)\big) = 0$ if the above s-volume sum converges. Actually, monotonicity on ψ can be removed (*exercise*) and we have proved the following Hausdorff measure analogue of Theorem 1.2.1. Recall, that $\mathcal{H}^1$ and one-dimensional Lebesgue measure m are comparable.

Theorem 1.3.1 *Let* $\psi : \mathbb{N} \to \mathbb{R}^+$ *be a function and* $s \geq 0$ *such that*

$$\sum_{q=1}^{\infty} q^{1-s}\psi^s(q) < \infty.$$

Then

$$\mathcal{H}^s\big(W(\psi)\big) = 0\,.$$

Now put $\psi(q) = q^{-\tau}$ $(\tau \geq 1)$ and notice that for $s > \frac{2}{\tau+1}$,

$$\sum_{q=1}^{\infty} q^{1-s}\psi^s(q) = \sum_{q=1}^{\infty} q^{-(\tau s + s - 1)} < \infty\,.$$

Then the following statement is a simple consequence of the above theorem and the definition of Hausdorff dimension.

Corollary 1.3.2 *For* $\tau \geq 2$, *we have that* $\dim W(\tau) \leq \frac{2}{\tau+1}$.

Note that the above convergence result and thus the upper bound dimension result simply exploit the natural cover associated with the lim sup set under consideration. The corollary constitutes the easy part of the famous Jarník–Besicovitch theorem.

Theorem 1.3.3 (The Jarník–Besicovitch Theorem) *Let* $\tau > 1$. *Then*

$$\dim\big(W(\tau)\big) = 2/(\tau+1).$$

Jarník proved the result in 1928. Besicovitch proved the same result in 1932 by completely different methods. The Jarník–Besicovitch theorem implies that

$$\dim W(2) = 2/3 \quad \text{and} \quad \dim W(2015) = 2/2016$$

and so $W(2015)$ is 'smaller' than $W(2)$ as expected. In view of Corollary 1.3.2, we need to establish the lower bound result $\dim\big(W(\tau)\big) \geq 2/(\tau+1)$ in order to complete the proof of Theorem 1.3.3. We will see that this is a consequence of Jarník's measure result discussed in the next section.

The dimension theorem is clearly an excellent result, but it gives no information regarding $\mathcal{H}^s$ at the critical exponent $d := 2/(\tau+1)$. By definition,

$$\mathcal{H}^s(W(\tau)) = \begin{cases} 0 & \text{if } s > d \\ \infty & \text{if } s < d, \end{cases}$$

but

$$\mathcal{H}^s(W(\tau)) = \textbf{?} \quad \text{if } s = d\,.$$

In short, it would be highly desirable to have a Hausdorff measure analogue of Khintchine's theorem.

1.3.3 Jarník's Theorem

Theorem 1.3.1 is the easy case of the following fundamental statement in metric Diophantine approximation. It provides an elegant criterion for the 'size' of the set $W(\psi)$ expressed in terms of Hausdorff measure.

Theorem 1.3.4 (Jarník's Theorem, 1931) *Let* $\psi : \mathbb{N} \to \mathbb{R}^+$ *be a monotonic function and* $s \in (0, 1)$. *Then*

$$\mathcal{H}^s\big(W(\psi)\big) = \begin{cases} 0 & \textit{if } \sum_{q=1}^{\infty} q^{1-s}\psi^s(q) < \infty \\ \infty & \textit{if } \sum_{q=1}^{\infty} q^{1-s}\psi^s(q) = \infty \end{cases}.$$

Remark 1.3.5 With $\psi(q) = q^{-\tau}$ $(\tau > 1)$, not only does the above theorem imply that $\dim W(\tau) = 2/(1+\tau)$ but it tells us that the Hausdorff measure at the critical exponent is infinite; i.e.

$$\mathcal{H}^s\big(W(\tau)\big) = \infty \quad \text{at} \quad s = 2/(1+\tau).$$

Remark 1.3.6 As in Khintchine's theorem, the assumption that ψ is monotonic is only required in the divergent case. In Jarník's original statement, apart from assuming stronger monotonicity conditions, various technical conditions on ψ and indirectly s were imposed, which prevented $s = 1$. Note that, even as stated, it is natural to exclude the case $s = 1$ since

$$\mathcal{H}^1\big(W(\psi)\big) \asymp m\big(W(\psi)\big) = 1.$$

The clear-cut statement without the technical conditions was established in [14] and it allows us to combine the theorems of Khintchine and Jarník into a unifying statement.

Theorem 1.3.7 (Khintchine–Jarník 2006) *Let $\psi : \mathbb{N} \to \mathbb{R}^+$ be a monotonic function and $s \in (0,1]$. Then*

$$\mathcal{H}^s\big(W(\psi)\big) = \begin{cases} 0 & \text{if } \sum_{q=1}^{\infty} q^{1-s}\psi^s(q) < \infty\,, \\[2ex] \mathcal{H}^s(I) & \text{if } \sum_{q=1}^{\infty} q^{1-s}\psi^s(q) = \infty\,. \end{cases}$$

Obviously, the Khintchine–Jarník theorem implies Khintchine's theorem.

In view of the Mass Transference Principle established in [21], one actually has that

$$\text{Khintchine's theorem} \quad \Longrightarrow \quad \text{Jarník's theorem.}$$

Thus, the Lebesgue theory of $W(\psi)$ underpins the general Hausdorff theory. At first glance this is rather surprising because the Hausdorff theory had previously been thought to be a subtle refinement of the Lebesgue theory. Nevertheless, the Mass Transference Principle allows us to transfer Lebesgue measure theoretic statements for lim sup sets to Hausdorff statements and naturally obtain a complete metric theory.

1.3.4 The Mass Transference Principle

Let (Ω, d) be a locally compact metric space and suppose there exist constants $\delta > 0$, $0 < c_1 < 1 < c_2 < \infty$ and $r_0 > 0$ such that

$$c_1\, r^\delta \le \mathcal{H}^\delta(B) \le c_2\, r^\delta\,, \tag{1.22}$$

for any ball $B = B(x, r)$ with $x \in \Omega$ and radius $r \leq r_0$. For the sake of simplicity, the definition of Hausdorff measure and dimension given in §1.3.1 is restricted to $\mathbb{R}^n$. Clearly, it can easily be adapted to the setting of arbitrary metric spaces – see [50, 72]. A consequence of (1.22) is that

$$0 < \mathcal{H}^\delta(\Omega) \leq \infty \quad \text{and} \quad \dim \Omega = \delta.$$

Next, given a dimension function f and a ball $B = B(x, r)$ we define the scaled ball

$$B^f := B\big(x, f(r)^{\frac{1}{\delta}}\big).$$

When $f(r) = r^s$ for some $s > 0$, we adopt the notation B^s, i.e.

$$B^s := B\big(x, r^{\frac{s}{\delta}}\big)$$

and so by definition $B^\delta = B$.

The Mass Transference Principle [21] allows us to transfer $\mathcal{H}^\delta$-measure theoretic statements for lim sup subsets of Ω to general $\mathcal{H}^f$-measure theoretic statements. Note that in the case where $\delta = k \in \mathbb{N}$, the measure $\mathcal{H}^\delta$ coincides with k-dimensional Lebesgue measure and the Mass Transference Principle allows us to transfer Lebesgue measure theoretic statements for lim sup subsets of $\mathbb{R}^k$ to Hausdorff measure theoretic statements.

Theorem 1.3.8 *Let $\{B_i\}_{i \in \mathbb{N}}$ be a sequence of balls in Ω with $r(B_i) \to 0$ as $i \to \infty$. Let f be a dimension function such that $x^{-\delta} f(x)$ is monotonic. For any ball $B \in \Omega$ with $\mathcal{H}^\delta(B) > 0$, if*

$$\mathcal{H}^\delta\big(B \cap \limsup_{i \to \infty} B_i^f\,\big) = \mathcal{H}^\delta(B)$$

then

$$\mathcal{H}^f\big(B \cap \limsup_{i \to \infty} B_i^\delta\,\big) = \mathcal{H}^f(B).$$

Remark 1.3.9 There is one point that is well worth making. The Mass Transference Principle is purely a statement concerning lim sup sets arising from a sequence of balls. There is absolutely no monotonicity assumption on the radii of the balls. Even the imposed condition that $r(B_i) \to 0$ as $i \to \infty$ is redundant, but is included to avoid unnecessary tedious discussion.

1.3.4.1 Khintchine's Theorem Implies Jarník's Theorem

First of all, let us dispose of the case that $\psi(r)/r \nrightarrow 0$ as $r \to \infty$. Then trivially, $W(\psi) = \mathrm{I}$ and the result is obvious. Without loss of generality, assume that $\psi(r)/r \to 0$ as $r \to \infty$. With respect to the Mass Transference Principle, let $\Omega = \mathrm{I}$, d be the supremum norm, $\delta = 1$ and $f(r) = r^s$ with $s \in (0,1)$. We are given that $\sum q^{1-s}\psi(q)^s = \infty$. Let $\theta(q) := q^{1-s}\psi(q)^s$. Then θ is an approximating function and $\sum \theta(q) = \infty$. Thus, assuming θ is monotonic, Khintchine's theorem implies that $\mathcal{H}^1(B \cap W(\theta)) = \mathcal{H}^1(B \cap \mathrm{I})$ for any ball B in $\mathbb{R}$. The condition that θ is monotonic can be relaxed by using the Duffin–Schaeffer Theorem instead of Khintchine's theorem. It now follows via the Mass Transference Principle that $\mathcal{H}^s(W(\psi)) = \mathcal{H}^s(\mathrm{I}) = \infty$ and this completes the proof of the divergence part of Jarník's theorem. As we have already seen, the convergence part is straightforward.

1.3.4.2 Dirichlet's Theorem Implies the Jarník–Besicovitch Theorem

Dirichlet's theorem (Theorem 1.1.2) states that for any irrational $x \in \mathbb{R}$, there exist infinitely many reduced rationals p/q $(q > 0)$ such that $|x - p/q| \leq q^{-2}$; i.e. $W(1) = \mathrm{I}$. Thus, with $f(r) := r^d$ $(d := 2/(1+\tau))$ the Mass Transference Principle implies that $\mathcal{H}^d(W(\tau)) = \infty$. Hence $\dim W(\tau) \geq d$. The upper bound is trivial. Note that we have actually proved a lot more than the Jarník–Besicovitch theorem. We have proved that the s-dimensional Hausdorff measure $\mathcal{H}^s$ of $W(\tau)$ at the critical exponent $s = d$ is infinite.

1.3.5 The Generalised Duffin–Schaeffer Conjecture

As with Khintchine's theorem, it is natural to seek an appropriate statement in which one removes the monotonicity condition in Jarník's theorem. In the case of Khintchine's theorem, the appropriate statement is the Duffin–Schaeffer conjecture – see §1.2.2.1. With this in mind, we work with the set $W'(\psi)$ in which the coprimeness condition $(p,q) = 1$ is imposed on the rational approximates p/q. For any function $\psi : \mathbb{N} \to \mathbb{R}^+$ and $s \in (0,1]$ it is easily verified that

$$\mathcal{H}^s\big(W(\psi)\big) = 0 \quad \text{if} \quad \sum_{q=1}^{\infty} \varphi(q) \left(\frac{\psi(q)}{q}\right)^s < \infty.$$

In the case the above s-volume sum diverges it is reasonable to believe in the truth of the following Hausdorff measure version of the Duffin–Schaeffer conjecture [21].

Conjecture 1.3.10 (Generalised Duffin–Schaeffer Conjecture, 2006) *For any function $\psi : \mathbb{N} \to \mathbb{R}^+$ and $s \in (0,1]$,*

$$\mathcal{H}^s\big(W'(\psi)\big) = \mathcal{H}^s(I) \quad \textit{if} \quad \sum_{q=1}^{\infty} \varphi(q) \left(\frac{\psi(q)}{q}\right)^s = \infty.$$

Remark 1.3.11 If $s = 1$, then $\mathcal{H}^1(\mathrm{I}) = m(\mathrm{I})$ and Conjecture 1.3.10 reduces to the Lebesgue measure conjecture of Duffin and Schaeffer (Conjecture 1.2.12).

Remark 1.3.12 In view of the Mass Transference Principle, it follows that

$$\text{Conjecture 1.2.12} \quad \Longrightarrow \quad \text{Conjecture 1.3.10}$$

Exercise: Prove the above implication.

1.4 The Higher-Dimensional Theory

We start with a generalisation of Theorem 1.1.1 to simultaneous approximation in $\mathbb{R}^n$.

Theorem 1.4.1 (Dirichlet in $\mathbb{R}^n$) *Let $(i_1, \ldots, i_n)$ be any n-tuple of numbers satisfying*

$$0 < i_1, \ldots, i_n < 1 \quad \text{and} \quad \sum_{t=1}^{n} i_t = 1. \tag{1.23}$$

Then, for any $\mathbf{x} = (x_1, \ldots, x_n) \in \mathbb{R}^n$ and $N \in \mathbb{N}$, there exists $q \in \mathbb{Z}$ such that

$$\max\{\|qx_1\|^{1/i_1}, \ldots, \|qx_n\|^{1/i_n}\} < N^{-1} \qquad \text{and} \qquad 1 \leq q \leq N. \tag{1.24}$$

Remark 1.4.2 The symmetric case corresponding to $i_1 = \ldots = i_n = 1/n$ is the more familiar form of the theorem. In this symmetric case, when N is an nth power, the one-dimensional proof using the Pigeonhole principle can easily be adapted to prove the associated statement (*exercise*). The above general form is a neat consequence of a fundamental theorem in the geometry of numbers; namely, Minkowski's theorem for systems of linear forms – see §1.4.1 below. At this point, simply observe that for a fixed q the first inequality in (1.24) corresponds to considering rectangles centred at rational points

$$\left(\frac{p_1}{q}, \ldots, \frac{p_n}{q}\right) \quad \text{of sidelength} \quad \frac{2}{qN^{i_1}}, \ldots, \frac{2}{qN^{i_n}} \quad \text{respectively.}$$

Now the shape of the rectangles are clearly governed by $(i_1, \ldots, i_n)$. However the volume is not. Indeed, for any $(i_1, \ldots, i_n)$ satisfying (1.23), the n-dimensional Lebesgue measure m_n of any rectangle centred at a rational point with denominator q is $2^n q^{-n} N^{-1}$.

1.4.1 Minkowski's Linear Forms Theorem

We begin by introducing various terminology and establishing Minkowski's theorem for convex bodies.

Definition 1.4.3 A subset B of $\mathbb{R}^n$ is said to be *convex* if, for any two points $\mathbf{x}, \mathbf{y} \in B$

$$\{\lambda \mathbf{x} + (1-\lambda)\mathbf{y} : 0 \leq \lambda \leq 1\} \subset B,$$

that is, the line segment joining $\mathbf{x}$ and $\mathbf{y}$ is contained in B. A *convex body* in $\mathbb{R}^n$ is a bounded convex set.

Definition 1.4.4 A subset B in $\mathbb{R}^n$ is said to be *symmetric about the origin* if, for every $\mathbf{x} \in B$, we have that $-\mathbf{x} \in B$.

The following is a simple but nevertheless powerful observation concerning symmetric convex bodies.

Theorem 1.4.5 (Minkowski's Convex Body Theorem) *Let B be a convex body in $\mathbb{R}^n$ symmetric about the origin. If* $\mathrm{vol}(B) > 2^n$, *then B contains a non-zero integer point.*

Proof The following proof is attributed to Mordell. For $m \in \mathbb{N}$ let $A(m, B) = \{\mathbf{a} \in \mathbb{Z}^m : \mathbf{a}/m \in B\}$. Then we have that

$$\lim_{m\to\infty} m^{-n} \# A(m, B) = \mathrm{vol}(B).$$

Since $\mathrm{vol}(B) > 2^n$, there is a sufficiently large m such that $m^{-n}\#A(m, B) > 2^n$, that is $\#A(m, B) > (2m)^n$. Since there are $2m$ different residue classes modulo $2m$ and each point in $A(Q, m)$ has n co-ordinates, there are two distinct points in $A(Q, m)$, say $\mathbf{a} = (a_1, \ldots, a_n)$ and $\mathbf{b} = (b_1, \ldots, b_n)$ such that

$$a_i \equiv b_i \pmod{2m} \quad \text{for each } i = 1, \ldots, n\,.$$

Hence

$$\mathbf{z} = \frac{1}{2}\frac{\mathbf{a}}{m} + \frac{1}{2}\left(-\frac{\mathbf{b}}{m}\right) = \frac{\mathbf{a}-\mathbf{b}}{2m} \in \mathbb{Z}^n \setminus \{\mathbf{0}\}\,.$$

Since B is symmetric about the origin, $-\mathbf{b}/m \in B$ and since B is convex $\mathbf{z} \in B$. The proof is complete. □

The above convex body result enables us to prove the following extremely useful statement.

Theorem 1.4.6 (Minkowski's Theorem for Systems of Linear Forms) *Let $\beta_{i,j} \in \mathbb{R}$, where $1 \le i, j \le k$, and let $C_1, \ldots, C_k > 0$. If*

$$|\det(\beta_{i,j})_{1\le i,j\le k}| \le \prod_{i=1}^{k} C_i, \tag{1.25}$$

then there exists a non-zero integer point $\mathbf{x} = (x_1, \ldots, x_k)$ such that

$$\begin{cases} |x_1\beta_{i,1} + \cdots + x_k\beta_{i,k}| < C_i & (1 \le i \le k-1) \\ |x_1\beta_{k,1} + \cdots + x_n\beta_{k,k}| \le C_k. \end{cases} \tag{1.26}$$

Proof The set of $(x_1, \ldots, x_k) \in \mathbb{R}^k$ satisfying (1.26) is a convex body symmetric about the origin. First consider the case when $\det(\beta_{i,j})_{1\le i,j\le k} \ne 0$ and (1.25) is strict. Then

$$\mathrm{vol}(B) = \frac{\prod_{i=1}^{k}(2C_i)}{|\det(\beta_{i,j})_{1\le i,j\le k}|} > 2^n ,$$

and the body contains a non-zero integer point $(x_1, \ldots, x_k)$, by Theorem 1.4.5, as required.

If $\det(\beta_{i,j})_{1\le i,j\le k} = 0$, then B is unbounded and has infinite volume. Then there exists a sufficiently large $m \in \mathbb{N}$ such that $B_m = B\cap[-m, m]$ has volume $\mathrm{vol}(B_m) > 2^n$. Next, B_m is convex and symmetric about the origin, since it is the intersection of 2 sets with these properties. Again, by Theorem 1.4.5, B_m contains a non-zero integer point $(x_1, \ldots, x_k)$. Since $B_m \subset B$ we again get the required statement.

Finally, consider the situation when (1.25) is an equation. In this case $\det(\beta_{i,j})_{1\le i,j\le k} \ne 0$. Define $C_k^{\varepsilon} = C_k + \varepsilon$ for some $\varepsilon > 0$. Then

$$|\det(\beta_{i,j})_{1\le i,j\le k}| < \prod_{i=1}^{k-1} C_i \times C_k^{\varepsilon} \tag{1.27}$$

and, by what we have already shown, there exists a non-zero integer solution $\mathbf{x}_{\varepsilon} = (x_1, \ldots, x_k)$ to the system

$$\begin{cases} |x_1\beta_{i,1} + \cdots + x_k\beta_{i,k}| < C_i & (1 \le i \le k-1) \\ |x_1\beta_{k,1} + \cdots + x_n\beta_{k,k}| \le C_k^{\varepsilon} . \end{cases} \tag{1.28}$$

For $\varepsilon \le 1$ all the points $\mathbf{x}_{\varepsilon}$ satisfy (1.28) with $\varepsilon = 1$. That is, they lie in a bounded body. Hence, there are only finitely many of them. Therefore there is a sequence ε_i tending to 0 such that $\mathbf{x}_{\varepsilon_i}$ are all the same, say $\mathbf{x}_0$. On letting $i \to \infty$ within (1.28) we get that (1.26) holds with $\mathbf{x} = \mathbf{x}_0$. □

It is easily verified that Theorem 1.4.1 (Dirichlet in $\mathbb{R}^n$) is an immediate consequence of Theorem 1.4.6 with $k = n + 1$ and

$$C_t = N^{-i_t} \quad (1 \le t \le k-1) \quad \text{and} \qquad C_k = N$$

and

$$(\beta_{i,j}) = \begin{pmatrix} -1 & 0 & 0 & \dots & \alpha_1 \\ 0 & -1 & 0 & \dots & \alpha_2 \\ 0 & 0 & -1 & \dots & \\ \vdots & & & \ddots & \alpha_n \\ 0 & 0 & 0 & \dots & 1 \end{pmatrix}.$$

Another elegant application of Theorem 1.4.6 is the following statement.

Corollary 1.4.7 *For any $(\alpha_1, \dots, \alpha_n) \in \mathbb{R}^n$ and any real $N > 1$, there exist $q_1, \dots, q_n, p \in \mathbb{Z}$ such that*

$$|q_1\alpha_1 + \cdots + q_n\alpha - p| < N^{-n} \qquad \text{and} \qquad 1 \le \max_{1\le i\le n} |q_i| \le N .$$

In particular, there exist infinitely many $((q_1, \dots, q_n), p) \in \mathbb{Z}^n \setminus \{0\} \times \mathbb{Z}$ such that

$$|q_1\alpha_1 + \cdots + q_n\alpha - p| < \Big(\max_{1\le i\le n} |q_i|\Big)^{-n} .$$

Proof Exercise. □

1.4.2 Bad in $\mathbb{R}^n$

An important consequence of Dirichlet's theorem (Theorem 1.4.1) is the following higher-dimensional analogue of Theorem 1.1.2.

Theorem 1.4.8 *Let $(i_1, \dots, i_n)$ be any n-tuple of real numbers satisfying (1.23). Let $\mathbf{x} = (x_1, \dots, x_n) \in \mathbb{R}^n$. Then there exist infinitely many integers $q > 0$ such that*

$$\max\{\|qx_1\|^{1/i_1}, \dots, \|qx_n\|^{1/i_n}\} < q^{-1}. \tag{1.29}$$

Now, just as in the one-dimensional set-up, we can ask the following natural question.

Question. Can we replace the right-hand side of (1.29) by ϵq^{-1} where $\epsilon > 0$ is arbitrary?

No. For any $(i_1, \ldots, i_n)$ satisfying (1.23), there exists $(i_1, \ldots, i_n)$ *badly approximable points.*

Denote by $\mathbf{Bad}(i_1, \ldots, i_n)$ the set of all $(i_1, \ldots, i_n)$ badly approximable points; that is the set of $(x_1, \ldots, x_n) \in \mathbb{R}^n$ such that there exists a positive constant $c(x_1, \ldots, x_n) > 0$ so that

$$\max\{\|qx_1\|^{1/i_1}, \ldots, \ \|qx_n\|^{1/i_n}\} > c(x_1, \ldots, x_n)\, q^{-1} \quad \forall q \in \mathbb{N}.$$

Remark 1.4.9 Let $n = 2$ and note that if $(x, y) \in \mathbf{Bad}(i, j)$ for some pair (i, j), then it would imply that

$$\liminf_{q \to \infty} q\,\|qx\|\,\|qy\| = 0.$$

Hence $\bigcap_{i+j=1} \mathbf{Bad}(i, j) = \emptyset$ would imply that Littlewood's conjecture is true. We will return to this famous conjecture in §1.4.4.

Remark 1.4.10 Geometrically speaking, $\mathbf{Bad}(i_1, \ldots, i_n)$ consists of points $\mathbf{x} \in \mathbb{R}^n$ that avoid all rectangles of size $c^{i_1} q^{-(1+i_1)} \times \ldots \times c^{i_n} q^{-(1+i_n)}$ centred at rational points $(p_1/q, \ldots, p_n/q)$ with $c = c(\mathbf{x})$ sufficiently small. Note that in the symmetric case $i_1 = \ldots = i_n = 1/n$, the rectangles are squares (or essentially balls), and this makes a profound difference when investigating the 'size' of $\mathbf{Bad}(i_1, \ldots, i_n)$ – it makes life significantly easier!

Perron [74] in 1921 observed that $(x, y) \in \mathbf{Bad}(\frac{1}{2}, \frac{1}{2})$ whenever x and y are linearly independent numbers in a cubic field; e.g. $x = \cos\frac{2\pi}{7}$, $y = \cos\frac{4\pi}{7}$. Thus, certainly $\mathbf{Bad}(\frac{1}{2}, \frac{1}{2})$ is not the empty set. It was shown by Davenport in 1954 that $\mathbf{Bad}(\frac{1}{2}, \frac{1}{2})$ is uncountable and later in [42] he gave a simple and more illuminating proof of this fact. Furthermore, the ideas in his 1964 paper show that $\mathbf{Bad}(i_1, \ldots, i_n)$ is uncountable. In 1966, Schmidt [85] showed that in the symmetric case the corresponding set $\mathbf{Bad}(\frac{1}{n}, \ldots, \frac{1}{n})$ is of full Hausdorff dimension. In fact, Schmidt proved the significantly stronger statement that the symmetric set is winning in the sense of his now famous (α, β)-games (see §1.7.2). Almost forty years later it was proved in [77] that

$$\dim \mathbf{Bad}(i_1, \ldots, i_n) = n.$$

Now let us return to the symmetric case of Theorem 1.4.8. It implies that every point $\mathbf{x} = (x_1, \ldots, x_n) \in \mathbb{R}^n$ can be approximated by rational points $(p_1/q, \ldots, p_n/q)$ with rate of approximation given by $q^{-(1+\frac{1}{n})}$. The above discussion shows that this rate of approximation cannot in general be improved by an arbitrary constant – $\mathbf{Bad}(\frac{1}{n}, \ldots, \frac{1}{n})$ is non-empty. However, if we exclude a set of real numbers of measure zero, then from a measure theoretic point of view the rate of approximation can be improved, just as in the one-dimensional set-up.

1.4.3 Higher-Dimensional Khintchine

Let $\mathrm{I}^n := [0,1)^n$ denote the unit cube in $\mathbb{R}^n$ and for $\mathbf{x} = (x_1, \ldots, x_n) \in \mathbb{R}^n$ let

$$\|q\mathbf{x}\| := \max_{1 \le i \le n} \|qx_i\|.$$

Given $\psi : \mathbb{N} \to \mathbb{R}^+$, let

$$W(n,\psi) := \{\mathbf{x} \in \mathrm{I}^n : \|q\mathbf{x}\| < \psi(q) \text{ for infinitely many } q \in \mathbb{N}\}$$

denote the set of *simultaneously ψ-well approximable* points $\mathbf{x} \in \mathrm{I}^n$. Thus, a point $\mathbf{x} \in \mathrm{I}^n$ is ψ-well approximable if there exist infinitely many rational points

$$\left(\frac{p_1}{q}, \ldots, \frac{p_n}{q}\right)$$

with $q > 0$, such that the inequalities

$$\left|x_i - \frac{p_i}{q}\right| < \frac{\psi(q)}{q}$$

are simultaneously satisfied for $1 \le i \le n$. For the same reason as in the $n = 1$ case there is no loss of generality in restricting our attention to the unit cube. In the case $\psi : q \to q^{-\tau}$ with $\tau > 0$, we write $W(n,\tau)$ for $W(n,\psi)$. The set $W(n,\tau)$ is the set of *simultaneously τ-well approximable numbers*. Note that in view of Theorem 1.4.8 we have that

$$W(n,\tau) = \mathrm{I}^n \ \text{ if } \ \tau \le \frac{1}{n}. \tag{1.30}$$

The following is the higher-dimensional generalisation of Theorem 1.2.5 to simultaneous approximation. Throughout, m_n will denote n-dimensional Lebesgue measure.

Theorem 1.4.11 (Khintchine's Theorem in $\mathbb{R}^n$) *Let $\psi : \mathbb{N} \to \mathbb{R}^+$ be a monotonic function. Then*

$$m_n(W(n,\psi)) = \begin{cases} 0 & \text{if } \sum_{q=1}^{\infty} \psi^n(q) < \infty\,, \\ 1 & \text{if } \sum_{q=1}^{\infty} \psi^n(q) = \infty\,. \end{cases}$$

Remark 1.4.12 The convergent case is a straightforward consequence of the Convergence Borel–Cantelli Lemma and does not require monotonicity.

Remark 1.4.13 The divergent case is the main substance of the theorem. When $n \geq 2$, a consequence of a theorem of Gallagher [54] is that the monotonicity condition can be dropped. Recall, that in view of the Duffin–Schaeffer counterexample (see §1.2.2.1) the monotonicity condition is crucial when $n = 1$.

Remark 1.4.14 Theorem 1.4.11 implies that

$$m_n(W(n, \psi)) = 1 \quad \text{if} \quad \psi(q) = 1/(q \log q)^{\frac{1}{n}}.$$

Thus, from a measure theoretic point of view the 'rate' of approximation given by Theorem 1.4.8 can be improved by $(\text{logarithm})^{\frac{1}{n}}$.

Remark 1.4.15 Theorem 1.4.11 implies that $m_n(\mathbf{Bad}(\frac{1}{n}, \ldots, \frac{1}{n})) = 0$.

Remark 1.4.16 For a generalisation of Theorem 1.4.11 to Hausdorff measures – that is, the higher-dimension analogue of Theorem 1.3.7 (Khintchine–Jarník theorem) – see Theorem 1.4.37 with $m = 1$ in §1.4.6. Also, see §1.5.3.1.

In view of Remark 1.4.13, one may think that there is nothing more to say regarding the Lebesgue theory of ψ-well approximable points in $\mathbb{R}^n$. After all, for $n \geq 2$ we do not even require monotonicity in Theorem 1.4.11. For ease of discussion let us restrict our attention to the plane $\mathbb{R}^2$ and assume that the n-volume sum in Theorem 1.4.11 diverges. So we know that almost all points (x_1, x_2) are ψ-well approximable but it tells us nothing for a given fixed x_1. For example, are there any points $(\sqrt{2}, x_2) \in \mathbb{R}^2$ that are ψ-well approximable? This will be discussed in §1.4.5 and the more general question of approximating points on a manifold will be the subject of §1.6.

1.4.4 Multiplicative Approximation: Littlewood's Conjecture

For any pair of real numbers $(\alpha, \beta) \in \mathrm{I}^2$, there exist infinitely many $q \in \mathbb{N}$ such that

$$\|q\alpha\| \, \|q\beta\| \leq q^{-1}.$$

This is a simple consequence of Theorem 1.4.8, or indeed the one-dimensional Dirichlet theorem, and the trivial fact that $\|x\| < 1$ for any x. For any arbitrary $\epsilon > 0$, the problem of whether or not the statement remains true by replacing the right-hand side of the inequality by $\epsilon\, q^{-1}$ now arises. This is precisely the content of Littlewood's conjecture.

Littlewood's Conjecture *For any pair* $(\alpha, \beta) \in I^2$,

$$\liminf_{q\to\infty} q\, ||q\alpha||\, ||q\beta|| = 0\,.$$

Equivalently, for any pair $(\alpha, \beta) \in \mathrm{I}^2$ there exist infinitely many rational points $(p_1/q, p_2/q)$ such that

$$\left|\alpha - \frac{p_1}{q}\right| \left|\beta - \frac{p_2}{q}\right| < \frac{\epsilon}{q^3} \quad (\epsilon > 0 \text{ arbitrary}).$$

Thus, geometrically, the conjecture states that every point in the (x, y)-plane lies in infinitely many hyperbolic regions given by $|x| \cdot |y| < \epsilon/q^3$ centred at rational points.

The analogous conjecture in the one-dimensional setting is false – Hurwitz's theorem tells us that the set **Bad** is non-empty. However, in the multiplicative situation the problem is still open.

We make various simple observations:

(i) *The conjecture is true for pairs* (α, β) *when either* α *or* β *are not in* **Bad**. Suppose $\beta \notin$ **Bad** and consider its convergents p_n/q_n. It follows from the right-hand side of inequality (1.8) that $q_n||q_n\alpha||\,||q_n\beta|| \leq 1/a_{n+1}$ for all n. Since β is not badly approximable the partial quotients a_i are unbounded and the conjecture follows. Alternatively, by definition if $\beta \notin$ **Bad**, then $\liminf_{q\to\infty} q\,||q\beta|| = 0$ and we are done. See also Remark 1.4.9.

(ii) *The conjecture is true for pairs* (α, β) *when either* α *or* β *lie in a set of full Lebesgue measure*. This follows at once from Khintchine's theorem. In fact, one has that for all α and almost all $\beta \in \mathrm{I}$,

$$q\, \log q\, \|q\alpha\|\, \|q\beta\| \leq 1 \quad \text{for infinitely many } q \in \mathbb{N} \tag{1.31}$$

or even

$$\liminf_{q\to\infty} q\, \log q\, ||q\alpha||\, ||q\beta|| = 0.$$

We now turn our attention to 'deeper' results regarding Littlewood.

Theorem (Cassels and Swinnerton-Dyer, 1955). *If* α, β *are both cubic irrationals in the same cubic field, then Littlewood's conjecture is true.*

This was subsequently strengthened by Peck [73].

Theorem (Peck, 1961). *If α, β are both cubic irrationals in the same cubic field, then (α, β) satisfy* (1.31) *with the constant* 1 *on the right-hand side replaced by a positive constant dependent on α and β.*

In view of (ii) above, when dealing with Littlewood we can assume without loss of generality that both α and β are in **Bad**. As mentioned in Chapter 1.1, it is conjectured (the Folklore Conjecture) that the only algebraic irrationals which are badly approximable are the quadratic irrationals. Of course, if this conjecture is true then the Cassels and Swinnerton-Dyer result follows immediately. On restricting our attention to just badly approximable pairs we have the following statement [76].

Theorem PV (2000). *Given $\alpha \in$ **Bad** we have that*

$$\dim\big(\{\beta \in \mathbf{Bad} : (\alpha, \beta) \text{ satisfy } (1.31)\}\big) = 1.$$

Regarding potential counterexamples to Littlewood, we have the following elegant statement [49].

Theorem EKL (2006).

$$\dim\big(\{(\alpha, \beta) \in I^2 : \liminf_{q\to\infty} q\,||q\alpha||\,||q\beta|| > 0\}\big) = 0.$$

Now let us turn our attention to non-trivial, purely metrical statements regarding Littlewood. The following result due to Gallagher [53] is the analogue of Khintchine's simultaneous approximation theorem (Theorem 1.4.11) within the multiplicative set-up. Given $\psi : \mathbb{N} \to \mathbb{R}^+$ let

$$W^{\times}(n, \psi) := \{\mathbf{x} \in \mathrm{I}^n : \|qx_1\| \ldots \|qx_n\| < \psi(q) \text{ for infinitely many } q \in \mathbb{N}\} \tag{1.32}$$

denote the set of multiplicative ψ-well approximable points $\mathbf{x} \in \mathrm{I}^n$.

Theorem 1.4.17 (Gallagher, 1962) *Let $\psi : \mathbb{N} \to \mathbb{R}^+$ be a monotonic function. Then*

$$m_n(W^{\times}(n, \psi)) = \begin{cases} 0 & \text{if } \sum_{q=1}^{\infty} \psi(q) \log^{n-1} q < \infty, \\ \\ 1 & \text{if } \sum_{q=1}^{\infty} \psi(q) \log^{n-1} q = \infty. \end{cases}$$

Remark 1.4.18 In the case of convergence, we can remove the condition that ψ is monotonic if we replace the above convergence condition by $\sum \psi(q)\,|\log \psi(q)|^{n-1} < \infty$; see [16] for more details.

An immediate consequence of Gallagher's theorem is that almost all (α, β) beat Littlewood's conjecture by 'log squared'; equivalently, almost surely Littlewood's conjecture is true with a 'log squared' factor to spare.

Corollary 1.4.19 *For almost all* $(\alpha, \beta) \in \mathbb{R}^2$

$$\liminf_{q\to\infty} q \, \log^2 q \, ||q\alpha|| \, ||q\beta|| = 0 \, . \tag{1.33}$$

Recall, that this is beyond the scope of what Khintchine's theorem can tell us; namely that

$$\liminf_{q\to\infty} q \, \log q \, ||q\alpha|| \, ||q\beta|| = 0 \quad \forall \, \alpha \in \mathbb{R} \quad \text{and} \quad \text{for almost all } \beta \in \mathbb{R}. \tag{1.34}$$

However, the extra log factor in the corollary comes at a cost of having to sacrifice a set of measure zero on the α side. As a consequence, unlike with (1.34) which is valid for any α, we are unable to claim that the stronger 'log squared' statement (1.33) is true for say when $\alpha = \sqrt{2}$. Obviously, the role of α and β in (1.34) can be reversed. This raises the natural question of whether (1.33) holds for every α. If true, it would mean that for any α we still beat Littlewood's conjecture by 'log squared' for almost all β.

1.4.4.1 Gallagher on Fibers

The following result is established in [17].

Theorem 1.4.20 *Let* $\alpha \in I$ *and* $\psi : \mathbb{N} \to \mathbb{R}^+$ *be a monotonic function such that*

$$\sum_{q=1}^{\infty} \psi(q) \, \log q = \infty \tag{1.35}$$

and such that

$$\exists \, \delta > 0 \qquad \liminf_{n\to\infty} q_n^{3-\delta} \psi(q_n) \geq 1 \, , \tag{1.36}$$

where q_n *denotes the denominators of the convergents of* α*. Then for almost every* $\beta \in I$*, there exists infinitely many* $q \in \mathbb{N}$ *such that*

$$\|q\alpha\| \, \|q\beta\| < \psi(q). \tag{1.37}$$

Remark 1.4.21 Condition (1.36) is not particularly restrictive. It holds for all α with Diophantine exponent $\tau(\alpha) < 3$. By definition,

$$\tau(x) = \sup\{\tau > 0 : \|q\alpha\| < q^{-\tau} \quad \text{for infinitely many } q \in \mathbb{N}\} \, .$$

Recall that by the Jarník–Besicovitch theorem (Theorem 1.3.3), the complement is of relatively small dimension; namely $\dim\{\alpha \in \mathbb{R} : \tau(\alpha) \geq 3\} = \frac{1}{2}$.

The theorem can be equivalently formulated as follows. Working within the (x, y)-plane, let L_x denote the line parallel to the y-axis passing through the point $(x, 0)$. Then, given $\alpha \in \mathrm{I}$, Theorem 1.4.20 simply states that

$$m_1(W^{\times}(2, \psi) \cap \mathrm{L}_\alpha) = 1 \quad \text{if} \quad \psi \text{ satisfies (1.35) and (1.36).}$$

An immediate consequence of the theorem is that (1.33) holds for every α as desired.

Corollary 1.4.22 *For every $\alpha \in \mathbb{R}$ one has that*

$$\liminf_{q\to\infty} q \log^2 q \, ||q\alpha|| \, ||q\beta|| = 0 \quad \textit{for almost all } \beta \in \mathbb{R}.$$

Pseudo sketch proof of Theorem 1.4.20 Given α and ψ, rewrite (1.37) as follows:

$$\|q\beta\| < \Psi_\alpha(q) \quad \textit{where} \quad \Psi_\alpha(q) := \frac{\psi(q)}{\|q\alpha\|}. \tag{1.38}$$

We are given (1.35) rather than the above divergent sum condition. So we need to show that

$$\sum_{q=1}^{\infty} \psi(q) \log q = \infty \quad \Longrightarrow \quad \sum_{q=1}^{\infty} \Psi_\alpha(q) = \infty. \tag{1.39}$$

This follows (*exercise*) on using partial summation together with the following fact established in [17]. For any irrational α and $Q \geq 2$

$$\sum_{q=1}^{Q} \frac{1}{\|q\alpha\|} \geq 2\, Q \log Q. \tag{1.40}$$

This lower bound estimate strengthens a result of Schmidt [84] – his result is for almost all α rather than all irrationals. Now, if $\Psi_\alpha(q)$ were a monotonic function of q we could have used Khintchine's theorem, which would then imply that

$$m_1(W(\Psi_\alpha)) = 1 \quad \text{if} \quad \sum_{q=1}^{\infty} \Psi_\alpha(q) = \infty. \tag{1.41}$$

Unfortunately, Ψ_α is not monotonic. Nevertheless, the argument given in [17] overcomes this difficulty. ⊠

It is worth mentioning that Corollary 1.4.22 together with Peck's theorem and Theorem PV adds weight to the argument made in [8] for the following strengthening of Littlewood's conjecture.

Conjecture 1.4.23 *For any pair $(\alpha, \beta) \in I^2$,*

$$\liminf_{q\to\infty} q \, \log q \, ||q\alpha|| \, ||q\beta|| < +\infty .$$

Furthermore, it is argued in [8] that the natural analogue of **Bad** within the multiplicative set-up is the set:

$$\mathbf{Mad} := \{(\alpha, \beta) \in \mathbb{R}^2 \; : \; \liminf_{q\to\infty} q \cdot \log q \cdot ||q\alpha|| \cdot ||q\beta|| > 0\}.$$

Note that Badziahin [4] has proven that there is a set of (α, β) of full Hausdorff dimension such that

$$\liminf_{q\to\infty} q \cdot \log q \cdot \log\log q \cdot ||q\alpha|| \cdot ||q\beta|| > 0 .$$

Regarding the convergence counterpart to Theorem 1.4.20, the following statement is established in [17].

Theorem 1.4.24 *Let $\alpha \in \mathbb{R}$ be any irrational real number and let $\psi : \mathbb{N} \to \mathbb{R}^+$ be such that*

$$\sum_{q=1}^{\infty} \psi(q) \, \log q \; < \; \infty .$$

Furthermore, assume either of the following two conditions:
(i) *$n \mapsto n\psi(n)$ is decreasing and*

$$\sum_{n=1}^{N} \frac{1}{n\|n\alpha\|} \ll (\log N)^2 \qquad \text{for all } N \geq 2; \tag{1.42}$$

(ii) *$n \mapsto \psi(n)$ is decreasing and*

$$\sum_{n=1}^{N} \frac{1}{\|n\alpha\|} \ll N \log N \qquad \text{for all } N \geq 2 . \tag{1.43}$$

Then for almost all $\beta \in \mathbb{R}$, there exist only finitely many $n \in \mathbb{N}$ such that

$$\|n\alpha\| \, \|n\beta\| < \psi(n). \tag{1.44}$$

The behaviour of the sums (1.42) and (1.43) is explicitly studied in terms of the continued fraction expansion of α. In particular, it is shown in [17] that

(1.42) holds for almost all real numbers α while (1.43) fails for almost all real numbers α. An intriguing question formulated in [17] concerns the behaviour of the above sums for algebraic α of degree ≥ 3. In particular, it is conjectured that (1.42) is true for any real algebraic number α of degree ≥ 3. As is shown in [17], this is equivalent to the following statement.

Conjecture 1.4.25 *For any algebraic $\alpha = [a_0; a_1, a_2, \dots] \in \mathbb{R}\setminus\mathbb{Q}$, we have that*

$$\sum_{k=1}^{n} a_k \ll n^2 .$$

Remark 1.4.26 Computational evidence for specific algebraic numbers does support this conjecture [34].

1.4.5 Khintchine on Fibers

In this section we look for a strengthening of Khintchine simultaneous theorem (Theorem 1.4.11) akin to the strengthening of Gallagher's multiplicative theorem described in §1.4.4.1. For ease of discussion, we begin with the case that $n = 2$ and whether or not Theorem 1.4.11 remains true if we fix $\alpha \in \mathrm{I}$. In other words, if L_α is the line parallel to the y-axis passing through the point $(\alpha, 0)$ and ψ is monotonic, then is it true that

$$m_1(W(2,\psi) \cap \mathrm{L}_\alpha) = \begin{cases} 0 & \text{if } \sum_{q=1}^{\infty} \psi^2(q) < \infty \\ \\ 1 & \text{if } \sum_{q=1}^{\infty} \psi^2(q) = \infty \end{cases} \qquad ????$$

The question marks are deliberate. They emphasise that the above statement is a question and not a fact or a claim. Indeed, it is easy to see that the convergent statement is false. Simply take α to be rational, say, $\alpha = \frac{a}{b}$. Then, by Dirichlet's theorem, for any β there exist infinitely many $q \in \mathbb{N}$ such that $\|q\beta\| < q^{-1}$ and so it follows that

$$\|bq\beta\| < \frac{b}{q} = \frac{b^2}{bq} \quad \text{and} \quad \|bq\alpha\| = 0 < \frac{b^2}{bq}.$$

This shows that every point on the rational vertical line L_α is $\psi(q) = b^2 q^{-1}$-approximable and so

$$m_1(W(2,\psi) \cap \mathrm{L}_\alpha) = 1 \quad \text{but} \quad \sum_{q=1}^{\infty} \psi^2(q) = \sum_{q=1}^{\infty} b^4 q^{-2} < \infty.$$

Now, concerning the divergent statement, we claim it is true.

Conjecture 1.4.27 *Let $\psi : \mathbb{N} \to \mathbb{R}^+$ be a monotonic function and $\alpha \in I$. Then*

$$m_1(W(2,\psi) \cap L_\alpha) = 1 \quad \textit{if} \quad \sum_{q=1}^{\infty} \psi^2(q) = \infty. \tag{1.45}$$

In order to state the current results, we need the notion of the Diophantine exponent of a real number. For $\mathbf{x} \in \mathbb{R}^n$, we let

$$\tau(\mathbf{x}) := \sup\{\tau : \mathbf{x} \in W(n,\tau)\} \tag{1.46}$$

denote the *Diophantine exponent of* $\mathbf{x}$. A word of warning: this notion of Diophantine exponent should not be confused with the Diophantine exponents introduced later in §1.4.6.1. Note that in view of (1.30), we always have that $\tau(\mathbf{x}) \geq 1/n$. In particular, for $\alpha \in \mathbb{R}$ we have that $\tau(\alpha) \geq 1$. The following result is established in [79].

Theorem 1.4.28 (F. Ramírez, D. Simmons, F. Süess) *Let $\psi : \mathbb{N} \to \mathbb{R}^+$ be a monotonic function and $\alpha \in I$.*

A. *If $\tau(\alpha) < 2$, then* (1.45) *is true.*
B. *If $\tau(\alpha) > 2$ and for some $\epsilon > 0$, $\psi(q) > q^{-\frac{1}{2}-\epsilon}$ for q large enough, then $W(2,\psi) \cap L_\alpha = I^2 \cap L_\alpha$. In particular, $m_1(W(2,\psi) \cap L_\alpha) = 1$.*

Remark 1.4.29 Though we have only stated it for lines in the plane, Theorem 1.4.28 is actually true for lines in $\mathbb{R}^n$. There, we fix an $(n-1)$-tuple of coordinates $\boldsymbol{\alpha} = (\alpha_1, \ldots, \alpha_{n-1})$, and consider the line $L_{\boldsymbol{\alpha}} \subset \mathbb{R}^n$. We obtain the same result, with a 'cut-off' at n in the *dual Diophantine exponent* of $\boldsymbol{\alpha} \in \mathbb{R}^{n-1}$. The dual Diophantine exponent $\tau^*(\mathbf{x})$ of a vector $\mathbf{x} \in \mathbb{R}^n$ is defined similarly to the (simultaneous) Diophantine exponent, defined above by (1.46), and in the case of numbers (i.e. one-dimensional vectors), the two notions coincide – see §1.4.6.1 for the formal definition of $\tau^*(\mathbf{x})$.

Remark 1.4.30 This cut-off in Diophantine exponent, which in Theorem 1.4.28 happens at $\tau(\alpha) = 2$, seems quite unnatural: why should real numbers with Diophantine exponent 2 be special? Still, such points are inaccessible to our methods. We will see the obstacle in the counting estimate (1.48) which is used for the proof of Part A and is unavailable for $\tau(\alpha) = 2$, and in our application of Khintchine's Transference Principle for the proof of Part B.

Remark 1.4.31 Note that in Part B, the 'in particular' full measure conclusion is immediate and does not even require the divergent sum condition associated with (1.45).

Regarding the natural analogous conjecture for higher-dimensional subspaces, we have the following statement from [79] which provides a complete solution in the case of affine co-ordinate subspaces of dimension at least two.

Theorem 1.4.32 *Let $\psi : \mathbb{N} \to \mathbb{R}^+$ be a monotonic function and given $\boldsymbol{\alpha} \in I^{n-d}$ where $2 \leq d \leq n-1$, let $L_{\boldsymbol{\alpha}} := \{\boldsymbol{\alpha}\} \times \mathbb{R}^d$. Then*

$$m_d(W(n,\psi) \cap L_{\boldsymbol{\alpha}}) = 1 \quad \text{if} \quad \sum_{q=1}^{\infty} \psi^n(q) = \infty. \tag{1.47}$$

Remark 1.4.33 Notice that Theorem 1.4.32 requires $d \geq 2$, thereby excluding lines in $\mathbb{R}^n$. In this case, the obstacle is easy to describe: the proof of Theorem 1.4.32 relies on Gallagher's extension of Khintchine's theorem, telling us that the monotonicity assumption can be dropped in higher dimensions (see Remark 1.4.13). In the proof of Theorem 1.4.32 we find a natural way to apply this directly to the fibers, therefore, we must require $d \geq 2$.

But this is again only a consequence of the chosen method of proof, and not necessarily a reflection of reality. Indeed, Theorem 1.4.28 (and its more general version for lines in $\mathbb{R}^n$) suggests that we should be able to relax Theorem 1.4.32 to include the case where $d = 1$.

Remark 1.4.34 The case when $d = n - 1$ was first treated in [78]. There, a number of results are proved in the direction of Theorem 1.4.32, but with various restrictions on Diophantine exponent, or on the approximating function.

Regarding the proof of Theorem 1.4.28, Part B makes use of Khintchine's Transference Principle (see §1.4.6.1) while the key to establishing Part A is the following measure theoretic statement (cf. Theorem 1.1.4) and ubiquity (see §1.5).

Proposition 1.1 *Let $\psi : \mathbb{N} \to \mathbb{R}^+$ be a monotonic function such that for all $\epsilon > 0$ we have $\psi(q) > q^{-\frac{1}{2}-\epsilon}$ for all q large enough. Let $\alpha \in \mathbb{R}$ be a number with Diophantine exponent $\tau(\alpha) < 2$. Then for any $0 < \epsilon < 1$ and integer $k \geq k_0(\epsilon)$, we have that*

$$m_1\left(\bigcup_{\substack{k^{n-1} < q \leq k^n : \\ \|q\alpha\| \leq \psi(k^n)}} \bigcup_{p=0}^{q} B\left(\frac{p}{q}, \frac{k}{k^{2n}\psi(k^n)}\right) \right) \geq 1 - \epsilon.$$

Remark 1.4.35 Note that within the context of Theorem 1.4.28, since α is fixed it is natural to consider only those $q \in \mathbb{N}$ for which $\|q\alpha\| \leq \psi(q)$ when considering solutions to the inequality $\|q\beta\| \leq \psi(q)$. In other words, if we let

$$\mathcal{A}_\alpha(\psi) := \{q \in \mathbb{N} : \|q\alpha\| \leq \psi(q)\}$$

then by definition

$$\begin{aligned} &W(2,\psi) \cap \mathrm{L}_\alpha \\ &\quad = \{(\alpha,\beta) \in \mathrm{L}_\alpha \cap \mathrm{I}^2 : \|q\beta\| \leq \psi(q) \text{ for infinitely many } q \in \mathcal{A}_\alpha(\psi)\}. \end{aligned}$$

It is clear that the one-dimensional Lebesgue measure m_1 of this set is the same as that of

$$\{\beta \in \mathrm{I} : \|q\beta\| \leq \psi(q) \text{ for infinitely many } q \in \mathcal{A}_\alpha(\psi)\}.$$

SKETCH PROOF OF PROPOSITION 1.1 In view of Minkowski's theorem for systems of linear forms, for any $(\alpha, \beta) \in \mathbb{R}^2$ and integer $N \geq 1$, there exists an integer $q \geq 1$ such that

$$\begin{aligned} \|q\alpha\| &\leq \psi(N) \\ \|q\beta\| &\leq \frac{1}{N\,\psi(N)} \\ q &\leq N. \end{aligned}$$

The desired statement follows on exploiting this with $N = k^n$ together with the following result which is a consequence of a general counting result established in [17]: given ψ and α satisfying the conditions imposed in Proposition 1.1, then for n sufficiently large

$$\#\{q \leq k^{n-1} : \|q\alpha\| \leq \psi(k^n)\} \ \leq\ 31\,\psi(k^n)\,k^{n-1}. \tag{1.48}$$

(An analogous count is established in [79] for vectors $\boldsymbol{\alpha} \in \mathbb{R}^{n-1}$.)
Exercise: Fill in the details of the above sketch. ⊠

1.4.6 Dual Approximation and Khintchine's Transference

Instead of simultaneous approximation by rational points as considered in the previous section, one can consider the closeness of the point $\mathbf{x} = (x_1, \ldots, x_m) \in \mathbb{R}^m$ to rational hyperplanes given by the equations $\mathbf{q} \cdot \mathbf{x} = p$ with $p \in \mathbb{Z}$ and $\mathbf{q} \in \mathbb{Z}^m$. The point $\mathbf{x} \in \mathbb{R}^n$ will be called *dually* ψ*-well approximable* if the inequality

$$|\mathbf{q} \cdot \mathbf{x} - p| < \psi(|\mathbf{q}|)$$

holds for infinitely many $(p, \mathbf{q}) \in \mathbb{Z} \times \mathbb{Z}^m$ with

$$|\mathbf{q}| := |\mathbf{q}|_\infty = \max\{|q_1|, \ldots, |q_m|\} > 0.$$

The set of dually ψ-approximable points in I^m is denoted by $W^*(m, \psi)$. In the case $\psi : q \to q^{-\tau}$ with $\tau > 0$, we write $W^*(m, \tau)$ for $W^*(m, \psi)$. The set $W^*(n, \tau)$ is the set of *dually τ-well approximable numbers*. Note that in view of Corollary 1.4.7 we have that

$$W^*(m, \tau) = \mathrm{I}^m \quad \text{if} \quad \tau \leq m. \tag{1.49}$$

The simultaneous and dual forms of approximation are special cases of a system of linear forms, covered by a general extension due to A. V. Groshev (see [90]). This treats real $m \times n$ matrices X, regarded as points in $\mathbb{R}^{mn}$, which are ψ-approximable. More precisely, $X = (x_{ij}) \in \mathbb{R}^{mn}$ is said to be ψ-approximable if the inequality

$$\|\mathbf{q}X\| < \psi(|\mathbf{q}|)$$

is satisfied for infinitely many $\mathbf{q} \in \mathbb{Z}^m$. Here $\mathbf{q}X$ is the system

$$q_1 x_{1j} + \cdots + q_m x_{m,j} \qquad (1 \leq j \leq n)$$

of n real linear forms in m variables and $\|\mathbf{q}X\| := \max_{1 \leq j \leq n} \|\mathbf{q} \cdot X^{(j)}\|$, where $X^{(j)}$ is the jth column vector of X. As the set of ψ-approximable points is translation invariant under integer vectors, we can restrict attention to the mn-dimensional unit cube I^{mn}. The set of ψ-approximable points in I^{mn} will be denoted by

$$W(m, n, \psi) := \{X \in \mathrm{I}^{mn} : \|\mathbf{q}X\| < \psi(|\mathbf{q}|) \text{ for infinitely many } \mathbf{q} \in \mathbb{Z}^m\}.$$

Thus, $W(n, \psi) = W(1, n, \psi)$ and $W^*(m, \psi) = W(m, 1, \psi)$. The following result naturally extends Khintchine's simultaneous theorem to the linear forms set-up. For obvious reasons, we write $|X|_{mn}$ rather than $m_{mn}(X)$ for mn-dimensional Lebesgue measure of a set $X \subset \mathbb{R}^{mn}$.

Theorem 1.4.36 (Khintchine–Groshev, 1938) *Let $\psi : \mathbb{N} \to \mathbb{R}^+$. Then*

$$|W(m, n, \psi)|_{mn} = \begin{cases} 0 & \text{if } \displaystyle\sum_{r=1}^{\infty} r^{m-1} \psi(r)^n < \infty, \\ 1 & \text{if } \displaystyle\sum_{r=1}^{\infty} r^{m-1} \psi(r)^n = \infty \text{ and } \psi \text{ is monotonic.} \end{cases}$$

The counterexample due to Duffin and Schaeffer mentioned in §1.2.2.1 means that the monotonicity condition cannot be dropped from Groshev's theorem when $m = n = 1$. To avoid this situation, let $mn > 1$. Then for $m = 1$, we have already mentioned (Remark 1.4.13) that the monotonicity condition can be removed. Furthermore, the monotonicity condition can also be removed for $m > 2$ – see [13, Theorem 8] and [90, Theorem 14]. The $m = 2$ situation was resolved only recently in [27], where it was shown that the monotonicity condition can be safely removed. The upshot of this discussion is that we only require the monotonicity condition in the Khintchine–Groshev theorem in the case when $mn = 1$.

Naturally, one can ask for a Hausdorff measure generalisation of the Khintchine–Groshev theorem. The following is such a statement and as one should expect it coincides with Theorem 1.3.7 when $m = n = 1$. In the simultaneous case ($m = 1$), the result was alluded to within Remark 1.4.16 following the simultaneous statement of Khintchine's theorem.

Theorem 1.4.37 *Let $\psi : \mathbb{N} \to \mathbb{R}^+$. Then*

$$\mathcal{H}^s(W(m,n,\psi)) = \begin{cases} 0 & \text{if } \sum_{r=1}^{\infty} r^{m(n+1)-1-s}\psi(r)^{s-n(m-1)} < \infty, \\ \mathcal{H}^s(\mathbb{I}^{mn}) & \text{if } \sum_{r=1}^{\infty} r^{m(n+1)-1-s}\psi(r)^{s-n(m-1)} = \infty \\ & \text{and } \psi \text{ is monotonic}. \end{cases}$$

This Hausdorff theorem follows from the corresponding Lebesgue statement in the same way that Khintchine's theorem implies Jarník's theorem via the Mass Transference Principle – see §1.3.4.1. The Mass Transference Principle introduced in §1.3.4 deals with lim sup sets which are defined by a sequence of balls. However, the 'slicing' technique introduced in [22] extends the Mass Transference Principle to deal with lim sup sets defined by a sequence of neighbourhoods of 'approximating' planes. This naturally enables us to generalise the Lebesgue measure statements for systems of linear forms to Hausdorff measure statements. The last sentence should come with a warning. It gives the impression that in view of the discussion preceding Theorem 1.4.36, one should be able to establish Theorem 1.4.37 directly, without the monotonicity assumption except when $m = n = 1$. However, as things currently stand we also need to assume monotonicity when $m = 2$. For further details see [13, §8].

Returning to Diophantine approximation in $\mathbb{R}^n$, we consider the following natural question.

Question. Is there a connection between the simultaneous ($m = 1$) and dual ($n = 1$) forms of approximating points in $\mathbb{R}^n$?

1.4.6.1 Khintchine's Transference

The simultaneous and dual forms of Diophantine approximation are related by a 'transference' principle in which a solution of one form is related to a solution of the other. In order to state the relationship we introduce the quantities ω^* and ω. For $\mathbf{x} = (x_1, \ldots, x_n) \in \mathbb{R}^n$, let

$$\omega^*(\mathbf{x}) := \sup\left\{\omega \in \mathbb{R} : \mathbf{x} \in W^*(n, n + \omega)\right\}$$

and

$$\omega(\mathbf{x}) := \sup\left\{\omega \in \mathbb{R} : \mathbf{x} \in W(n, \tfrac{1+\omega}{n})\right\}.$$

Note that

$$\tau(\mathbf{x}) = \frac{1 + \omega(\mathbf{x})}{n}$$

where $\tau(\mathbf{x})$ is the Diophantine exponent of $\mathbf{x}$ as defined by (1.46). For the sake of completeness we mention that the quantity

$$\tau^*(\mathbf{x}) = n + \omega^*(\mathbf{x})$$

is called the dual Diophantine exponent. The following statement provides a relationship between the dual and simultaneous Diophantine exponents.

Theorem 1.4.38 (Khintchine's Transference Principle) *For $\mathbf{x} \in \mathbb{R}^n$, we have that*

$$\frac{\omega^*(\mathbf{x})}{n^2 + (n-1)\omega^*(\mathbf{x})} \leq \omega(\mathbf{x}) \leq \omega^*(\mathbf{x})$$

with the left-hand side being interpreted as $1/(n-1)$ if $\omega^*(\mathbf{x})$ is infinite.

Remark 1.4.39 The transference principle implies that given any $\epsilon > 0$, if $\mathbf{x} \in W(n, \frac{1+\epsilon}{n})$ then $\mathbf{x} \in W^*(n, n + \epsilon^*)$ for some ϵ^* comparable to ϵ, and vice versa.

Proof of Part B of Theorem 1.4.28

Part B of Theorem 1.4.28 follows by plugging $n = 2$ and $d = 1$ into the following proposition, which is in turn a simple consequence of Khintchine's Transference Principle.

Proposition 1.2 *Let $\psi : \mathbb{N} \to \mathbb{R}^+$ be a monotonic function and given $\boldsymbol{\alpha} \in I^{n-d}$ where $1 \leq d \leq n-1$, let $L_{\boldsymbol{\alpha}} := \{\boldsymbol{\alpha}\} \times \mathbb{R}^d$. Assume that $\tau(\boldsymbol{\alpha}) > \frac{1+d}{n-d}$ and for $\epsilon > 0$, $\psi(q) > q^{-\frac{1}{n}-\epsilon}$ for q large enough. Then*

$$W(n, \psi) \cap L_{\boldsymbol{\alpha}} = I^n \cap L_{\boldsymbol{\alpha}}.$$

In particular, $m_d(W(n, \psi) \cap L_{\boldsymbol{\alpha}}) = 1$.

Proof We are given that $\tau(\boldsymbol{\alpha}) > \frac{1+d}{n-d}$ and so by definition $\omega(\boldsymbol{\alpha}) > d$. Thus, by Khintchine's Transference Principle, it follows that $\omega^*(\boldsymbol{\alpha}) > d$ and so $\omega^*(\mathbf{x}) > 0$ for any point $\mathbf{x} = (\boldsymbol{\alpha}, \boldsymbol{\beta}) \in \mathbb{R}^n$; i.e. $\boldsymbol{\beta} \in \mathbb{R}^d$ and $\mathbf{x}$ is a point on the d-dimensional plane $L_{\boldsymbol{\alpha}}$. On applying Khintchine's Transference Principle again, we deduce that $\omega(\mathbf{x}) > 0$ which together with the growth condition imposed on ψ implies the desired conclusion. □

1.5 Ubiquitous Systems of Points

In [14], a general framework is developed for establishing divergent results analogous to those of Khintchine and Jarník for a natural class of lim sup sets. The framework is based on the notion of 'ubiquity', which goes back to [10] and [46] and captures the key measure theoretic structure necessary to prove such measure theoretic laws. The 'ubiquity' introduced below is a much simplified version of that in [14]. In particular, we make no attempt to incorporate the linear forms theory of metric Diophantine approximation. However, this does have the advantage of making the exposition more transparent and also leads to cleaner statements which are more than adequate for the application we have in mind; namely to systems of points.

1.5.1 The General Framework and Fundamental Problem

The *general framework of ubiquity* considered within is as follows:

- (Ω, d) is a compact metric space.
- μ is a Borel probability measure supported on Ω.
- There exist positive constants δ and r_o such that for any $x \in \Omega$ and $r \leq r_0$,

$$a\, r^{\delta} \;\leq\; \mu(B(x, r)) \;\leq\; b\, r^{\delta}. \tag{1.50}$$

  The constants a and b are independent of the ball $B(x, r) := \{y \in \Omega : d(x, y) < r\}$.

- $\mathcal{R} = (R_\alpha)_{\alpha \in J}$ a sequence of points R_α in Ω indexed by an infinite countable set J. The points R_α are commonly referred to as *resonant points*.
- $\beta : J \to \mathbb{R}^+ : \alpha \mapsto \beta_\alpha$ is a positive function on J. It attaches a 'weight' β_α to the resonant point R_α.
- To avoid pathological situations:

$$\#\{\alpha \in J : \beta_\alpha \leq x\} < \infty \quad \text{for any } x \in \mathbb{R}. \tag{1.51}$$

Remark 1.5.1 The measure condition (1.50) on the *ambient measure* μ implies that μ is non-atomic; that is, $\mu(\{x\}) = 0$ for any $x \in \Omega$, and that

$$\mu(\Omega) := 1 \asymp \mathcal{H}^\delta(\Omega) \quad \text{and} \quad \dim \Omega = \delta.$$

Indeed, μ is comparable to δ-dimensional Hausdorff measure $\mathcal{H}^\delta$.

Given a decreasing function $\Psi : \mathbb{R}^+ \to \mathbb{R}^+$ let

$$\Lambda(\Psi) = \{x \in \Omega : x \in B(R_\alpha, \Psi(\beta_\alpha)) \text{ for infinitely many } \alpha \in J\}.$$

The set $\Lambda(\Psi)$ is a 'lim sup' set; it consists of points in Ω which lie in infinitely many of the balls $B(R_\alpha, \Psi(\beta_\alpha))$ centred at resonant points. As in the classical setting introduced in §1.2, it is natural to refer to the function Ψ as the *approximating function*. It governs the 'rate' at which points in Ω must be approximated by resonant points in order to lie in $\Lambda(\Psi)$. In view of the finiteness condition (1.51), it follows that for any fixed $k > 1$, the number of α in J with $k^{t-1} < \beta_\alpha \leq k^t$ is finite regardless of the value of $t \in \mathbb{N}$. Therefore $\Lambda(\Psi)$ can be rewritten as the lim sup set of

$$\Upsilon(\Psi, k, t) := \bigcup_{\alpha \in J : k^{t-1} < \beta_\alpha \leq k^t} B(R_\alpha, \Psi(\beta_\alpha))\,;$$

that is,

$$\Lambda(\Psi) = \limsup_{t \to \infty} \Upsilon(\Psi, k, t) := \bigcap_{m=1}^{\infty} \bigcup_{t=m}^{\infty} \Upsilon(\Psi, k, t).$$

It is reasonably straightforward to determine conditions under which $\mu(\Lambda(\Psi)) = 0$. In fact, this is implied by the convergence part of the Borel–Cantelli lemma from probability theory whenever

$$\textstyle\sum_{t=1}^{\infty} \mu(\Upsilon(\Psi, k, t)) < \infty. \tag{1.52}$$

In view of this it is natural to consider the following **fundamental problem**:

Under what conditions is $\mu(\Lambda(\psi)) > 0$ and more generally $\mathcal{H}^s(\Lambda(\Psi)) > 0$?

Ideally, we would like to be able to conclude the full measure statement $\mathcal{H}^s(\Lambda(\Psi)) = \mathcal{H}^s(\Omega)$. Recall that when $s = \delta$, the ambient measure μ coincides with $\mathcal{H}^\delta$. Also, if $s < \delta$ then $\mathcal{H}^s(\Omega) = \infty$.

1.5.1.1 The Basic Example

In order to illustrate and clarify the above general set-up, we show that the set $W(n, \psi)$ of simultaneously ψ-well approximable points $\mathbf{x} \in \mathrm{I}^n := [0, 1]^n$ can be expressed in the form of $\Lambda(\Psi)$. With this in mind, let:

- $\Omega := \mathrm{I}^n$ and $d(\mathbf{x}, \mathbf{y}) := \max_{1 \le i \le n} |x_i - y_i|$,
- μ be Lebesgue measure restricted to I^n and $\delta := n$,
- $J := \{(\mathbf{p}, q) \in \mathbb{Z}^n \times \mathbb{N} : \mathbf{p}/q \in \mathrm{I}^n\}$ and $\alpha := (\mathbf{p}, q) \in J$,
- $\mathcal{R} := (\mathbf{p}/q)_{(\mathbf{p},q) \in J}$ and $\beta_{(\mathbf{p},q)} := q$.

Thus, the resonant points R_α are simply rational points $\mathbf{p}/q := (p_1/q, \ldots, p_n/q)$ in the unit cube I^n. It is readily verified that the measure condition (1.50) and the finiteness condition (1.51) are satisfied and moreover that for any decreasing function $\psi : \mathbb{N} \to \mathbb{R}^+$,

$$\Lambda(\Psi) = W(n, \psi) \quad \text{with} \quad \Psi(q) := \psi(q)/q \, .$$

For this basic example, the solution to the fundamental problem is given by the simultaneous Khintchine–Jarník theorem (see Theorem 1.4.37 with $m = 1$ in §1.4.6).

1.5.2 The Notion of Ubiquity

The following 'system' contains the key measure theoretic structure necessary for our attack on the fundamental problem.

Let $\rho : \mathbb{R}^+ \to \mathbb{R}^+$ be a function with $\rho(r) \to 0$ as $r \to \infty$ and let

$$\Delta(\rho, k, t) := \bigcup_{\alpha \in J \,:\, \beta_\alpha \le k^t} B(R_\alpha, \rho(k^t)) \, ,$$

where $k > 1$ is a fixed real number. Note that when $\rho = \Psi$ the composition of $\Delta(\rho, k, t)$ is very similar to that of $\Upsilon(\Psi, k, t)$.

Definition (Ubiquitous System) Let $B = B(x, r)$ denote an arbitrary ball with centre x in Ω and radius $r \le r_0$. Suppose there exists a function ρ and absolute constants $\kappa > 0$ and $k > 1$ such that for any ball B as above

$$\mu\,(B \cap \Delta(\rho, k, t)) \ \ge\ \kappa\ \mu(B) \quad \text{for } t \ge t_0(B). \tag{1.53}$$

Then the pair $(\mathcal{R}, \beta)$ is said to be a *local μ-ubiquitous system relative to* (ρ, k). If (1.53) does not hold for arbitrary balls with centre x in Ω and radius $r \leq r_0$, but does hold with $B = \Omega$, the pair $(\mathcal{R}, \beta)$ is said to be a *global μ-ubiquitous system relative to* (ρ, k).

Loosely speaking, the definition of local ubiquity says that the underlying space Ω is locally 'approximated' by the set $\Delta(\rho, k, t)$ in terms of the measure μ. By 'locally' we mean balls centred at points in Ω. The function ρ is referred to as the *ubiquitous function*. The actual values of the constants κ and k in the above definition are irrelevant – it is their existence that is important. In practice, the μ-ubiquity of a system can be established using standard arguments concerning the distribution of the resonant points in Ω, from which the function ρ arises naturally. To illustrate this, we return to the basic example of §1.5.1.1.

Proposition 1.3 *There is a constant $k > 1$ such that the pair $(\mathcal{R}, \beta)$ defined in §1.5.1.1 is a local μ-ubiquitous system relative to (ρ, k) where $\rho : r \mapsto$ const $\times\, r^{-(n+1)/n}$.*

The one-dimensional case of this proposition follows from Theorem 1.1.4.

Exercise: Prove the above proposition for arbitrary n. Hint: you will need to use the multidimensional version of Dirichlet's theorem, or Minkowski's theorem.

1.5.3 The Ubiquity Statements

Before stating the main results regarding ubiquity we introduce one last notion. Given a real number $k > 1$, a function $h : \mathbb{R}^+ \to \mathbb{R}^+$ will be said to be *k-regular* if there exists a strictly positive constant $\lambda < 1$ such that for t sufficiently large

$$h(k^{t+1}) \leq \lambda\, h(k^t). \tag{1.54}$$

The constant λ is independent of t but may depend on k. A consequence of local ubiquity is the following result.

Theorem 1.5.2 (Ubiquity: The Hausdorff Measure Case) *Let (Ω, d) be a compact metric space equipped with a probability measure μ satisfying condition* (1.50) *and such that any open subset of Ω is μ-measurable. Suppose that $(\mathcal{R}, \beta)$ is a locally μ-ubiquitous system relative to (ρ, k) and that Ψ is an approximating function. Furthermore, suppose that $s \in (0, \delta]$, that ρ is k-regular and that*

$$\sum_{t=1}^{\infty} \frac{\Psi(k^t)^s}{\rho(k^t)^\delta} = \infty . \tag{1.55}$$

Then

$$\mathcal{H}^s\left(\Lambda(\Psi)\right) = \mathcal{H}^s\left(\Omega\right).$$

As already mentioned, if $s < \delta$ then $\mathcal{H}^s(\Omega) = \infty$. On the other hand, if $s = \delta$, the Hausdorff measure $\mathcal{H}^\delta$ is comparable to the ambient measure μ and the theorem implies that

$$\mu\left(\Lambda(\Psi)\right) = \mu(\Omega) := 1.$$

Actually, the notion of global ubiquity has implications in the ambient measure case.

Theorem 1.5.3 (Ubiquity: The Ambient Measure Case) *Let (Ω, d) be a compact metric space equipped with a measure μ satisfying condition* (1.50) *and such that any open subset of Ω is μ-measurable. Suppose that $(\mathcal{R}, \beta)$ is a globally μ-ubiquitous system relative to (ρ, k) and that Ψ is an approximating function. Furthermore, suppose that either ρ or Ψ is k-regular and that*

$$\sum_{t=1}^{\infty} \left(\frac{\Psi(k^t)}{\rho(k^t)}\right)^{\delta} = \infty . \tag{1.56}$$

Then

$$\mu\left(\Lambda(\Psi)\right) > 0.$$

If in addition $(\mathcal{R}, \beta)$ is a locally μ-ubiquitous system relative to (ρ, k), then

$$\mu\left(\Lambda(\Psi)\right) = 1.$$

Remark 1.5.4 Note that in Theorem 1.5.3 we can get away with either ρ or Ψ being k-regular. In the ambient measure case, it is also possible to weaken the measure condition (1.50) (see Theorem 1 in [14, §3]).

Remark 1.5.5 If we know via some other means that $\Lambda(\Psi)$ satisfies a zero-full law (as indeed is the case for the classical set of $W(n, \psi)$ of ψ-well approximable points), then it is enough to show that $\mu\left(\Lambda(\Psi)\right) > 0$ in order to conclude full measure.

The above results constitute the main theorems appearing in [14] tailored to the set-up considered here. In fact, Theorem 1.5.2 as stated appears in [25] for the first time. Previously, the Hausdorff and ambient measure cases had been thought of and stated separately.

The concept of ubiquity was originally formulated by Dodson, Rynne and Vickers [46] to obtain lower bounds for the Hausdorff dimension of lim sup sets. Furthermore, the ubiquitous systems of [46] essentially coincide with the regular systems of Baker and Schmidt [10] and both have proved very useful in obtaining lower bounds for the Hausdorff dimension of lim sup sets. However, unlike the framework developed in [14], both [10] and [46] fail to shed any light on establishing the more desirable divergent Khintchine and Jarník type results. The latter clearly implies lower bounds for the Hausdorff dimension. For further details regarding regular systems and the original formulation of ubiquitous systems see [14, 31].

1.5.3.1 The Basic Example and the Simultaneous Khintchine–Jarník Theorem

Regarding the basic example of §1.5.1.1, recall that

$$\Lambda(\Psi) = W(n, \psi) \quad \text{with} \quad \Psi(q) := \psi(q)/q$$

and that Proposition 1.3 states that for k large enough, the pair $(\mathcal{R}, \beta)$ is a local μ-ubiquitous system relative to (ρ, k) where

$$\rho : r \mapsto \text{const} \times r^{-(n+1)/n}.$$

Now, clearly the function ρ is k-regular. Also note that the divergence sum condition (1.55) associated with Theorem 1.5.2 becomes

$$\sum_{t=1}^{\infty} k^{t(n+1-s)} \psi(k^t)^s = \infty.$$

If ψ is monotonic, this is equivalent to

$$\sum_{q=1}^{\infty} q^{n-s} \psi(q)^s = \infty,$$

and Theorem 1.5.2 implies that

$$\mathcal{H}^s(W(n, \psi)) = \mathcal{H}^s(\mathrm{I}^n).$$

The upshot is that Theorem 1.5.2 implies the divergent case of the simultaneous Khintchine–Jarník theorem; namely, Theorem 1.4.37 with $m = 1$ in §1.4.6.

Remark 1.5.6 It is worth standing back a little and thinking about what we have actually used in establishing the classical results – namely, local ubiquity. Within the classical set-up, local ubiquity is a simple measure theoretic statement concerning the distribution of rational points with respect to Lebesgue measure – the natural measure on the unit interval. From this we are able to obtain the divergent parts of both Khintchine's theorem (a Lebesgue measure statement) and Jarník's theorem (a Hausdorff measure statement). In other words, the Lebesgue measure statement of local ubiquity underpins the general Hausdorff measure theory of the lim sup set $W(n,\psi)$. This of course is very much in line with the subsequent discovery of the Mass Transference Principle discussed in §1.3.4.

The applications of ubiquity are widespread, as demonstrated in [14, §12]. We now consider a more recent application of ubiquity to the 'fibers' strengthening of Khintchine's simultaneous theorem described in §1.4.5.

1.5.3.2 Proof of Theorem 1.4.28: Part A

Let $\psi : \mathbb{N} \to \mathbb{R}^+$ be a monotonic function and $\alpha \in \mathrm{I}$ such that it has Diophantine exponent $\tau(\alpha) < 2$. In view of Remark 1.4.35 in §1.4.5, establishing Theorem 1.4.28 is equivalent to showing that

$$m(\Pi(\psi,\alpha)) = 1 \quad \text{if} \quad \sum_{q=1}^{\infty} \psi^2(q) = \infty$$

where

$$\Pi(\psi,\alpha) := \{\beta \in \mathrm{I} : \|q\beta\| \le \psi(q) \text{ for infinitely many } q \in \mathcal{A}_\alpha(\psi)\}.$$

Recall,

$$\mathcal{A}_\alpha(\psi) := \{q \in \mathbb{N} : \|q\alpha\| \le \psi(q)\}.$$

Remark 1.5.7 Without loss of generality, we can assume that

$$q^{-\frac{1}{2}}(\log q)^{-1} \le \psi(q) \le q^{-\frac{1}{2}} \quad \forall\, q \in \mathbb{N}. \tag{1.57}$$

Exercise: Verify that this is indeed the case. For the right-hand side of (1.57), consider the auxiliary function

$$\tilde{\psi} : q \to \tilde{\psi} := \min\{q^{-\frac{1}{2}}, \psi(q)\}$$

and show that $\sum_{q=1}^{\infty} \tilde{\psi}^2(q) = \infty$. For the left-hand side of (1.57), consider the auxiliary function

$$\tilde{\psi} : q \to \tilde{\psi}(q) := \max\{\hat{\psi}(q) := q^{-\frac{1}{2}}(\log q)^{-1}, \psi(q)\}$$

and show that $m(\Pi(\hat{\psi},\alpha)) = 0$ by making use of the counting estimate (1.48) and the convergence Borel–Cantelli lemma.

We now show that the set $\Pi(\psi,\alpha)$ can be expressed in the form of $\Lambda(\Psi)$. With this in mind, let:

- $\Omega := [0, 1]$ and $d(x, y) := |x - y|$;
- μ be Lebesgue measure restricted to I and $\delta := 1$;
- $J := \{(p, q) \in \mathbb{Z} \times \mathcal{A}_\alpha(\psi) : p/q \in \mathrm{I}\}$ and $\alpha := (p, q) \in J$;
- $\mathcal{R} := (p/q)_{(p,q)\in J}$ and $\beta_{(p,q)} := q$.

Thus, the resonant points R_α are simply rational points p/q in the unit interval I with denominators q restricted to the set $\mathcal{A}_\alpha(\psi)$. It is readily verified that the measure condition (1.50) and the finiteness condition (1.51) are satisfied and moreover that for any decreasing function $\psi : \mathbb{N} \to \mathbb{R}^+$,

$$\Lambda(\Psi) = \Pi(\psi, \alpha) \quad \text{with} \quad \Psi(q) := \psi(q)/q\,.$$

Note that since ψ is decreasing, the function Ψ is k-regular. Now, in view of Remark 1.5.7, the conditions of Proposition 1.1 are satisfied and we conclude that for k large enough, the pair $(\mathcal{R}, \beta)$ is a global m-ubiquitous system relative to (ρ, k) where

$$\rho : r \mapsto \frac{k}{r^2\psi(r)}.$$

Now, since ψ is monotonic

$$\sum_{t=1}^{\infty} \frac{\Psi(k^t)}{\rho(k^t)} = \sum_{t=1}^{\infty} k^{t-1}\psi^2(k^t) = \infty \iff \sum_{q=1}^{\infty} \psi^2(q) = \infty$$

and Theorem 1.5.3 implies that

$$\mu\Big(\Pi(\psi, \alpha)\Big) > 0\,.$$

Now observe that $\Pi(\psi, \alpha)$ is simply the set $W(\bar{\psi})$ of $\bar{\psi}$-well approximable numbers with $\bar{\psi}(q) := \psi(q)$ if $q \in \mathcal{A}_\alpha(\psi)$ and zero otherwise. Thus, Cassels' zero-full law [38] implies the desired statement; namely that

$$\mu\Big(\Pi(\psi, \alpha)\Big) = 1.$$

1.6 Diophantine Approximation on Manifolds

Diophantine approximation on manifolds (as coined by Bernik and Dodson in their Cambridge Tract [31]) or *Diophantine approximation of dependent quantities* (as coined by Sprindžuk in his monograph [90]) refers to the study of Diophantine properties of points in $\mathbb{R}^n$ whose coordinates are confined by functional relations or equivalently are restricted to a sub-manifold $\mathcal{M}$ of $\mathbb{R}^n$. Thus, in the case of simultaneous Diophantine approximation one studies sets such as

$$\mathcal{M} \cap W(n, \psi).$$

To some extent, we have already touched upon the theory of Diophantine approximation on manifolds when we considered Gallagher multiplicative theorem on fibers in §1.4.4.1 and Khintchine simultaneous theorem on fibers in §1.4.5. In these sections, the points of interest are confined to an affine co-ordinate subspace of $\mathbb{R}^n$; namely the manifold

$$\mathrm{L}_{\boldsymbol{\alpha}} := \{\boldsymbol{\alpha}\} \times \mathbb{R}^d, \text{ where } 1 \leq d \leq n-1 \text{ and } \boldsymbol{\alpha} \in \mathrm{I}^{n-d}.$$

In general, a manifold $\mathcal{M}$ can locally be given by a system of equations, for instance, the unit sphere in $\mathbb{R}^3$ is given by the equation

$$x^2 + y^2 + z^2 = 1;$$

or it can be immersed into $\mathbb{R}^n$ by a map $\mathbf{f} : \mathbb{R}^d \to \mathbb{R}^n$ (the actual domain of $\mathbf{f}$ can be smaller than $\mathbb{R}^d$); for example, the Veronese curve is given by the map

$$x \mapsto (x, x^2, \ldots, x^n).$$

Such a map $\mathbf{f}$ is often referred to as a parameterisation and *without loss of generality we will assume that the domain of* $\mathbf{f}$ *is* I^d *and that the manifold* $\mathcal{M} \subseteq I^n$. Locally, a manifold given by a system of equations can be parameterised by some map $\mathbf{f}$ and, conversely, if a manifold is immersed by a map $\mathbf{f}$, it can be written using a system of $n-d$ equations, where d is the dimension of the manifold.

Exercise: Parameterise the upper hemisphere $x^2 + y^2 + z^2 = 1$, $z > 0$, and also write the Veronese curve (see above) by a system of equations.

In these notes we will mainly concentrate on the simultaneous (rather than dual) theory of Diophantine approximation on manifolds. In particular, we consider the following two natural problems:

Problem 1. To develop a Lebesgue theory for $\mathcal{M} \cap W(n, \psi)$;

Problem 2. To develop a Hausdorff theory for $\mathcal{M} \cap W(n, \psi)$.

In short, the aim is to establish analogues of the two fundamental theorems of Khintchine and Jarník, and thereby provide a complete measure theoretic description of the sets $\mathcal{M} \cap W(n, \psi)$. The fact that the points $\mathbf{x} \in \mathbb{R}^n$ of interest are of dependent variables, which reflects the fact that $\mathbf{x} \in \mathcal{M}$, introduces major difficulties in attempting to describe the measure theoretic structure of $\mathcal{M} \cap W(n, \psi)$. This is true even in the specific case that $\mathcal{M}$ is a planar curve. More to the point, even for seemingly simple curves such as the unit circle or the parabola, the above problems are fraught with difficulties. In these notes we will concentrate mainly on describing the Lebesgue theory.

Unless stated otherwise, the approximating function $\psi : \mathbb{N} \to \mathbb{R}^+$ *throughout this section is assumed to be monotonic.*

1.6.1 The Lebesgue Theory for Manifolds

The goal is to obtain a Khintchine-type theorem that describes the Lebesgue measure of the set $\mathcal{M} \cap W(n, \psi)$ of simultaneously ψ-approximable points lying on $\mathcal{M}$. First of all, notice that if the dimension d of the manifold $\mathcal{M}$ is strictly less than n then $m_n(\mathcal{M} \cap W(n, \psi)) = 0$ irrespective of the approximating function ψ. Thus, in attempting to develop a Lebesgue theory for $\mathcal{M} \cap W(n, \psi)$ it is natural to use the induced d-dimensional Lebesgue measure on $\mathcal{M}$. Alternatively, if $\mathcal{M}$ is immersed by a map $\mathbf{f} : \mathrm{I}^d \to \mathbb{R}^n$ we use the d-dimensional Lebesgue measure m_d on the set of parameters of $\mathbf{f}$; namely I^d. In either case, the measure under consideration will be denoted by $|.|_{\mathcal{M}}$.

Remark 1.6.1 Notice that for $\tau \leq 1/n$, we have that $|\mathcal{M} \cap W(n, \tau)|_{\mathcal{M}} = |\mathcal{M}|_{\mathcal{M}} :=$ FULL as it should be since, by Dirichlet's theorem, we have that $W(n, \tau) = \mathrm{I}^n$.

The two-dimension fiber problem considered in §1.4.5, in which the manifold $\mathcal{M}$ is a vertical line L_α, shows that it is not possible to obtain a Khintchine-type theorem (both the convergence and divergence aspects) for all manifolds. Indeed, the convergence statement fails for vertical lines. Thus, in a quest for developing a general Khintchine-type theory for manifolds (cf. Problem 1 above), it is natural to avoid lines and more generally hyperplanes. In short, we insist that the manifold under consideration is 'sufficiently' curved.

1.6.1.1 Non-Degenerate Manifolds

In order to make any reasonable progress with Problems 1 and 2 above, we assume that the manifolds $\mathcal{M}$ under consideration are **non-degenerate** [67]. Essentially, these are smooth sub-manifolds of $\mathbb{R}^n$ which are sufficiently curved so as to deviate from any hyperplane. Formally, a manifold $\mathcal{M}$ of dimension d embedded in $\mathbb{R}^n$ is said to be non-degenerate if it arises from a non-degenerate map $\mathbf{f} : U \to \mathbb{R}^n$ where U is an open subset of $\mathbb{R}^d$ and $\mathcal{M} := \mathbf{f}(U)$. The map $\mathbf{f} : U \to \mathbb{R}^n : \mathbf{x} \mapsto \mathbf{f}(\mathbf{x}) = (f_1(\mathbf{x}), \ldots, f_n(\mathbf{x}))$ is said to be *non-degenerate at* $\mathbf{x} \in U$ if there exists some $l \in \mathbb{N}$ such that $\mathbf{f}$ is l times continuously differentiable on some sufficiently small ball centred at $\mathbf{x}$

and the partial derivatives of $\mathbf{f}$ at $\mathbf{x}$ of orders up to l span $\mathbb{R}^n$. The map $\mathbf{f}$ is *non-degenerate* if it is non-degenerate at almost every (in terms of d-dimensional Lebesgue measure) point in U; in turn the manifold $\mathcal{M} = \mathbf{f}(U)$ is also said to be non-degenerate. Any real, connected analytic manifold not contained in any hyperplane of $\mathbb{R}^n$ is non-degenerate. Indeed, if $\mathcal{M}$ is immersed by an analytic map $\mathbf{f} = (f_1, \ldots, f_n) : U \to \mathbb{R}^n$ defined on a ball $U \subset \mathbb{R}^d$, then $\mathcal{M}$ is non-degenerate if and only if the functions $1, f_1, \ldots, f_n$ are linearly independent over $\mathbb{R}$.

Without loss of generality, we will assume that U is I^d and that the manifold $\mathcal{M} \subseteq I^n$

Note that in the case the manifold $\mathcal{M}$ is a planar curve $\mathcal{C}$, a point on $\mathcal{C}$ is non-degenerate if the curvature at that point is non-zero. Thus, $\mathcal{C}$ is a non-degenerate planar curve if the set of points on $\mathcal{C}$ at which the curvature vanishes is a set of one-dimensional Lebesgue measure zero. Moreover, it is not difficult to show that the set of points on a planar curve at which the curvature vanishes but the curve is non-degenerate is at most countable. In view of this, the curvature completely describes the non-degeneracy of planar curves. Clearly, a straight line is degenerate everywhere.

The claim is that the notion of non-degeneracy is the right description for a manifold $\mathcal{M}$ to be 'sufficiently' curved in order to develop a general Khintchine-type theory (both convergent and divergent cases) for $\mathcal{M} \cap W(n, \psi)$. With this in mind, the key then lies in understanding the distribution of rational points 'close' to such manifolds.

1.6.1.2 Rational Points Near Manifolds: The Heuristics

Given a point $\mathbf{x} = (x_1, \ldots, x_n) \in \mathbb{R}^n$ and a set $A \subseteq \mathbb{R}^n$, let

$$\operatorname{dist}(\mathbf{x}, A) := \inf\{d(\mathbf{x}, \mathbf{a}) : \mathbf{a} \in A\}$$

where, as usual, $d(\mathbf{x}, \mathbf{a}) := \max_{1 \le i \le n} |x_i - a_i|$. Now let $\mathbf{x} \in \mathcal{M} \cap W(n, \psi)$. Then, by definition, there exist infinitely many $q \in \mathbb{N}$ and $\mathbf{p} \in \mathbb{Z}^n$ such that

$$\operatorname{dist}\Big(\mathcal{M}, \frac{\mathbf{p}}{q}\Big) \le d\Big(\mathbf{x}, \frac{\mathbf{p}}{q}\Big) < \frac{\psi(q)}{q}.$$

This means that the rational points

$$\frac{\mathbf{p}}{q} := \Big(\frac{p_1}{q}, \ldots, \frac{p_n}{q}\Big)$$

of interest must lie within the $\frac{\psi(q)}{q}$-neighbourhood of $\mathcal{M}$. In particular, assuming that ψ is decreasing, we have that the points $\mathbf{p}/q$ of interest with $k^{t-1} < q \le k^t$ are contained in the $\frac{\psi(k^{t-1})}{k^{t-1}}$-neighbourhood of $\mathcal{M}$. Let us denote this neighbourhood by $\Delta_k^+(t,\psi)$ and by $N_k^+(t,\psi)$ the set of rational points with $k^{t-1} < q \le k^t$ contained in $\Delta_k^+(t,\psi)$. In other words,

$$N_k^+(t,\psi) := \left\{\mathbf{p}/q \in \mathrm{I}^n : k^{t-1} < q \le k^t \text{ and } \operatorname{dist}\big(\mathcal{M}, \mathbf{p}/q\big) \le \tfrac{\psi(k^{t-1})}{k^{t-1}}\right\}. \tag{1.58}$$

Recall, that $\mathcal{M} \subseteq \mathrm{I}^n$. Hence, regarding the n-dimensional volume of the neighbourhood $\Delta_k^+(t,\psi)$, it follows that

$$m_n\Big(\Delta_k^+(t,\psi)\Big) \asymp \left(\frac{\psi(k^{t-1})}{k^{t-1}}\right)^{n-d}.$$

Now let $Q_k(t)$ denote the set of rational points with $k^{t-1} < q \le k^t$ lying in the unit cube I^n. Then,

$$\#Q_k(t) \asymp (k^t)^{n+1}$$

and if we assume that the points in $Q_k(t)$ are 'fairly' distributed within I^n, we would expect that

the number of these points that fall into $\Delta_k^+(t,\psi)$
is proportional to the measure of $\Delta_k^+(t,\psi)$.

In other words, and more formally, under the above distribution assumption, we would expect that

$$\#\{Q_k(t) \cap \Delta_k^+(t,\psi)\} \asymp \#Q_k(t) \times m_n\Big(\Delta_k^+(t,\psi)\Big) \tag{1.59}$$

and since the left-hand side is $\#N_k^+(t,\psi)$, we would be able to conclude that

$$\#N_k^+(t,\psi) \asymp (k^t)^{n+1}\left(\frac{\psi(k^{t-1})}{k^{t-1}}\right)^{n-d} \asymp (k^{t-1})^{d+1}\psi(k^{t-1})^{n-d}. \tag{1.60}$$

For the moment, let us assume that (1.59) and hence (1.60) are fact. Now

$$\begin{aligned}\mathcal{M} \cap W(n,\psi) &= \bigcap_{m=1}^{\infty}\bigcup_{t=m}^{\infty}\ \bigcup_{k^{t-1}<q\le k^t}\ \bigcup_{\mathbf{p}\in\mathbb{Z}^n:\mathbf{p}/q\in\mathrm{I}^n} B\Big(\tfrac{\mathbf{p}}{q}, \tfrac{\psi(q)}{q}\Big) \cap \mathcal{M} \\ &\subset \bigcap_{m=1}^{\infty}\bigcup_{t=m}^{\infty} A_k^+(t,\psi,\mathcal{M}),\end{aligned}$$

where

$$A_k^+(t,\psi,\mathcal{M}) := \bigcup_{k^{t-1}<q\le k^t}\ \bigcup_{\mathbf{p}\in\mathbb{Z}^n:\mathbf{p}/q\in\mathrm{I}^n} B\Big(\tfrac{\mathbf{p}}{q}, \tfrac{\psi(k^{t-1})}{k^{t-1}}\Big) \cap \mathcal{M}.$$

It is easily verified that

$$|A_k^+(t,\psi,\mathcal{M})|_{\mathcal{M}} \le \sum_{k^{t-1}<q\le k^t}\ \sum_{\mathbf{p}\in\mathbb{Z}^n:\mathbf{p}/q\in\mathrm{I}^n} \underbrace{\left|B\left(\tfrac{\mathbf{p}}{q},\tfrac{\psi(k^{t-1})}{k^{t-1}}\right)\cap\mathcal{M}\right|_{\mathcal{M}}}_{\ll(\psi(k^{t-1})/k^{t-1})^d}$$

$$\ll \#N_k^+(t,\psi)\ (\psi(k^{t-1})/k^{t-1})^d$$

$$\overset{(1.60)}{\asymp} (k^{t-1})^{d+1}\psi(k^{t-1})^{n-d}(\psi(k^{t-1})/k^{t-1})^d$$

$$\asymp k^{t-1}\psi(k^{t-1})^n .$$

Hence

$$\sum_{t=1}^{\infty}|A_k^+(t,\psi,\mathcal{M})|_{\mathcal{M}} \ll \sum_{t=1}^{\infty}k^t\psi(k^t)^n \asymp \sum_{q=1}^{\infty}\psi(q)^n . \tag{1.61}$$

All the steps in the above argument apart from (1.59) and hence (1.60), can be turned into a rigorous proof. Indeed, the estimate (1.60) is not always true.

Exercise. Consider the circle $\mathcal{C}_{\sqrt{3}}$ in $\mathbb{R}^2$ given by the equation $x^2+y^2=3$. Prove that $\mathcal{C}$ does not contain any rational points. Next let $\psi(q)=q^{-1-\varepsilon}$ for some $\varepsilon>0$. Prove that

$$\mathcal{C}_{\sqrt{3}}\cap W(2,\psi)=\varnothing .$$

The upshot is that even for non-degenerate manifolds, we cannot expect the heuristic estimate (1.60) to hold for any decreasing ψ – some restriction on the rate at which ψ decreases to zero is required. On the other hand, affine subspaces of $\mathbb{R}^n$ may contain too many rational points; for instance, if $\mathcal{M}$ is a linear subspace of $\mathbb{R}^n$ with a basis of rational vectors. Of course, such manifolds are not non-degenerate.

However, *whenever the upper bound associated with the heuristic estimate* (1.60) *is true*, inequality (1.61) together with the convergence Borel–Cantelli lemma implies that

$$|\mathcal{M}\cap W(n,\psi)|_{\mathcal{M}}=0 \quad\text{if}\quad \sum_{q=1}^{\infty}\psi(q)^n<\infty .$$

This statement represents the convergent case of the 'Dream Theorem' for manifolds – see §1.6.1.3 immediately below. Note that the associated sum

$\sum \psi(q)^n$ coincides with the sum appearing in Theorem 1.4.11 (Khintchine in $\mathbb{R}^n$) but the associated measure $|.|_{\mathcal{M}}$ is d-dimensional Lebesgue measure (induced on $\mathcal{M}$) rather than n-dimensional Lebesgue measure.

1.6.1.3 The Dream Theorem and Its Current Status

The Dream Theorem. Let $\mathcal{M}$ be a non-degenerate sub-manifold of $\mathbb{R}^n$. Let $\psi : \mathbb{N} \to \mathbb{R}^+$ be a monotonic function. Then

$$|\mathcal{M} \cap W(n, \psi)|_{\mathcal{M}} = \begin{cases} 0 & \text{if } \sum_{q=1}^{\infty} \psi(q)^n < \infty \, , \\ \\ 1 & \text{if } \sum_{q=1}^{\infty} \psi(q)^n = \infty \, . \end{cases} \tag{1.62}$$

We emphasise that the Dream Theorem is a desired statement rather than an established fact.

As we have already demonstrated, the convergence case of the Dream Theorem would follow on establishing the upper bound estimate

$$\#N_k^+(t, \psi) \ll (k^{t-1})^{d+1} \psi(k^{t-1})^{n-d} \tag{1.63}$$

for non-degenerate manifolds. Recall that the rational points of interest are given by the set

$$N_k(t, \psi) := \left\{ \mathbf{p}/q \in \mathrm{I}^n : k^{t-1} < q \le k^t \text{ and } \operatorname{dist}(\mathcal{M}, \mathbf{p}/q) \le \tfrac{\psi(q)}{q} \right\} ,$$

and that $\#N_k^+(t, \psi)$ is an upper bound for $\#N_k(t, \psi)$. Obviously, a lower bound for $\#N_k(t, \psi)$ is given by $\#N_k^-(t, \psi)$ where

$$N_k^-(t, \psi) := \left\{ \mathbf{p}/q \in \mathrm{I}^n : k^{t-1} < q \le k^t \text{ and } \operatorname{dist}(\mathcal{M}, \mathbf{p}/q) \le \tfrac{\psi(k^t)}{k^t} \right\} ,$$

and if ψ is k-regular (see (1.54)) then $N_k^+(t, \psi) \asymp N_k^-(t, \psi)$. In particular, whenever we are able to establish the heuristic estimate (1.60) or equivalently the upper bound estimate (1.63) together with the lower bound estimate

$$\#N_k^-(t, \psi) \gg (k^{t-1})^{d+1} \psi(k^{t-1})^{n-d} , \tag{1.64}$$

we would have that

$$\#N_k(t, \psi) \asymp (k^{t-1})^{d+1} \psi(k^{t-1})^{n-d} . \tag{1.65}$$

It is worth stressing that the lower bound estimate (1.64) is by itself not enough to prove the divergence case of the Dream Theorem. Loosely speaking, we also need to know that rational points associated with $N_k^-(t, \psi)$ are 'ubiquitous' within the $\frac{\psi(k^t)}{k^t}$–neighbourhood of $\mathcal{M}$. Indeed, when establishing the divergence case of Khintchine's theorem (Theorem 1.2.5), we trivially have the right count of k^{2t} for the number of rational points $p/q \in \mathrm{I}$ with $k^{t-1} < q \leq k^t$. The crux is to establish the associated distribution type result given by Theorem 1.1.4. This in turn implies that the rational points under consideration give rise to a ubiquitous system – see §1.5.3.1.

We now turn our attention to reality and describe various 'general' contributions towards the Dream Theorem.

- *Extremal manifolds.* A sub-manifold $\mathcal{M}$ of $\mathbb{R}^n$ is called *extremal* if

$$\left|\mathcal{M} \cap W(n, \tfrac{1+\varepsilon}{n})\right|_{\mathcal{M}} = 0 \qquad \forall\, \varepsilon > 0\,.$$

Note that $\mathcal{M} \cap W(n, \frac{1}{n}) = \mathcal{M}$ – see Remark 1.6.1. In their pioneering work [67] published in 1998, Kleinbock and Margulis proved that any non-degenerate sub-manifold $\mathcal{M}$ of $\mathbb{R}^n$ is extremal. It is easy to see that this implies the convergence case of the Dream Theorem for functions of the shape

$$\psi_\varepsilon(q) := q^{-\frac{1+\varepsilon}{n}}\,.$$

Indeed,

$$\textstyle\sum_{q=1}^{\infty} \psi_\varepsilon(q)^n = \sum_{q=1}^{\infty} q^{-(1+\varepsilon)} < \infty$$

and so whenever the convergent case of (1.62) is fulfilled, the corresponding manifold is extremal.

- *Planar curves.* The Dream Theorem is true when $n = 2$; that is, when $\mathcal{M}$ is a non-degenerate planar curve. The convergence case of (1.62) for planar curves was established in [91] and subsequently strengthened in [30]. The divergence case of (1.62) for planar curves was established in [15].
- *Beyond planar curves.* The divergence case of the Dream Theorem is true for analytic non-degenerate sub-manifolds of $\mathbb{R}^n$ [11]. Recently, the divergence case of (1.62) has been shown to be true for non-degenerate curves and manifolds that can be 'fibred' into such curves [20]. The latter includes C^∞ non-degenerate sub-manifolds of $\mathbb{R}^n$ which are not necessarily analytic. The convergence case of the Dream Theorem is true for a large subclass of 2-non-degenerate sub-manifolds of $\mathbb{R}^n$ with dimension d strictly greater than $(n+1)/2$ [19]. Earlier, manifolds satisfying a geometric (curvature) condition were shown to satisfy the convergence case of the Dream Theorem [47].

The upshot of the above is that the Dream Theorem is in essence fact for a fairly generic class of non-degenerate sub-manifolds $\mathcal{M}$ of $\mathbb{R}^n$ apart from the case of convergence when $n \geq 3$ and $d \leq (n+1)/2$.

Remark 1.6.2 The theory of Diophantine approximation stems from Mahler's problem (1932) regarding the *extremality* of the Veronese curve

$$\mathcal{V} := \{(x, x^2, \ldots, x^n) : x \in \mathbb{R}\}.$$

Following a substantial number of partial results (initially for $n = 2$, then $n = 3$ and some for higher n), a complete solution to the problem was given by Sprindžuk in 1965. For a historical account of the manifold theory we refer the reader to the monographs [31, 90] and the introduction given in the paper [15].

Remark 1.6.3 Note that, in view of Khintchine's Transference Principle, we could have easily defined extremality via the dual form of Diophantine approximation (see Remark 1.4.39); namely, $\mathcal{M}$ is extremal if

$$\left|\mathcal{M} \cap W^*(n, n+\varepsilon)\right|_{\mathcal{M}} = 0 \qquad \forall\, \varepsilon > 0\,.$$

The point is that both definitions are equivalent. This is not the case in the inhomogeneous set-up considered in §1.6.3.1.

Remark 1.6.4 It is worth mentioning that in [67], Kleinbock and Margulis established a stronger (multiplicative) form of extremality (see §1.6.4.1) that settled the Baker–Sprindžuk conjecture from the eighties. Not only did their work solve a long-standing fundamental problem, but it also developed new techniques utilising the link between Diophantine approximation and homogeneous dynamics. Without doubt the work of Kleinbock and Margulis has been the catalyst for the subsequent contributions towards the Dream Theorem described above.

1.6.2 The Hausdorff Theory for Manifolds

The goal is to obtain a Jarník-type theorem that describes the Hausdorff measure $\mathcal{H}^s$ of the set $\mathcal{M} \cap W(n, \psi)$ of simultaneously ψ-approximable points lying on $\mathcal{M}$. In other words, we wish to obtain a Hausdorff measure version of the Dream Theorem. In view of this, by default, we consider approximating functions ψ which decrease sufficiently rapidly so that the d-dimensional Lebesgue measure of $\mathcal{M} \cap W(n, \psi)$ is zero. Now, as the example in §1.6.1.2 demonstrates, in order to obtain a coherent Hausdorff measure theory we must impose some restriction on the rate at which ψ decreases. Indeed, with reference to that example, the point is that $\mathcal{H}^s(\mathcal{C}_{\sqrt{3}} \cap W(2, 1+\varepsilon)) = 0$ irrespective of $\varepsilon > 0$ and the measure $\mathcal{H}^s$. On the other hand, for the unit circle $\mathcal{C}_1$ in $\mathbb{R}^2$ given by the equation $x^2 + y^2 = 1$, it can be shown [14, Theorem 19] that for any $\varepsilon > 0$,

$$\mathcal{H}^s(\mathcal{C}_1 \cap W(2, 1+\varepsilon)) = \infty \quad \text{with} \quad s = \tfrac{1}{2+\varepsilon}.$$

Nevertheless, it is believed that if the rate of decrease of ψ is 'close' to the approximating function $q^{-1/n}$ associated with Dirichlet's theorem, then the behaviour of $\mathcal{H}^s(\mathcal{M} \cap W(n,\psi))$ can be captured by a single, general criterion. In the following statement, the condition on ψ is captured in terms of the deviation of $\mathcal{H}^s$ from d-dimensional Lebesgue measure.

The Hausdorff Dream Theorem. Let $\mathcal{M}$ be a non-degenerate sub-manifold of $\mathbb{R}^n$, $d := \dim \mathcal{M}$ and $m := \operatorname{codim} \mathcal{M}$. Thus, $d + m = n$. Let $\psi : \mathbb{N} \to \mathbb{R}^+$ be a monotonic function. Then, for any $s \in (\frac{m}{m+1}d, d)$

$$\mathcal{H}^s(\mathcal{M} \cap W(n,\psi)) = \begin{cases} 0 & \text{if } \sum_{q=1}^{\infty} \psi^{s+m}(q) q^{-s+d} < \infty\,, \\ \infty & \text{if } \sum_{q=1}^{\infty} \psi^{s+m}(q) q^{-s+d} = \infty. \end{cases} \tag{1.66}$$

We emphasize that the above is a desired statement rather than an established fact.

We now turn our attention to reality and describe various 'general' contributions towards the Hausdorff Dream Theorem.

- *Planar curves.* As with the Dream Theorem, the convergence case of (1.66) for planar curves ($n = 2, d = m = 1$) was established in [91] and subsequently strengthened in [30]. The divergence case of (1.66) for planar curves was established in [15].
- *Beyond planar curves.* The divergence case of the Hausdorff Dream Theorem is true for analytic non-degenerate sub-manifolds of $\mathbb{R}^n$ [11]. The convergence case is rather fragmented. To the best of our knowledge, the partial results obtained in [19, Corollaries 3 & 5] for 2-non-degenerate sub-manifolds of $\mathbb{R}^n$ with dimension d strictly greater than $(n+1)/2$, represent the first significant coherent contribution towards the convergence case.

Exercise. Prove the convergent case of (1.66) assuming the heuristic estimate (1.60) for the number of rational points near $\mathcal{M}$ – see §1.6.1.2.

Remark 1.6.5 Regarding the divergence case of (1.66), it is tempting to claim that it follows from the divergence case of the (Lebesgue) Dream Theorem via the Mass Transference Principle introduced in §1.3.4. After all, this is true when $\mathcal{M} = \mathrm{I}^n$; namely that Khintchine's theorem implies Jarník's theorem as demonstrated in §1.3.4.1. However, this is far from the truth within the context of manifolds. The reason for this is simple. With respect to the set-up of the

Mass Transference Principle, the set Ω that supports the $\mathcal{H}^{\delta}$-measure (with $\delta = \dim \mathcal{M}$) is the manifold $\mathcal{M}$ itself and is embedded in $\mathbb{R}^n$. The set $\mathcal{M} \cap W(n, \psi) \subset \Omega$ of interest can be naturally expressed as the intersection with $\mathcal{M}$ of the lim sup set arising from balls $B(\frac{\mathbf{p}}{q}, \frac{\psi(q)}{q})$ centred at rational points $\mathbf{p}/q \in \mathbb{R}^n$. However, the centre of these balls do not necessarily lie in the support of the measure $\Omega = \mathcal{M}$ and this is where the problem lies. A prerequisite for the framework of the Mass Transference Principle is that $\{B_i\}_{i\in\mathbb{N}}$ is a sequence of balls in Ω.

1.6.3 Inhomogeneous Diophantine Approximation

When considering the well approximable sets $W(n, \psi)$ or indeed the badly approximable sets $\mathbf{Bad}(i_1, \ldots, i_n)$, we are in essence investigating the behaviour of the fractional part of $q\mathbf{x}$ about the origin as q runs through $\mathbb{N}$. Clearly, we could consider the set-up in which we investigate the behaviour of the orbit of $\{q\mathbf{x}\}$ about some other point. With this in mind, given $\psi : \mathbb{N} \to \mathbb{R}^+$ and a fixed point $\boldsymbol{\gamma} = (\gamma_1, \ldots, \gamma_n) \in \mathbb{R}^n$, let

$$W_{\boldsymbol{\gamma}}(n, \psi) := \{\mathbf{x} \in \mathrm{I}^n : \|q\mathbf{x} - \boldsymbol{\gamma}\| < \psi(q) \text{ for infinitely many } q \in \mathbb{N}\}$$

denote the *inhomogeneous* set of all *simultaneously ψ-well approximable* points $\mathbf{x} \in \mathrm{I}^n$. Thus, a point $\mathbf{x} \in W_{\boldsymbol{\gamma}}(n, \psi)$ if there exist infinitely many 'shifted' rational points

$$\left(\frac{p_1 - \gamma_1}{q}, \ldots, \frac{p_n - \gamma_n}{q}\right)$$

with $q > 0$, such that the inequalities

$$|x_i - (p_i - \gamma_i)/q| < \psi(q)/q$$

are simultaneously satisfied for $1 \leq i \leq n$. The following is the natural generalisation of the simultaneous Khintchine–Jarník theorem to the inhomogeneous set-up. For further details, see [13, 14] and references within.

Theorem 1.6.6 (Inhomogeneous Khintchine–Jarník) *Let $\psi : \mathbb{N} \to \mathbb{R}^+$ be a monotonic function, $\boldsymbol{\gamma} \in \mathbb{R}^n$ and $s \in (0, n]$. Then*

$$\mathcal{H}^s(W_{\boldsymbol{\gamma}}(n, \psi)) = \begin{cases} 0 & \text{if } \sum_{r=1}^{\infty} r^{n-s}\psi(r)^s < \infty\,, \\[2ex] \mathcal{H}^s(I^n) & \text{if } \sum_{r=1}^{\infty} r^{n-s}\psi(r)^s = \infty\,. \end{cases}$$

Remark 1.6.7 For the sake of completeness we state the inhomogeneous analogue of Hurwitz's Theorem due to Khintchine [62, §10.10]: *for any irrational* $x \in \mathbb{R}$, $\gamma \in \mathbb{R}$ *and* $\varepsilon > 0$, *there exist infinitely many integers* $q > 0$ *such that*

$$q\,\|qx - \gamma\| \leq (1+\varepsilon)/\sqrt{5}.$$

Note that the presence of the ε term means that the inhomogeneous statement is not quite as sharp as the homogeneous one (i.e. when $\gamma = 0$). Also, for obvious reasons, in the inhomogeneous situation it is necessary to exclude the case that x is rational.

We now swiftly move on to the inhomogeneous theory for manifolds. In short, the heuristics of §1.6.1.2, adapted to the inhomogeneous set-up, gives evidence towards the following natural generalisation of the Dream Theorem.

The Inhomogeneous Dream Theorem. Let $\mathcal{M}$ be a non-degenerate submanifold of $\mathbb{R}^n$. Let $\psi : \mathbb{N} \to \mathbb{R}^+$ be a monotonic function and $\boldsymbol{\gamma} \in \mathbb{R}^n$. Then

$$|\mathcal{M} \cap W_{\boldsymbol{\gamma}}(n,\psi)|_{\mathcal{M}} = \begin{cases} 0 & \text{if } \sum_{q=1}^{\infty} \psi(q)^n < \infty\,, \\[2ex] 1 & \text{if } \sum_{q=1}^{\infty} \psi(q)^n = \infty\,. \end{cases}$$

Regarding what is known, the current state of knowledge is absolutely in line with the homogeneous situation. The inhomogeneous analogue of the extremality result of Kleinbock and Margulis [67] is established in [24, 26]. We will return to this in §1.6.3.1 below. For planar curves, the Inhomogeneous Dream Theorem is established in [18]. Beyond planar curves, the results in [19, 20] are obtained within the inhomogeneous framework. So in summary, the Inhomogeneous Dream Theorem is in essence fact for non-degenerate submanifolds $\mathcal{M}$ of $\mathbb{R}^n$ apart from the case of convergence when $n \geq 3$ and $d \leq (n+1)/2$.

1.6.3.1 Inhomogeneous Extremality and a Transference Principle

First we need to decide on what precisely we mean by inhomogeneous extremality. With this in mind, a manifold $\mathcal{M}$ is said to be *simultaneously inhomogeneously extremal* (SIE for short) if, for every $\boldsymbol{\gamma} \in \mathbb{R}^n$,

$$\Big|\mathcal{M} \cap W_{\boldsymbol{\gamma}}(n, \tfrac{1+\varepsilon}{n})\Big|_{\mathcal{M}} = 0 \qquad \forall\, \varepsilon > 0\,. \tag{1.67}$$

On the other hand, a manifold $\mathcal{M}$ is said to be *dually inhomogeneously extremal* (DIE for short) if, for every $\gamma \in \mathbb{R}$,

$$\Big|\mathcal{M} \cap W^*_{\gamma}(n, n+\varepsilon)\Big|_{\mathcal{M}} = 0 \qquad \forall\, \varepsilon > 0\,.$$

Here, given $\tau > 0$ and a fixed point $\gamma \in \mathbb{R}$, $W^*_\gamma(n, \tau)$ is the *inhomogeneous* set of *dually τ-well approximable* points consisting of points $\mathbf{x} \in \mathrm{I}^n$ for which the inequality

$$\|\mathbf{q} \cdot \mathbf{x} - \gamma\| < |\mathbf{q}|^{-\tau}$$

holds for infinitely many $\mathbf{q} \in \mathbb{Z}^n$. Moreover, a manifold $\mathcal{M}$ is simply said to be *inhomogeneously extremal* if it is both SIE and DIE.

As mentioned in Remark 1.6.3, in the homogeneous case ($\boldsymbol{\gamma}$=0) the simultaneous and dual forms of extremality are equivalent. Recall that this is a simply consequence of Khintchine's Transference Principle (Theorem 1.4.38). However, in the inhomogeneous case, there is no classical transference principle that allows us to deduce SIE from DIE and vice versa. The upshot is that the two forms of inhomogeneous extremality have to be treated separately. It turns out that establishing the dual form of inhomogeneous extremality is technically far more complicated than establishing the simultaneous form [26]. The framework developed in [24] naturally incorporates both forms of inhomogeneous extremality and indeed other stronger (multiplicative) notions associated with the inhomogeneous analogue of the Baker–Sprindžuk conjecture.

Conjecture. *Let $\mathcal{M}$ be a non-degenerate sub-manifold of $\mathbb{R}^n$. Then $\mathcal{M}$ is inhomogeneously extremal.*

The proof given in [24] of this inhomogeneous conjecture relies very much on the fact that we know that the homogeneous statement is true. In particular, the general inhomogeneous transference principle of [24, §5] enables us to establish the following transference for non-degenerate manifolds:

$$\mathcal{M} \text{ is extremal} \iff \mathcal{M} \text{ is inhomogeneously extremal.} \tag{1.68}$$

Clearly, this enables us to conclude that:

$$\mathcal{M} \text{ is SIE} \iff \mathcal{M} \text{ is DIE.}$$

In other words, a transference principle between the two forms of inhomogeneous extremality does exist at least for the class of non-degenerate manifolds.

Trivially, inhomogeneous extremality implies (homogeneous) extremality. Thus, the main substance of (1.68) is the reverse implication. This rather surprising fact relies on the fact that the inhomogeneous lim sup sets $\mathcal{M} \cap W_{\boldsymbol{\gamma}}(n, \frac{1+\varepsilon}{n})$ and the induced measure $|.|_{\mathcal{M}}$ on non-degenerate manifolds satisfy the *intersection property* and the *contracting property* described in [24,

§5]. These properties are at the heart of the Inhomogeneous Transference Principle [24, Theorem 5] that enables us to transfer zero measure statements for homogeneous lim sup sets to inhomogeneous lim sup sets. The general set-up, although quite natural, is rather involved and will not be reproduced in these notes. Instead, we refer the reader to the papers [24, 26]. We advise the reader to first look at [26] in which the easier statement

$$\mathcal{M} \text{ is extremal} \quad \Longrightarrow \quad \mathcal{M} \text{ is SIE} \tag{1.69}$$

is established. This has the great advantage of bringing to the forefront the main ideas of [24] while omitting the abstract and technical notions that come with describing the inhomogeneous transference principle in all its glory. In order to illustrate the basic line of thinking involved in establishing (1.69) and indeed (1.68) we shall prove the following statement concerning extremality on $\mathrm{I} = [0, 1]$:

$$m(W(1+\varepsilon)) = 0 \forall \varepsilon > 0 \quad \Longrightarrow \quad m(W_\gamma(1+\varepsilon)) = 0 \ \forall\, \varepsilon > 0. \tag{1.70}$$

Of course, it is easy to show that the inhomogeneous set $W_\gamma(1+\varepsilon)$ is of zero Lebesgue measure m by using the convergence Borel–Cantelli lemma. However, the point here is to develop an argument that exploits the fact that we know the homogeneous set $W_0(1+\varepsilon) := W(1+\varepsilon)$ is of zero Lebesgue measure.

To prove (1.70), we make use of the fact that $W_\gamma(1+\varepsilon)$ is a lim sup set given by

$$W_\gamma(1+\varepsilon) = \bigcap_{s=1}^{\infty} \bigcup_{q=s}^{\infty} \bigcup_{p\in\mathbb{Z}} B^\gamma_{p,q}(\varepsilon) \cap \mathrm{I}, \tag{1.71}$$

where, given $q \in \mathbb{N}$, $p \in \mathbb{Z}$, $\gamma \in \mathbb{R}$ and $\varepsilon > 0$

$$B^\gamma_{p,q}(\varepsilon) := \{\, y \in \mathbb{R} : |qy + p + \gamma| < |q|^{-1-\varepsilon} \,\}\,.$$

As usual, if $B = B(x, r)$ denotes the ball (interval) centred at x and of radius $r > 0$, then it is easily seen that

$$B^\gamma_{p,q}(\varepsilon) = B\Big(-\frac{p+\gamma}{q}, |q|^{-2-\varepsilon}\Big).$$

Now we consider 'blown up' balls $B^\gamma_{p,q}(\varepsilon/2)$ and observe that Lebesgue measure m satisfies the following contracting property: for any choice $q \in \mathbb{N}$, $p \in \mathbb{Z}$, $\gamma \in \mathbb{R}$ and $\varepsilon > 0$ we have that

$$m\Big(B^\gamma_{p,q}(\varepsilon)\Big) = \frac{2}{q^{2+\varepsilon}} = q^{-\frac{\varepsilon}{2}} \frac{2}{q^{2+(\varepsilon/2)}} = q^{-\frac{\varepsilon}{2}}\, m\Big(B^\gamma_{p,q}(\varepsilon/2)\Big). \tag{1.72}$$

Next we separate the balls $B^{\gamma}_{p,q}(\varepsilon)$ into classes of disjoint and non-disjoint balls. Fix $q \in \mathbb{N}$ and $p \in \mathbb{Z}$. Clearly, there exists a unique integer $t = t(q)$ such that $2^t \leq q < 2^{t+1}$. The ball $B^{\gamma}_{p,q}(\varepsilon)$ is said to be *disjoint* if for every $q' \in \mathbb{N}$ with $2^t \leq q' < 2^{t+1}$ and every $p' \in \mathbb{Z}$

$$B^{\gamma}_{p,q}(\varepsilon/2) \cap B^{\gamma}_{p',q'}(\varepsilon/2) \cap \mathrm{I} = \varnothing .$$

Otherwise, the ball $B^{\gamma}_{p,q}(\varepsilon/2)$ is said to be *non-disjoint*. This notion of disjoint and non-disjoint balls enables us to decompose the $W_\gamma(1+\varepsilon)$ into the two lim sup subsets:

$$D^{\gamma}(\varepsilon) := \bigcap_{s=0}^{\infty} \bigcup_{t=s}^{\infty} \bigcup_{2^t \leq |q| < 2^{t+1}} \bigcup_{\substack{p \in \mathbb{Z} \\ B^{\gamma}_{p,q}(\varepsilon) \text{ is disjoint}}} B^{\gamma}_{p,q}(\varepsilon) \cap \mathrm{I},$$

and

$$N^{\gamma}(\varepsilon) := \bigcap_{s=0}^{\infty} \bigcup_{t=s}^{\infty} \bigcup_{2^t \leq |q| < 2^{t+1}} \bigcup_{\substack{p \in \mathbb{Z} \\ B^{\gamma}_{p,q}(\varepsilon) \text{ is non-disjoint}}} B^{\gamma}_{p,q}(\varepsilon) \cap \mathrm{I}.$$

Formally,

$$W_\gamma(1+\varepsilon) = \bigcap_{s=1}^{\infty} \bigcup_{q=s}^{\infty} \bigcup_{p \in \mathbb{Z}} B^{\gamma}_{p,q}(\varepsilon) \cap \mathrm{I} = D^{\gamma}(\varepsilon) \cup N^{\gamma}(\varepsilon).$$

We now show that $m(D^{\gamma}(\varepsilon)) = 0 = m(N^{\gamma}(\varepsilon))$. This would clearly imply (1.70). Naturally, we deal with the disjoint and non-disjoint sets separately.

The disjoint case: By the definition of disjoint balls, for every fixed t we have that

$$\sum_{2^t \leq q < 2^{t+1}} \sum_{\substack{p \in \mathbb{Z} \\ B^{\gamma}_{p,q}(\varepsilon) \text{ is disjoint}}} m(B^{\gamma} p, q(\varepsilon/2) \cap \mathrm{I})$$
$$= m\Big(\bigcup_{2^t \leq q < 2^{t+1}} \bigcup_{\substack{p \in \mathbb{Z} \\ B^{\gamma}_{p,q}(\varepsilon) \text{ is disjoint}}} B^{\gamma}_{p,q}(\varepsilon/2) \cap \mathrm{I}\Big) \leq m(\mathrm{I}) = 1.$$

This, together with the contracting property (1.72) of the measure m, implies that

$$m\Big(\bigcup_{2^t \leq q < 2^{t+1}} \bigcup_{\substack{p \in \mathbb{Z} \\ B^{\gamma}_{p,q}(\varepsilon) \text{ is disjoint}}} B^{\gamma}_{p,q}(\varepsilon) \cap \mathrm{I}\Big) = \sum_{2^t \leq q < 2^{t+1}} \sum_{\substack{p \in \mathbb{Z} \\ B^{\gamma}_{p,q}(\varepsilon) \text{ is disjoint}}} m(B^{\gamma}_{p,q}(\varepsilon) \cap \mathrm{I})$$

$$\leq \sum_{2^t \leq q < 2^{t+1}} \sum_{\substack{p \in \mathbb{Z} \\ B^{\gamma}_{p,q}(\varepsilon) \text{ is disjoint}}} q^{-\frac{\varepsilon}{2}} \; m(B^{\gamma}_{p,q}(\varepsilon/2) \cap \mathrm{I})$$

$$\leq 2^{-t\frac{\varepsilon}{2}} \sum_{2^t \leq q < 2^{t+1}} \sum_{\substack{p \in \mathbb{Z} \\ B^{\gamma}_{p,q}(\varepsilon) \text{ is disjoint}}} m(B^{\gamma}_{p,q}(\varepsilon/2) \cap \mathrm{I})$$

$$\leq 2^{-t\frac{\varepsilon}{2}}.$$

Since $\sum_{t=1}^{\infty} 2^{-t\frac{\varepsilon}{2}} < \infty$, the convergence Borel–Cantelli lemma implies that

$$m(D^{\gamma}(\varepsilon)) = 0.$$

The non-disjoint case: Let $B^{\gamma}_{p,q}(\varepsilon)$ be a non-disjoint ball and let $t = t(q)$ be as above. Clearly

$$B^{\gamma}_{p,q}(\varepsilon) \subset B^{\gamma}_{p,q}(\varepsilon/2)\,.$$

By the definition of non-disjoint balls, there is another ball $B^{\gamma}_{p',q'}(\varepsilon/2)$ with $2^t \leq q < 2^{t+1}$ such that

$$B^{\gamma}_{p,q}(\varepsilon/2) \cap B^{\gamma}_{\mathbf{p}',q'}(\varepsilon/2) \cap \mathrm{I} \neq \varnothing\,. \tag{1.73}$$

It is easily seen that $q' \neq q$, as otherwise we would have that $B^{\gamma}_{p,q}(\varepsilon/2) \cap B^{\gamma}_{p',q}(\varepsilon/2) = \varnothing$. The point here is that rationals with the same denominator q are separated by $1/q$. Take any point y in the non-empty set appearing in (1.73). By the definition of $B^{\gamma}_{p,q}(\varepsilon/2)$ and $B^{\gamma}_{p',q'}(\varepsilon/2)$, it follows that

$$|qy + p + \gamma| \;<\; q^{-1-\frac{\varepsilon}{2}} \;\leq\; 2^{t(-1-\frac{\varepsilon}{2})}$$

and

$$|q'y + p' + \gamma| \;<\; (q')^{-1-\frac{\varepsilon}{2}} \;\leq\; 2^{t(-1-\frac{\varepsilon}{2})}\,.$$

On combining these inequalities in the obvious manner and assuming without loss of generality that $q > q'$, we deduce that

$$|\underbrace{(q-q')}_{q''}y + \underbrace{(p-p')}_{p''}| \;<\; 2 \cdot 2^{t(-1-\frac{\varepsilon}{2})} \;<\; 2^{(t+2)(-1-\frac{\varepsilon}{3})} \tag{1.74}$$

for all t sufficiently large. Furthermore, $0 < q'' \leq 2^{t+2}$, which together with (1.74) yields that

$$|q''y + p''| \;<\; (q'')^{-1-\frac{\varepsilon}{3}}\,.$$

If the latter inequality holds for infinitely many different $q'' \in \mathbb{N}$, then $y \in W(1+\varepsilon/3)$. Otherwise, there is a fixed pair $(p'', q'') \in \mathbb{Z} \times \mathbb{N}$ such that (1.74) is satisfied for infinitely many t. Thus, we must have that $q''y + p'' = 0$ and so y is a rational point. The upshot of the non-disjoint case is that

$$N^{\gamma}(\varepsilon) \subset W(1+\varepsilon/3) \cup \mathbb{Q}.$$

However, we are given that the homogeneous set $W(1 + \varepsilon/3)$ is of measure zero and since $\mathbb{Q}$ is countable, it follows that

$$m(N^{\gamma}(\varepsilon)) = 0.$$

This completes the proof of (1.70).

1.6.4 The Inhomogeneous Multiplicative Theory

For completeness, we include a short section surveying recent striking developments in the theory of inhomogeneous multiplicative Diophantine approximation. Nevertheless, we start by highlighting the fact that there remain gaping holes in the theory.

Given $\psi : \mathbb{N} \to \mathbb{R}^+$ and a fixed point $\boldsymbol{\gamma} = (\gamma_1, \ldots, \gamma_n) \in \mathbb{R}^n$, let

$$W_{\boldsymbol{\gamma}}^{\times}(n,\psi) := \{\mathbf{x} \in \mathrm{I}^n : \|qx_1 - \gamma_1\| \ldots \|qx_n - \gamma_n\| < \psi(q) \quad \text{for infinitely many } q \in \mathbb{N}\} \tag{1.75}$$

denote the *inhomogeneous* set of all *multiplicatively ψ-well approximable* points $\mathbf{x} \in \mathrm{I}^n$. When $\boldsymbol{\gamma} = \{\mathbf{0}\}$, the corresponding set $W_{\boldsymbol{\gamma}}^{\times}(n,\psi)$ naturally coincides with the homogeneous set $W^{\times}(n,\psi)$ given by (1.32) in §1.4.4. It is natural to ask for an inhomogeneous generalisation of Gallagher's theorem (§1.4.4, Theorem 1.4.17). A straightforward 'volume' argument making use of the lim sup nature of $W_{\boldsymbol{\gamma}}^{\times}(n,\psi)$, together with the convergence Borel–Cantelli lemma implies the following statement.

Lemma 1.6.8 (Inhomogeneous Gallagher: convergence) *Let $\psi : \mathbb{N} \to \mathbb{R}^+$ be a monotonic function and $\boldsymbol{\gamma} \in \mathbb{R}^n$. Then*

$$m_n(W_{\boldsymbol{\gamma}}^{\times}(n,\psi)) = 0 \quad \textit{if} \quad \sum_{q=1}^{\infty} \psi(q)\log^{n-1} q < \infty.$$

The context of Remark 1.4.18 remains valid in the inhomogeneous set-up; namely, we can remove the condition that ψ is monotonic if we replace the above convergence sum condition by $\sum \psi(q)|\log \psi(q)|^{n-1} < \infty$.

Surprisingly, the divergence counterpart of Lemma 1.6.8 is not known.

Conjecture 1.6.9 (Inhomogeneous Gallagher: divergence) *Let* $\psi : \mathbb{N} \to \mathbb{R}^+$ *be a monotonic function and* $\boldsymbol{\gamma} \in \mathbb{R}^n$. *Then*

$$m_n(W_{\boldsymbol{\gamma}}^{\times}(n,\psi)) = 1 \quad if \quad \sum_{q=1}^{\infty} \psi(q)\log^{n-1} q = \infty.$$

Restricting our attention to $n = 2$, it is shown in [17, Theorem 13] that the conjecture is true if given $\boldsymbol{\gamma} = (\gamma_1, \gamma_2) \in \mathbb{R}^2$, either $\gamma_1 = 0$ or $\gamma_2 = 0$. In other words, we are able to deal with the situation in which one of the two 'approximating quantities' is inhomogeneous but not both. For further details see [17, §2.2].

We now turn our attention to the Hausdorff theory. Given that the Lebesgue theory is so incomplete, it would be reasonable to have low expectations for a coherent Hausdorff theory. However, when $n = 2$, we are bizarrely in pretty good shape. To begin with, note that

$$\text{if } s \leq 1 \text{ then } \mathcal{H}^s(W_{\boldsymbol{\gamma}}^{\times}(2,\psi)) = \infty \text{ irrespective of the approximating function } \psi. \tag{1.76}$$

To see this, given $\boldsymbol{\gamma} = (\gamma_1, \gamma_2) \in \mathbb{R}^2$, we observe that for any $\alpha \in W_{\gamma_1}(1,\psi)$ the whole line $x_1 = \alpha$ within the unit interval is contained in $W_{\boldsymbol{\gamma}}^{\times}(2,\psi)$. Hence,

$$W_{\gamma_1}(1,\psi) \times \mathrm{I} \subset W_{\boldsymbol{\gamma}}^{\times}(2,\psi). \tag{1.77}$$

It is easy to verify that $W_{\gamma_1}(1,\psi)$ is an infinite set for any approximating function ψ and so (1.77) implies (1.76). Thus, when considering the s-dimensional Hausdorff measure of $W_{\boldsymbol{\gamma}}^{\times}(2,\psi)$, there is no loss of generality in assuming that $s \in (1,2]$. The following inhomogeneous multiplicative analogue of Jarník's theorem is established in [28, Theorem 1].

Theorem 1.6.10 *Let* $\psi : \mathbb{N} \to \mathbb{R}^+$ *be a monotonic function,* $\boldsymbol{\gamma} \in \mathbb{R}^2$ *and* $s \in (1,2)$. *Then*

$$\mathcal{H}^s\big(W_{\boldsymbol{\gamma}}^{\times}(2,\psi)\big) = \begin{cases} 0 & if \ \sum_{q=1}^{\infty} q^{2-s}\psi^{s-1}(q) < \infty\,, \\ \infty & if \ \sum_{q=1}^{\infty} q^{2-s}\psi^{s-1}(q) = \infty\,. \end{cases} \tag{1.78}$$

Remark 1.6.11 Recall that Gallagher's multiplicative statement and its conjectured inhomogeneous generalisation (Conjecture 1.6.9) have the extra 'log factor' in the Lebesgue 'volume' sum compared to Khintchine's simultaneous statement (Theorem 1.6.6 with $s = n = 2$). A priori, it is natural to expect the log factor to appear in one form or another when determining the Hausdorff measure $\mathcal{H}^s$ of $W_\gamma^\times(2, \psi)$ for $s \in (1, 2)$. This, as we see from Theorem 1.6.10, is very far from the truth. The 'log factor' completely disappears. Thus, genuine 'fractal' Hausdorff measures are insensitive to the multiplicative nature of $W_\gamma^\times(2, \psi)$.

Remark 1.6.12 Note that, in view of the previous remark, even if we had written $\mathcal{H}^s(\mathrm{I}^2)$ instead of ∞ in the divergence case of Theorem 1.6.10, it is still necessary to exclude the case $s = 2$.

For $n > 2$, the proof given in [28] of Theorem 1.6.10 can be adapted to show that for any $s \in (n-1, n)$

$$\mathcal{H}^s\big(W_\gamma^\times(n, \psi)\big) = 0 \qquad \text{if} \qquad \sum_{q=1}^{\infty} q^{n-s}\psi^{s+1-n}(q)\log^{n-2} q < \infty .$$

Thus, for convergence in higher dimensions we lose a log factor from the Lebesgue volume sum appearing in Gallagher's homogeneous result and indeed Lemma 1.6.8. This of course is absolutely consistent with the $n = 2$ situation given by Theorem 1.6.10. Regarding a divergent statement, the arguments used in proving Theorem 1.6.10 can be adapted to show that, for any $s \in (n-1, n)$,

$$\mathcal{H}^s\big(W_\gamma^\times(n, \psi)\big) = \infty \qquad \text{if} \qquad \sum_{q=1}^{\infty} q^{n-s}\psi^{s+1-n}(q) = \infty .$$

Thus, there is a discrepancy in the above 's-volume' sum conditions for convergence and divergence when $n > 2$. In view of this, it remains an interesting open problem to determine the necessary and sufficient condition for $\mathcal{H}^s\big(W_\gamma^\times(n, \psi)\big)$ to be zero or infinite in higher dimensions.

1.6.4.1 The Multiplicative Theory for Manifolds

Let $\mathcal{M}$ be a non-degenerate sub-manifold of $\mathbb{R}^n$. In a nutshell, as in the simultaneous case, the overarching problem is to develop a Lebesgue and Hausdorff theory for $\mathcal{M} \cap W_\gamma^\times(n, \psi)$. Given that our current knowledge for the independent theory (i.e. when $\mathcal{M} = \mathbb{R}^n$) is pretty poor, we should not expect too much in terms of the dependent (manifold) theory. We start by describing coherent aspects of the Lebesgue theory. The following is the multiplicative analogue of the statement that $\mathcal{M}$ is inhomogeneously extremal. Given $\tau > 0$ and a fixed point $\gamma \in \mathbb{R}^n$, we write $W_\gamma^\times(n, \tau)$ for the set $W_\gamma^\times(n, \psi)$ with $\psi(q) = q^{-\tau}$.

Theorem 1.6.13 *Let $\mathcal{M}$ be a non-degenerate sub-manifold of $\mathbb{R}^n$. Then*

$$\left|\mathcal{M} \cap W_{\gamma}^{\times}(n, 1+\varepsilon)\right|_{\mathcal{M}} = 0 \qquad \forall\, \varepsilon > 0\,.$$

In the homogeneous case, the above theorem is due to Kleinbock and Margulis [67] and implies that non-degenerate manifolds are *strongly extremal* (by definition). It is easily seen that strongly extremal implies extremal. The inhomogeneous statement is established via the general Inhomogeneous Transference Principle developed in [24].

Beyond strong extremality, we have the following convergent statement for the Lebesgue measure of $\mathcal{M} \cap W_{\gamma}^{\times}(n, \psi)$ in the case $\mathcal{M}$ is a planar curve $\mathcal{C}$.

Theorem 1.6.14 *Let $\psi : \mathbb{N} \to \mathbb{R}^+$ be a monotonic function and $\gamma \in \mathbb{R}^2$. Let $\mathcal{C}$ be a non-degenerate planar curve. Then*

$$\left|\mathcal{C} \cap W_{\gamma}^{\times}(2, \psi)\right|_{\mathcal{C}} = 0 \quad \text{if} \quad \sum_{q=1}^{\infty} \psi(q) \log q \ < \infty\,. \tag{1.79}$$

The homogeneous case is established in [5, Theorem 1]. However, on making use of the upper bound counting estimate appearing within Theorem 2 of [18], it is easy to adapt the homogeneous proof to the inhomogeneous set-up. The details are left as an *exercise*. Just as in the homogeneous theory, obtaining the counterpart divergent statement for the Lebesgue measure of $\mathcal{C} \cap W_{\gamma}^{\times}(2, \psi)$ remains a stubborn problem. However, for genuine fractal Hausdorff measures $\mathcal{H}^s$ we have a complete convergence/divergence result [28, Theorem 2].

Theorem 1.6.15 *Let $\psi : \mathbb{N} \to \mathbb{R}^+$ be a monotonic function, $\gamma \in \mathbb{R}^2$ and $s \in (0, 1)$. Let $\mathcal{C}$ be a $C^{(3)}$-planar curve with non-zero curvature everywhere apart from a set of s-dimensional Hausdorff measure zero. Then*

$$\mathcal{H}^s\left(\mathcal{C} \cap W_{\gamma}^{\times}(2, \psi)\right) = \begin{cases} 0 & \text{if } \sum_{q=1}^{\infty} q^{1-s}\psi^s(q) \ < \infty, \\ \infty & \text{if } \sum_{q=1}^{\infty} q^{1-s}\psi^s(q) = \infty. \end{cases}$$

It is evident from the proof of the divergence case of the above theorem [28, §2.1.3], that imposing the condition that $\mathcal{C}$ is a $C^{(1)}$-planar curve suffices.

Beyond planar curves, the following lower bound dimension result represents the current state of knowledge.

Theorem 1.6.16 *Let $\mathcal{M}$ be an arbitrary Lipschitz manifold in $\mathbb{R}^n$ and $\boldsymbol{\gamma} \in \mathbb{R}^n$. Then, for any $\tau \geq 1$*

$$\dim\left(\mathcal{M} \cap W_{\boldsymbol{\gamma}}^{\times}(n,\tau)\right) \geq \dim \mathcal{M} - 1 + \frac{2}{1+\tau}. \tag{1.80}$$

The homogeneous case is established in [23, Theorem 5]. The homogeneous proof [23, §6.2] rapidly reduces to the inequality

$$\dim\left(\mathcal{M} \cap W_{\mathbf{0}}^{\times}(n,\tau)\right) \geq \dim \mathcal{M} - 1 + \dim W_0^{\times}(1,\tau).$$

But $W_0^{\times}(1,\tau) := W(1,\tau)$ and the desired statement follows on applying the Jarník–Besicovitch theorem (Theorem 1.3.3). Now, Theorem 1.6.6 implies that the inhomogeneous generalisation of the Jarník–Besicovitch theorem is valid; namely that, for any $\gamma \in \mathbb{R}$ and $\tau \geq 1$,

$$\dim W_{\gamma}(1,\tau) = \frac{2}{1+\tau}.$$

Thus, the short argument given in [23, §6.2] can be adapted in the obvious manner to establish Theorem 1.6.16.

1.6.4.2 Cassels' Problem

A straightforward consequence of Theorem 1.6.6 with $s = 2$ (inhomogeneous Khintchine) is that for any $\boldsymbol{\gamma} = (\gamma_1, \gamma_2) \in \mathbb{R}^2$, the set

$$W_{\boldsymbol{\gamma}}^{\times} := \{\mathbf{x} \in \mathrm{I}^2 \colon \liminf_{q\to\infty} q\,\|qx_1 - \gamma_1\|\,\|qx_2 - \gamma_2\| = 0\} \tag{1.81}$$

is of full Lebesgue measure; i.e. for any $\boldsymbol{\gamma} \in \mathbb{R}^2$, we have that

$$m_2(W_{\boldsymbol{\gamma}}^{\times}) = 1.$$

Of course, one can actually deduce the stronger 'fibre' statement that for any $x \in \mathrm{I}$ and $\boldsymbol{\gamma} = (\gamma_1, \gamma_2) \in \mathbb{R}^2$, the set

$$\{y \in \mathrm{I} \colon \liminf_{q\to\infty} q\,\|qx - \gamma_1\|\,\|qy - \gamma_2\| = 0\}$$

is of full Lebesgue measure. In a beautiful paper [89], Shapira establishes the following statement which solves a problem of Cassels dating back to the fifties.

Theorem 1.6.17 (U. Shapira)

$$m_2\Big(\bigcap_{\boldsymbol{\gamma}\in\mathbb{R}^2} W_{\boldsymbol{\gamma}}^{\times}\Big) = 1.$$

Thus, almost any pair of real numbers $(x_1, x_2) \in \mathbb{R}^2$ satisfies

$$\forall\,(\gamma_1,\gamma_2)\in\mathbb{R}^2 \qquad \liminf_{q\to\infty} q\,\|qx_1-\gamma_1\|\,\|qx_2-\gamma_2\| = 0. \tag{1.82}$$

In fact, Cassels asked for the existence of just one pair (x_1, x_2) satisfying (1.82). Furthermore, Shapira showed that if $1, x_1, x_2$ form a basis for a totally real cubic number field, then (x_1, x_2) satisfies (1.82). On the other hand, if $1, x_1, x_2$ are linearly dependent over $\mathbb{Q}$, then (x_1, x_2) cannot satisfy (1.82).

Most recently, Gorodnik and Vishe [55] have strengthened Shapira's result in the following manner: *almost any pair of real numbers* $(x_1, x_2) \in \mathbb{R}^2$ *satisfies*

$$\forall\,(\gamma_1,\gamma_2)\in\mathbb{R}^2 \qquad \liminf_{q\to\infty} q\,\log_5 q\,\|qx_1-\gamma_1\|\,\|qx_2-\gamma_2\| = 0,$$

where $\log_5$ is the fifth iterate of log. This 'rate' result makes a contribution towards the following open problem.

Conjecture 1.6.18 *Almost any pair of real numbers* $(x_1, x_2) \in \mathbb{R}^2$ *satisfies*

$$\forall\,(\gamma_1,\gamma_2)\in\mathbb{R}^2 \qquad \liminf_{q\to\infty} q\,\log q\,\|qx_1-\gamma_1\|\,\|qx_2-\gamma_2\| < \infty. \tag{1.83}$$

Remark 1.6.19 It is relatively straightforward to show (*exercise*) that for any $\tau > 2$

$$\left\{\mathbf{x}\in\mathrm{I}^2 : \forall\,(\gamma_1,\gamma_2)\in\mathbb{R}^2\ \liminf_{q\to\infty} q\,\log^{\tau} q\,\|qx_1-\gamma_1\|\,\|qx_2-\gamma_2\| = 0\right\} = \emptyset.$$

We end this section by mentioning Cassels' problem within the context of Diophantine approximation on manifolds. By exploiting the work of Shah [88], it is shown in [56] that for any non-degenerate planar curve $\mathcal{C}$

$$\left|\,\mathcal{C}\cap\bigcap\nolimits_{\gamma\in\mathbb{R}^2} W_{\gamma}^{\times}\,\right|_{\mathcal{C}} = 1.$$

1.7 The Badly Approximable Theory

We have had various discussions regarding badly approximable points in earlier sections, in particular within §1.1.3 and §1.4.2. We mentioned that the badly approximable set **Bad** and its higher-dimensional generalisation $\mathbf{Bad}(i_1, \ldots, i_n)$ are small in the sense that they are of zero Lebesgue measure but are nevertheless large in the sense that they have full Hausdorff dimension. In this section we outline the basic techniques used in establishing the dimension results. For transparency and simplicity, we shall concentrate on the one-dimensional case. We begin with the classical nearly-100-year-old result due to Jarník.

1.7.1 Bad Is of Full Dimension

The key purpose of this section is to introduce a basic Cantor set construction and show how it can be utilised to show that **Bad** is of maximal dimension – a result first established by Jarník in [59]. Towards the end we shall mention the additional ideas required in higher dimensions.

Theorem 1.7.1 (Jarník, 1928) *The Hausdorff dimension of* **Bad** *is one; that is*

$$\dim \mathbf{Bad} = 1 .$$

The proof utilises the following simple **Cantor set construction.** Let $R, M \in \mathbb{N}$ and $M \leq R - 1$. Let $E_0 = [0, 1]$. Partition the interval E_0 into R equal close subintervals and remove any M of them. This gives E_1 – the union of $(R - M)$ closed intervals $\{I_{1,j}\}_{1\leq j\leq R-M}$ of length $|I_{1,j}| = R^{-1}$. Then repeat the procedure: partition each interval $I_{1,j}$ within E_1 into R equal close subinterval and remove any M intervals of the partitioning of each $I_{1,j}$. This procedure gives rise to E_2 – the union of $(R - M)^2$ closed intervals $\{I_{2,j}\}_{1\leq j\leq(R-M)^2}$ of length $|I_{2,j}| = R^{-2}$. The process goes on recurrently/inductively as follows: for $n \geq 1$, given that E_{n-1} is constructed and represents the union of $(R - M)^{n-1}$ closed intervals $\{I_{n-1,j}\}_{1\leq j\leq(R-M)^{n-1}}$ of length $|I_{n-1,j}| = R^{-(n-1)}$, to construct E_n we

(i) partition each interval $I_{n-1,j}$ within E_{n-1} into R equal closed subintervals, and
(ii) remove any M of the R intervals of the above partitioning of each $I_{n-1,j}$.

Observe that E_n will be the union of exactly $(R - M)^n$ closed intervals $\{I_{n,j}\}_{1\leq j\leq(R-M)^n}$ of length $|I_{n,j}| = R^{-n}$. The corresponding Cantor set is defined to be

$$\mathcal{K} := \bigcap_{n=0}^{\infty} E_n .$$

Remark 1.7.2 Of course, the Cantor set constructed above is not unique and depends on the specific choices of M intervals being removed in each case. Indeed, there are continuum many possibilities for the resulting set $\mathcal{K}$. For example, if $R = 3$, $M = 1$ and we always remove the middle interval of the partitioning, the set $\mathcal{K}$ is the famous middle third Cantor set as described in Example 1.3.1 of §1.3.1.

Trivially, the Cantor set $\mathcal{K}$ is non-empty since it is the intersection of a nested sequence of closed intervals within $[0, 1]$. Indeed, if $0 \leq M \leq R - 2$ then we have that $\mathcal{K}$ is uncountable. The following result relates the Hausdorff dimension of $\mathcal{K}$ to the parameters R and M associated with $\mathcal{K}$.

Lemma 1.7.3 *Let $\mathcal{K}$ be the Cantor set constructed above. Then*

$$\dim \mathcal{K} = \frac{\log(R-M)}{\log R}. \tag{1.84}$$

Proof Let $\{I_{n,j}\}_{1\leq j\leq (R-M)^n}$ be the collection of intervals within E_n associated with the construction of $\mathcal{K}$. Recall that this is a collection of $(R-M)^n$ closed intervals, each of length R^{-n}. Naturally, $\{I_{n,j}\}_{1\leq j\leq (R-M)^n}$ is a cover of $\mathcal{K}$. Furthermore, for every $\rho > 0$ there is a sufficiently large n such that $\{I_{n,j}\}_{1\leq j\leq (R-M)^n}$ is a ρ-cover of $\mathcal{K}$ – simply make sure that $R^{-n} < \rho$. Observe that

$$\sum_j \operatorname{diam}(I_{n,j})^s = (R-M)^n R^{-ns} = 1 \qquad \text{where} \qquad s := \frac{\log(R-M)}{\log R}.$$

Hence, by definition, $\mathcal{H}^s_\rho(\mathcal{K}) \leq 1$ for all sufficiently small $\rho > 0$. Consequently, $\mathcal{H}^s(\mathcal{K}) \leq 1$ and it follows that

$$\dim \mathcal{K} \leq s.$$

For the lower bound, let $0 < \rho < 1$ and $\{B_i\}$ be an arbitrary ρ-cover of $\mathcal{K}$. We show that

$$\sum_i \operatorname{diam}(B_i)^s \geq \kappa,$$

where s is as above and the constant $\kappa > 0$ is independent of the cover. Without loss of generality, we will assume that each B_i is an open interval. Since $\mathcal{K}$ is the intersection of closed subsets of $[0, 1]$, it is bounded and closed and hence compact. Therefore, $\{B_i\}$ contains a finite subcover. Thus, without loss of generality, we can assume that $\{B_i\}$ is a finite ρ-cover of $\mathcal{K}$. For each B_i, let $k \in \mathbb{Z}$ be the unique integer such that

$$R^{-(k+1)} \leq \operatorname{diam}(B_i) < R^{-k}.$$

Then B_i intersects at most two intervals of E_k as the intervals in E_k are R^{-k} in length. If $j \geq k$, then B_i intersects at most

$$2(R-M)^{j-k} = 2(R-M)^j R^{-sk} \leq 2(R-M)^j R^s \operatorname{diam}(B_i)^s \tag{1.85}$$

intervals within E_j. These are the intervals that are contained in the (at most) two intervals of E_k that intersect B_i. Now choose j large enough so that

$$R^{-(j+1)} \leq \operatorname{diam}(B_i) \qquad \forall\, B_i\,.$$

This is possible since the cover $\{B_i\}$ is finite. Since $\{B_i\}$ is a cover of $\mathcal{K}$, it must intersect every interval of E_j. There are $(R-M)^j$ intervals within E_j. Hence, by (1.85) it follows that

$$(R-M)^j \leq \sum_i 2(R-M)^j R^s \operatorname{diam}(B_i)^s.$$

The upshot of this is that for any ρ-cover $\{B_i\}$ of $\mathcal{K}$, we have that

$$\sum_i \operatorname{diam}(B_i)^s \geq \tfrac{1}{2} R^{-s} = \frac{1}{2(R-M)}\,.$$

Hence, by definition, we have that $\mathcal{H}^s_\rho(\mathcal{K}) \geq \frac{1}{2(R-M)}$ for all sufficiently small $\rho > 0$. Therefore, $\mathcal{H}^s(\mathcal{K}) \geq \frac{1}{2(R-M)} > 0$ and it follows that

$$\dim \mathcal{K} \geq s = \frac{\log(R-M)}{\log R}$$

as required. □

Armed with Lemma 1.7.3, it is relatively straightforward to prove Jarník's full dimension result.

Proof of Theorem 1.7.1 Let $R \geq 4$ be an integer. For $n \in \mathbb{Z}$, $n \geq 0$ let

$$Q_n = \{p/q \in \mathbb{Q} : R^{\frac{n-3}{2}} \leq q < R^{\frac{n-2}{2}}\} \subset \mathbb{Q}\,, \tag{1.86}$$

where p/q is a reduced fraction of integers. Observe that $Q_0 = Q_1 = Q_2 = \varnothing$, that the sets Q_n are disjoint and that

$$\mathbb{Q} = \bigcup_{n=3}^{\infty} Q_n\,. \tag{1.87}$$

Furthermore, note that

$$\left|\frac{p}{q} - \frac{p'}{q'}\right| \geq \frac{1}{q'q} > R^{-n+2} \qquad \text{for different } p/q \text{ and } p'/q' \text{ in } Q_n. \tag{1.88}$$

Fix $0 < \delta \leq \frac{1}{2}$. Then for $p/q \in Q_n$, define the *dangerous interval* $\Delta(p/q)$ as follows:

$$\Delta(p/q) := \left\{x \in [0,1] : \left|x - \frac{p}{q}\right| < \delta R^{-n}\right\}. \tag{1.89}$$

The goal is to construct a Cantor set $\mathcal{K} = \bigcap_{n=0}^{\infty} E_n$ such that for every $n \in \mathbb{N}$

$$E_n \cap \Delta(p/q) = \varnothing \qquad \text{for all } p/q \in Q_n \,. \tag{1.90}$$

To this end, let $E_0 = [0, 1]$ and suppose that E_{n-1} has already been constructed. Let I be any of the intervals $I_{n-1,j}$ within E_{n-1}. Then $|I| = R^{-n+1}$ and, by (1.88) and (1.89), there is at most one dangerous interval $\Delta(p_I/q_I)$ with $p_I/q_I \in Q_n$ that intersects I. Partition I into R closed subintervals of length $R^{-n} = R^{-1}|I|$. Note that since $\delta \leq \frac{1}{2}$, the dangerous interval $\Delta(p_I/q_I)$, if it exists, can intersect at most two intervals of the partitioning of I. Hence, by removing $M = 2$ intervals of the partitioning of each I within E_{n-1} we construct E_n while ensuring that (1.90) is satisfied. By Lemma 1.7.3, it follows that for any $R \geq 4$

$$\dim \mathcal{K} \geq \frac{\log(R-2)}{\log R} \,.$$

Now take any $x \in \mathcal{K}$ and any $p/q \in \mathbb{Q}$. Then $p/q \in Q_n$ for some $n \in \mathbb{N}$ and since $\mathcal{K} \subset E_n$ we have that $x \in E_n$. Then, by (1.90), we have that $x \notin \Delta(p/q)$, which implies that

$$\left| x - \frac{p}{q} \right| \geq \delta R^{-n} \geq \delta R^{-3} q^{-2} \,. \tag{1.91}$$

Since $p/q \in \mathbb{Q}$ is arbitrary and R and δ are fixed, we have that $x \in \mathbf{Bad}$. That is, $\mathcal{K} \subset \mathbf{Bad}$ and thus it follows that

$$\dim \mathbf{Bad} \geq \dim \mathcal{K} \geq \frac{\log(R-2)}{\log R} .$$

This is true for any $R \geq 4$ and so on letting $R \to \infty$, it follows that $\dim \mathbf{Bad} \geq 1$. The complementary upper bound statement $\dim \mathbf{Bad} \leq 1$ is trivial since $\mathbf{Bad} \subset \mathbb{R}$. □

Remark 1.7.4 The crucial property underpinning the proof of Theorem 1.7.1 is the separation property (1.88) of rationals. Indeed, without appealing to Lemma 1.7.3, the above proof based on (1.88) alone shows that **Bad** is uncountable. The construction of the Cantor set $\mathcal{K}$ as well as the proof of Theorem 1.7.1 can be generalised to higher dimensions in order to show that

$$\dim \mathbf{Bad}(i_1, \ldots, i_n) = n.$$

Regarding the higher-dimensional generalisation of the proof of Theorem 1.7.1, the appropriate analogue of (1.88) is the following elegant Simplex Lemma – see for example [69, Lemma 4].

Lemma 1.7.5 (Simplex Lemma) *Let $m \geq 1$ be an integer and $Q > 1$ be a real number. Let $E \subseteq \mathbb{R}^m$ be a convex set of m-dimensional Lebesgue measure*

$$|E| \leq (m!)^{-1} Q^{-(m+1)}.$$

Suppose that E contains $m+1$ rational points $(p_i^{(1)}/q_i, \ldots, p_i^{(m)}/q_i)$ with $1 \leq q_i < Q$, where $0 \leq i \leq m$. Then these rational points lie in some hyperplane of $\mathbb{R}^m$.

1.7.2 Schmidt's Games

In his pioneering work [85], Wolfgang M. Schmidt introduced the notion of (α, β)-games which now bear his name. These games are an extremely powerful tool for investigating badly approximable sets. The simplified account which we are about to present is sufficient to bring out the main features of the games.

Suppose that $0 < \alpha < 1$ and $0 < \beta < 1$. Consider the following game involving the two arch rivals **Ayesha** and **Bhupen** – often simply referred to as players **A** and **B**. First, **B** chooses a closed ball $\mathbf{B}_0 \subset \mathbb{R}^m$. Next, **A** chooses a closed ball $\mathbf{A}_0$ contained in $\mathbf{B}_0$ of diameter $\alpha\, \rho(\mathbf{B}_0)$ where $\rho(\,.\,)$ denotes the diameter of the ball under consideration. Then, **B** chooses at will a closed ball $\mathbf{B}_1$ contained in $\mathbf{A}_0$ of diameter $\beta\, \rho(\mathbf{A}_0)$. Alternating in this manner between the two players, generates a nested sequence of closed balls in $\mathbb{R}^m$:

$$\mathbf{B}_0 \supset \mathbf{A}_0 \supset \mathbf{B}_1 \supset \mathbf{A}_1 \supset \ldots \supset \mathbf{B}_n \supset \mathbf{A}_n \supset \ldots \tag{1.92}$$

with diameters

$$\rho(\mathbf{B}_n) = (\alpha\, \beta)^n\, \rho(\mathbf{B}_0) \quad \text{and} \quad \rho(\mathbf{A}_n) = \alpha\, \rho(\mathbf{B}_n).$$

A subset X of $\mathbb{R}^m$ is said to be (α, β)-*winning* if **A** can play in such a way that the unique point of the intersection

$$\bigcap_{n=0}^{\infty} \mathbf{B}_n = \bigcap_{n=0}^{\infty} \mathbf{A}_n$$

lies in X, regardless of how **B** plays. The set X is called α-*winning* if it is (α, β)-winning for all $\beta \in (0, 1)$. Finally, X is simply called *winning* if it is α-winning for some α. Informally, player **B** tries to stay away from the 'target' set X whilst player **A** tries to land on X. As shown by Schmidt in [85], the following are the key consequences of winning.

- *If $X \subset \mathbb{R}^m$ is a winning set, then* $\dim X = m$.
- *The intersection of countably many α-winning sets is α-winning.*

Schmidt [85] proved the following fundamental result for the symmetric case of the higher-dimensional analogue of **Bad** which, given the above properties, has implications well beyond simply full dimension.

Theorem 1.7.6 (Schmidt, 1966) *For any $m \in \mathbb{N}$, the set $\mathbf{Bad}(\frac{1}{m}, \ldots, \frac{1}{m})$ is winning.*

Proof To illustrate the main ideas involved in proving the theorem we shall restrict our attention to when $m = 1$. In this case, we are able to establish the desired winning statement by naturally modifying the proof of Theorem 1.7.1. Without loss of generality, we can restrict $\mathbf{Bad} := \mathbf{Bad}(1)$ to the unit interval $[0, 1]$. Let $0 < \alpha < \frac{1}{2}$ and $0 < \beta < 1$. Let $R = (\alpha\beta)^{-1}$ and define Q_n by (1.86). Again $Q_0 = Q_1 = Q_2 = \varnothing$; the sets Q_n are disjoint; (1.87) and (1.88) are both true. Furthermore, for $p/q \in Q_n$ the corresponding dangerous interval $\Delta(p/q)$ is defined by (1.89), where $0 < \delta < 1$ is to be specified below and will be dependent on α and the first move made by **Bhupen** .
Our goal is to show that **Ayesha** has a strategy to ensure that sequence (1.92) satisfies

$$\mathbf{A}_n \cap \Delta(p/q) = \varnothing \qquad \text{for all } \ p/q \in Q_n \, . \tag{1.93}$$

Then the single point x corresponding to the intersection over all the closed and nested intervals $\mathbf{A}_n$ would satisfy (1.91) for all $p/q \in \mathbb{Q}$ meaning that x is badly approximable. By definition, this would imply that **Bad** is α-winning as desired.

Let $\mathbf{B}_0 \subset [0, 1]$ be any closed interval. Now we set

$$\delta := \rho(\mathbf{B}_0)(\tfrac{1}{2} - \alpha).$$

Suppose that

$$\mathbf{B}_0 \supset \mathbf{A}_0 \supset \mathbf{B}_1 \supset \mathbf{A}_1 \supset \ldots \supset \mathbf{B}_{n-1} \supset \mathbf{A}_{n-1}$$

are already chosen and satisfy the required properties; namely (1.93). Suppose that $\mathbf{B}_n \subset \mathbf{A}_{n-1}$ is any closed interval of length

$$\rho(\mathbf{B}_n) = \beta\rho(\mathbf{A}_{n-1}) = (\alpha\,\beta)^n\,\rho(\mathbf{B}_0) = R^{-n}\rho(\mathbf{B}_0).$$

Next, **A** has to choose a closed interval $\mathbf{A}_n$ contained in $\mathbf{B}_n$ of diameter

$$\rho(\mathbf{A}_n) = \alpha\,\rho(\mathbf{B}_n) = \alpha R^{-n}\rho(\mathbf{B}_0)$$

and satisfying (1.93). If (1.93) is satisfied with $\mathbf{A}_n$ replaced by $\mathbf{B}_n$, then choosing $\mathbf{A}_n$ obviously represents no problem. Otherwise, using (1.88) one readily verifies that there is exactly one point $p_n/q_n \in Q_n$ such that the interval $\Delta(p_n/q_n)$ intersects $\mathbf{B}_n$. In this case $\mathbf{B}_n \setminus \Delta(p_n/q_n)$ is either the union of two closed intervals, the larger one being of length

$$\geq \tfrac{1}{2}\Big(\rho(\mathbf{B}_n) - \rho(\Delta(p_n/q_n))\Big) = \tfrac{1}{2}R^{-n}\Big(\rho(\mathbf{B}_0) - 2\delta\Big)$$
$$= \alpha R^{-n}\rho(\mathbf{B}_0) = \alpha\rho(\mathbf{B}_n)$$

or a single closed interval of even greater length. Hence, it is possible to choose a closed interval $\mathbf{A}_n \subset \mathbf{B}_n \setminus \Delta(p_n/q_n)$ of length $\rho(\mathbf{A}_n) = \alpha\rho(\mathbf{B}_n)$. By construction, (1.93) is satisfied, thus proving the existence of a winning strategy for **A**. □

Remark 1.7.7 For various reasons, over the last decade or so there has been an explosion of interest in Schmidt's games. This has given rise to several ingenious generalisations of the original game leading to stronger notions of winning, such as modified winning, absolute winning, hyperplane winning and potential winning. For details see [51, 68] and references within.

The framework of Schmidt games and thus the notion of winning is defined in terms of balls. Thus, it is naturally applicable when considering the symmetric case ($i_1 = \ldots = i_n = 1/n$) of the badly approximable sets $\mathbf{Bad}(i_1 \ldots, i_n)$. Recall, that in the symmetric case, points in $\mathbf{Bad}(\frac{1}{n}, \ldots, \frac{1}{n})$ avoid squares (which are essentially balls) centred around rational points were as in the general case the points avoiding rectangles (far from being balls). We now turn our attention to the general case. Naturally, it would be desirable to be able to show that the general set $\mathbf{Bad}(i_1 \ldots, i_n)$ is winning.

1.7.3 Properties of General $\mathbf{Bad}(i_1 \ldots, i_n)$ Sets Beyond Full Dimension

Despite the fact that the sets $\mathbf{Bad}(i_1, \ldots, i_n)$ have long been known to be uncountable and indeed of full dimension, see [42, 68, 69, 77], the following conjecture of Schmidt dating back to 1982 remained unresolved until reasonably recently.

Schmidt's Conjecture

$$\mathbf{Bad}(\tfrac{1}{3}, \tfrac{2}{3}) \cap \mathbf{Bad}(\tfrac{2}{3}, \tfrac{1}{3}) \neq \emptyset.$$

As is already highlighted in Remark 1.4.9, if false then it would imply that Littlewood's conjecture is true.

Schmidt's conjecture was proved in [7] by establishing the following stronger statement regarding the intersection of $\mathbf{Bad}(i,j)$ sets with vertical lines $L_\alpha := \{(\alpha, y) : y \in \mathbb{R}\} \subset \mathbb{R}^2$. To some extent it represents the badly approximable analogue of the 'fiber' results that appeared in §1.4.5.

Theorem 1.7.8 *Let* (i_k, j_k) *be a countable sequence of non-negative reals such that* $i_k + j_k = 1$ *and let* $i := \sup\{i_k : k \in \mathbb{N}\}$. *Suppose that*

$$\liminf_{k\to\infty} \min\{i_k, j_k\} > 0. \tag{1.94}$$

Then, for any $\alpha \in \mathbb{R}$ *such that* $\liminf_{q\to\infty} q^{1/i} \|q\alpha\| > 0$, *we have that*

$$\dim \bigcap_k \mathbf{Bad}(i_k, j_k) \cap L_\alpha = 1. \tag{1.95}$$

Remark 1.7.9 The Diophantine condition imposed on α associated with the vertical line L_α is easily seen to be necessary – see [7, §1.3]. Note that the condition is automatically satisfied if $\alpha \in \mathbf{Bad}$. On the other hand, condition (1.94) is present for technical reason and can be removed – see Theorem 1.7.15 and discussion below. At this point, simply observe that it is automatically satisfied for any finite collection of pairs (i_k, j_k) and thus Theorem 1.7.8 implies Schmidt's conjecture. Indeed, together with a standard 'slicing' result from fractal geometry one obtains the following full dimension statement – see [7, §1.2] for details.

Corollary 1.7.10 *Let* (i_k, j_k) *be a countable sequence of non-negative reals such that* $i_k + j_k = 1$ *and satisfying condition* (1.94). *Then,*

$$\dim \bigcap_k \mathbf{Bad}(i_k, j_k) = 2. \tag{1.96}$$

At the heart of establishing Theorem 1.7.8 is the 'raw' construction of the generalised Cantor sets framework formulated in [8]. For the purposes of these notes, we opt to follow the framework of *Cantor rich sets* introduced in [12], which is a variation of the aforementioned generalised Cantor sets.

Let $R \geq 3$ be an integer. Given a collection $\mathcal{I}$ of compact intervals in $\mathbb{R}$, let $\frac{1}{R}\mathcal{I}$ denote the collection of intervals obtained by splitting each interval in $\mathcal{I}$ into R equal closed subintervals with disjoint interiors. Given a compact interval $I_0 \subset \mathbb{R}$, a sequence $(\mathcal{I}_q)_{q\geq 0}$ such that

$$\mathcal{I}_0 = \{I_0\} \qquad \text{and} \qquad \mathcal{I}_q \subset \tfrac{1}{R}\mathcal{I}_{q-1} \quad \text{for } q \geq 1$$

is called *an R-sequence in I_0*. It defines the corresponding *generalised Cantor set*:

$$\mathcal{K}((\mathcal{I}_q)_{q\geq 0}) := \bigcap_{q\geq 0} \bigcup_{I_q\in\mathcal{I}_q} I_q. \tag{1.97}$$

Given $q \in \mathbb{N}$ and any interval J, let

$$\widehat{\mathcal{I}}_q := \left(\tfrac{1}{R}\mathcal{I}_{q-1}\right)\setminus\mathcal{I}_q \qquad \text{and} \qquad \widehat{\mathcal{I}}_q \sqcap J := \{I_q \in \widehat{\mathcal{I}}_q : I_q \subset J\}.$$

Furthermore, define

$$d_q(\mathcal{I}_q) := \min_{\{\widehat{\mathcal{I}}_{q,p}\}} \sum_{p=0}^{q-1} \left(\frac{4}{R}\right)^{q-p} \max_{I_p\in\mathcal{I}_p} \#\big(\widehat{\mathcal{I}}_{q,p} \sqcap I_p\big)\,, \tag{1.98}$$

where the minimum is taken over all partitions $\{\widehat{\mathcal{I}}_{q,p}\}_{p=0}^{q-1}$ of $\widehat{\mathcal{I}}_q$; that is $\widehat{\mathcal{I}}_q = \bigcup_{p=0}^{q-1}\widehat{\mathcal{I}}_{q,p}$.

The following dimension statement was established in [8, Theorem 4], see also [12, Theorem 5].

Lemma 1.7.11 *Let $\mathcal{K}((\mathcal{I}_q)_{q\geq 0})$ be the Cantor set given by* (1.97)*. Suppose that*

$$d_q(\mathcal{I}_q) \leq 1 \tag{1.99}$$

for all $q \in \mathbb{N}$. Then

$$\dim \mathcal{K}((\mathcal{I}_q)_{q\geq 0}) \geq 1 - \frac{\log 2}{\log R}\,.$$

Although the lemma can be viewed as a generalisation of Lemma 1.7.3, we stress that its proof is substantially more involved and requires new ideas. At the heart of the proof is the 'extraction' of a 'local' Cantor type subset $\mathcal{K}$ of $\mathcal{K}((\mathcal{I}_q)_{q\geq 0})$. By a local Cantor set we mean a set arising from a construction as described in §1.7.1. The parameter M associated with the extracted local Cantor set $\mathcal{K}$ is essentially $\frac{1}{2}R$.

It is self-evident from Lemma 1.7.11 that if a given set $X \subset \mathbb{R}$ contains a generalised Cantor set given by (1.97) with arbitrarily large R, then $\dim X = 1$. The following definition of Cantor-rich [12] imposes a stricter requirement than (1.99) in order to ensure that the countable intersection of generalised Cantor sets is of full dimension. To some extent, building upon the raw construction of [7, §7.1], the full dimension aspect for countable intersections had previously been investigated in [8, §7].

Definition 1.7.12 Let $M > 1$, $X \subset \mathbb{R}$ and I_0 be a compact interval. The set X is said to be *M-Cantor-rich in I_0* if for any $\varepsilon > 0$ and any integer $R \geq M$ there exists an R-sequence $(\mathcal{I}_q)_{q\geq 0}$ in I_0 such that $\mathcal{K}((\mathcal{I}_q)_{q\geq 0}) \subset X$ and

$$\sup_{q\in\mathbb{N}} d_q(\mathcal{I}_q) \leq \varepsilon.$$

The set X is said to be *Cantor-rich in I_0* if it is *M-Cantor-rich in I_0* for some M, and it is said to be *Cantor-rich* if it is *Cantor-rich in I_0* for some compact interval I_0.

The following summarises the key properties of Cantor-rich sets:

(i) Any Cantor-rich set X in $\mathbb{R}$ satisfies $\dim X = 1$*;*
(ii) For any given compact interval I_0 and any given fixed $M \in \mathbb{N}$, any countable intersection of M-Cantor-rich sets in I_0 is M-Cantor-rich in I_0.

The framework of Cantor-rich sets was utilised in the same paper [12] to establish the following result concerning badly approximable points on manifolds.

Theorem 1.7.13 *For any non-degenerate analytic sub-manifold $\mathcal{M} \subset \mathbb{R}^n$ and any sequence $(i_{1,k}, \ldots, i_{n,k})$ of non-negative reals such that $i_{1,k} + \cdots + i_{n,k} = 1$ and*

$$\inf\{i_{j,k} > 0 : 1 \leq j \leq n,\ k \in \mathbb{N}\} > 0, \tag{1.100}$$

one has that

$$\dim \textstyle\bigcap_k \mathbf{Bad}(i_{1,k}, \ldots, i_{n,k}) \cap \mathcal{M} = \dim \mathcal{M}. \tag{1.101}$$

The condition of analyticity from Theorem 1.7.13 can be omitted in the case the sub-manifold $\mathcal{M} \subset \mathbb{R}^n$ is a curve. Indeed, establishing the theorem for curves is very much the crux since any manifold can be 'fibred' into an appropriate collection of curves – see [12, §2.1] for details. In the case $n = 2$, so that $\mathcal{M}$ is a non-degenerate planar curve, the theorem was previously established in [9] and provides a solution to an explicit problem of Davenport dating back to the swinging sixties concerning the existence of badly approximable pairs on the parabola. Furthermore, in [9] partial results for lines (degenerate curves) with slopes satisfying certain Diophantine constraints are also obtained. Although not optimal, they naturally extend Theorem 1.7.8 beyond vertical lines. As already mentioned, Theorem 1.7.13

as stated for general n was established in [12] and it settles the natural generalisations of Schmidt's conjecture and Davenport's problem in arbitrary dimensions.

Remark 1.7.14 Building upon the one-dimensional, generalised Cantor sets framework formulated in [8], an abstract 'metric space' framework of higher-dimensional generalised Cantor sets, branded as 'Cantor-winning sets', has recently been introduced in [6]. Projecting this framework onto the specific one-dimensional construction of Cantor-rich sets given above, the definition of Cantor-winning sets reads as follows. Let $\varepsilon_0 > 0$, $X \subset \mathbb{R}$ and I_0 be a compact interval. Then the set X is ε_0-*Cantor-winning in* I_0 if for any positive $\varepsilon < \varepsilon_0$ there exists a positive integer R_ε such that for any integer $R \geq R_\varepsilon$ there exists an R-sequence $(\mathcal{I}_q)_{q\geq 0}$ in I_0 such that $\mathcal{K}((\mathcal{I}_q)_{q\geq 0}) \subset X$ and

$$\max_{I_p \in \mathcal{I}_p} \#\big(\widehat{\mathcal{I}}_{q,p} \sqcap I_p\big) \leq R^{(q-p)(1-\varepsilon)} .$$

The latter key condition implies that $d_q(\mathcal{I}_q)$ is no more than $8R^{-\varepsilon}$ provided that $8R^{-\varepsilon} < 1$. Most recently, David Simmons has shown that the notion of Cantor-winning as defined in [6] is equivalent to the notion of potential winning as defined in [51].

The use of Cantor-rich sets in establishing statements such as Theorems 1.7.8 and 1.7.13 comes at a cost of having to impose, seemingly for technical reasons, conditions such as (1.94) and (1.100). Although delivering some additional benefits, unfortunately the framework of Cantor-winning sets described above does not seem to resolve this issue. However, if, for example, we could show that $\mathbf{Bad}(i_1, \ldots, i_n)$ is (Schmidt) winning, then we would be able to intersect countably many such sets without imposing any technical conditions. When $n = 2$, this has been successfully accomplished by Jinpeng An in his elegant paper [2].

Theorem 1.7.15 (J. An) *For any pair of non-negative reals* (i, j) *such that* $i + j = 1$, *the two-dimensional set* $\mathbf{Bad}(i, j)$ *is winning.*

A simple consequence of this is that we can remove condition (1.94) from the statement of Corollary 1.7.10. Prior to [2], it is important to note that An in [1] had shown that $\mathbf{Bad}(i, j) \cap \mathrm{L}_\alpha$ is winning, where L_α is a vertical line as in Theorem 1.7.8. Of course, this implies that Theorem 1.7.8 is true without imposing condition (1.94). On combining the ideas and techniques introduced in the papers [1, 9, 12], it is shown in [3] that $\mathbf{Bad}(i, j) \cap \mathcal{C}$ is winning, where

$\mathcal{C}$ is a non-degenerate planar curve. This implies that we can remove condition (1.100) from the $n = 2$ statement of Theorem 1.7.13. In higher dimensions ($n > 2$), removing condition (1.100) remains very much a key open problem. The recent work of Guan and Yu [44] makes a contribution towards this problem. Building upon the work of An [2], they show that the set $\mathbf{Bad}(i_1, \ldots, i_n)$ is winning whenever $i_1 = \cdots = i_{n-1} \geq i_n$.

So far we have discussed the homogeneous theory of badly approximable sets. We now turn our attention to the inhomogeneous theory.

1.7.4 Inhomogeneous Badly Approximable Points

Given $\theta \in \mathbb{R}$ the natural inhomogeneous generalisation of the one-dimensional set **Bad** is the set

$$\mathbf{Bad}(\theta) := \{x \in \mathbb{R} : \exists\, c(x) > 0 \ \text{ so that } \ \|qx - \theta\| \ > \ c(x)\, q^{-1} \ \forall\, q \in \mathbb{N}\}.$$

Within these notes we shall prove the following inhomogeneous strengthening of Theorem 1.7.1.

Theorem 1.7.16 *For any $\theta \in \mathbb{R}$, we have that*

$$\dim \mathbf{Bad}(\theta) = 1\,.$$

The basic philosophy behind the proof is simple and exploits the already discussed homogeneous 'intervals construction'; namely

$$\text{(homogeneous construction)} \quad + \quad (\boldsymbol{\theta} - \boldsymbol{\theta} = \mathbf{0})$$
$$\Longrightarrow \quad \text{(inhomogeneous statement).}$$

Remark 1.7.17 Recall that we have already made use of this type of philosophy in establishing the inhomogeneous extremality conjecture stated in §1.6.3.1, where the proof very much relies on the fact that we already know that any non-degenerate manifold is (homogeneously) extremal.

Proof of Theorem 1.7.16 Let $R \geq 4$ be an integer and $\delta = \frac{1}{2}$. For $n \in \mathbb{Z}$, $n \geq 0$, define the sets Q_n by (1.86) and additionally define the following sets of 'shifted' rational points

$$Q_n(\theta) = \{(p+\theta)/q \in \mathbb{R} : p, q \in \mathbb{Z},\ R^{\frac{n-5}{2}} \leq q < R^{\frac{n-4}{2}}\}. \tag{1.102}$$

Clearly, $Q_0(\theta) = \cdots = Q_4(\theta) = \varnothing$ and the union $Q(\theta) := \bigcup_{n=5}^{\infty} Q_n(\theta)$ contains all the possible points $(p+\theta)/q$ with $p, q \in \mathbb{Z}, q > 0$.

Next, for $p/q \in Q_n$ define the dangerous interval $\Delta(p/q)$ by (1.89) and additionally define the inhomogeneous family of dangerous intervals given by

$$\Delta((p+\theta)/q) := \left\{x \in [0,1] : \left|x - \frac{p+\theta}{q}\right| < \delta R^{-n}\right\}, \tag{1.103}$$

where $(p+\theta)/q \in Q(\theta)$. With reference to the Cantor construction of §1.7.1, our goal is to construct a Cantor set $\mathcal{K} = \bigcap_{n=0}^{\infty} E_n$ such that for every $n \in \mathbb{N}$

$$E_n \cap \Delta(p/q) = \varnothing \qquad \text{for all } p/q \in Q_n \tag{1.104}$$

and simultaneously

$$E_n \cap \Delta((p+\theta)/q) = \varnothing \qquad \text{for all } (p+\theta)/q \in Q_n(\theta)\,. \tag{1.105}$$

To this end, let $E_0 = [0,1]$ and suppose that E_{n-1} has been constructed as required. Let I be any interval within E_{n-1}. Then $|I| = R^{-n+1}$. When constructing E_n, I is partitioned into R subintervals. We need to decide how many of these subintervals have to be removed in order to satisfy (1.104) and (1.105). As was argued in the proof of Theorem 1.7.1, removing two intervals of the partitioning of I ensures that (1.104) is satisfied. We claim that the same applies to (1.105), that is removing two intervals of the partitioning of I ensures (1.105). Indeed, since the length of $\Delta((p+\theta)/q)$ is no more that R^{-n}, to verify this claim it suffices to show that there is only one point $(p+\theta)/q \in Q_n(\theta)$ such that

$$\Delta((p+\theta)/q) \cap I \neq \varnothing.$$

This condition implies that

$$|qx - p - \theta| < R^{\frac{n-4}{2}}(\delta R^{-n} + R^{-n+1}) \qquad \text{for any } x \in I\,. \tag{1.106}$$

For a contradiction, suppose there are two distinct points $(p_1+\theta)/q_1$ and $(p_2+\theta)/q_2$ in $Q_n(\theta)$ satisfying (1.106). Then, by (1.106) and the triangle inequality, we get that

$$|(q_1-q_2)x-(p_1-p_2)| < 2R^{\frac{n-4}{2}}(\delta R^{-n}+R^{-n+1}) \qquad \text{for any } x \in I\,. \tag{1.107}$$

Clearly $q_1 \neq q_2$ as otherwise we would have that

$$|p_1 - p_2| < 2R^{\frac{n-4}{2}}(\delta R^{-n} + R^{-n+1}) < 1,$$

implying that $p_1 = p_2$ and contradicting to the fact that $(p_1+\theta)/q_1$ and $(p_2+\theta)/q_2$ are distinct. In the above we have used that $n \geq 5$. Also without loss of generality we assume that $q_1 > q_2$. Then define $d = \gcd(q_1 - q_2, p_1 - p_2)$,

$q = (q_1 - q_2)/d$, $p = (p_1 - p_2)/d$ and let m be the unique integer such that

$$p/q \in Q_m .$$

Thus, $R^{\frac{m-3}{2}} \leq q < R^{\frac{m-2}{2}}$. Since $q < q_1 < R^{\frac{n-4}{2}}$ we have that $m \leq n-2$. Then, by (1.107),

$$\left| x - \frac{p}{q} \right| < R^{-\frac{m-3}{2}} 2R^{\frac{n-4}{2}} (\delta R^{-n} + R^{-n+1}) \leq \delta R^{-m} \tag{1.108}$$

for any $x \in I$ provided that $R \geq 36$ (recall that $\delta = \frac{1}{2}$). It means that $\Delta(p/q) \cap I \neq \varnothing$. But this is impossible since (1.104) is valid with n replaced by m and $I \subset E_{n-1} \subset E_m$. This proves our above claim. The upshot is that by removing $M = 4$ intervals of the partitioning of each I within E_{n-1} we construct E_n while ensuring that the desired conditions (1.104) and (1.105) are satisfied. The finale of the proof makes use of Lemma 1.7.3 and is almost identical to that of the proof of Theorem 1.7.1. We leave the details to the reader. □

Remark 1.7.18 Note that in the above proof of Theorem 1.7.16, we actually show that

$$\dim \mathbf{Bad} \cap \mathbf{Bad}(\theta) = 1 .$$

It seems that proving this stronger statement is simpler than any potential 'direct' proof of the implied fact that $\dim \mathbf{Bad}(\theta) = 1$.

Remark 1.7.19 In the same way that the proof of Theorem 1.7.1 can be modified to show that **Bad** is winning (see the proof of Theorem 1.7.6 for the details), the proof of Theorem 1.7.16 can be adapted to show that $\mathbf{Bad}(\theta)$ is winning.

In higher dimensions, the natural generalisation of the one-dimensional set $\mathbf{Bad}(\theta)$ is the set $\mathbf{Bad}(i_1, \ldots, i_n; \boldsymbol{\theta})$ defined in the following manner. For any $\boldsymbol{\theta} = (\theta_1, \ldots, \theta_n) \in \mathbb{R}^n$ and n-tuple of real numbers $i_1, \ldots, i_n \geq 0$ such that $i_1 + \cdots + i_n = 1$, we let $\mathbf{Bad}(i_1, \ldots, i_n; \boldsymbol{\theta})$ to be the set of points $(x_1, \ldots, x_n) \in \mathbb{R}^n$ for which there exists a positive constant $c(x_1, \ldots, x_n)$ such that

$$\max\{ \, ||qx_1 - \theta_1||^{1/i_1} \, , \ldots , \, ||qx_n - \theta_n||^{1/i_n} \, \} \; > \; c(x_1, \ldots, x_n) \, q^{-1} \forall q \in \mathbb{N}.$$

The ideas used in the proof of Theorem 1.7.16 can be naturally generalised to show that

$$\dim \mathbf{Bad}(i_1, \ldots, i_n; \boldsymbol{\theta}) = n.$$

In the case that $n = 2$, the details of the proof are explicitly given in [28, §3]. Indeed, as mentioned in [28, Remark 3.4], in the symmetric case $i_1 =$

$\ldots = i_n = 1/n$, we actually have that $\mathbf{Bad}(\frac{1}{n}, \ldots, \frac{1}{n}; \boldsymbol{\theta})$ is winning; i.e. the inhomogeneous strengthening of Theorem 1.7.6.

Remark 1.7.20 The basic philosophy exploited in proving Theorem 1.7.16 has been successfully incorporated within the context of Schmidt games to establish the inhomogeneous generalisation of the homogeneous winning statements discussed at the end of §1.7.3. In particular, let $\boldsymbol{\theta} \in \mathbb{R}^2$ and (i, j) be a pair of non-negative real numbers such that $i + j = 1$. Then, it is shown in [3] that (i) the set $\mathbf{Bad}(i, j; \boldsymbol{\theta})$ is winning and (ii) for any non-degenerate planar curve $\mathcal{C}$, the set $\mathbf{Bad}(i, j; \boldsymbol{\theta}) \cap \mathcal{C}$ is winning. Also, in [3] the following almost optimal winning result for the intersection of $\mathbf{Bad}(i, j)$ sets with arbitrary lines (degenerate curves) is obtained. It substantially extends and generalises the previous 'line' result obtained in [9].

Theorem 1.7.21 *Let (i, j) be a pair of non-negative real numbers such that $i + j = 1$ and given $a, b \in \mathbb{R}$ with $a \neq 0$, let $L_{a,b}$ denote the line defined by the equation $y = ax + b$. Suppose there exists $\epsilon > 0$ such that*

$$\liminf_{q \to \infty} q^{\frac{1}{\sigma} - \epsilon} \max\{\|qa\|, \|qb\|\} > 0 \qquad \text{where} \quad \sigma := \min\{i, j\}. \tag{1.109}$$

Then, for any $\boldsymbol{\theta} \in \mathbb{R}^2$ we have that $\mathbf{Bad}_{\boldsymbol{\theta}}(i, j) \cap L_{a,b}$ is winning. Moreover, if $a \in \mathbb{Q}$ the statement is true with $\epsilon = 0$ in (1.109).

The condition (1.109) is optimal up to the ϵ – see [3, Remark 4]. It is indeed both necessary and sufficient in the case $a \in \mathbb{Q}$. Note that the argument presented in [3, Remark 4] showing the necessity of (1.109) with $\epsilon = 0$ only makes use of the assumption that $\mathbf{Bad}(i, j) \cap \mathrm{L}_{a,b} \neq \emptyset$. It is plausible to suggest that this latter assumption is a necessary and sufficient condition for the conclusion of Theorem 1.7.21 to hold.

Conjecture 1.7.22 *Let (i, j) be a pair of non-negative real numbers such that $i + j = 1$ and given $a, b \in \mathbb{R}$ with $a \neq 0$, let $L_{a,b}$ denote the line defined by the equation $y = ax + b$. Then*

$$\mathbf{Bad}(i, j) \cap L_{a,b} \neq \emptyset$$

if and only if

$$\forall\, \boldsymbol{\theta} \in \mathbb{R}^2 \qquad \mathbf{Bad}_{\boldsymbol{\theta}}(i, j) \cap L_{a,b} \quad \text{is winning.}$$

Observe that the conjecture is true in the case $a \in \mathbb{Q}$ and when the line $\mathrm{L}_{a,b}$ is horizontal or vertical in the homogeneous case.

Acknowledgements. SV would like to thank the organisers of the 2014 Durham Easter School 'Dynamics and Analytic Number Theory' for giving him the opportunity to give a mini-course – it was a stimulating and enjoyable experience. Subsequently, the subject matter of that mini-course formed the foundations for a MAGIC graduate lecture course on metric number theory given jointly by VB and SV at the University of York in Spring 2015. We would like to thank the participants of these courses for providing valuable feedback on both the lectures and the accompanying notes. In particular, we thank Demi Allen and Henna Koivusalo for their detailed comments (well beyond the call of duty) on earlier drafts of this end product. For certain, their input has improved the clarity and the accuracy of the exposition. Of course, any remaining typos and mathematical errors are absolutely their fault!

Victor Beresnevich
Department of Mathematics, University of York
York, Y010 5DD, UK.
email: `victor.beresnevich@york.ac.uk`
`maths.york.ac.uk/www/vb8`

Felipe Ramírez
Department of Mathematics and Computer Science, Wesleyan University
265 Church Street, Middletown, CT 06459, USA.
email: `framirez@wesleyan.edu`
`framirez.web.wesleyan.edu`

Sanju Velani
Department of Mathematics, University of York
York, Y010 5DD, UK.
email: `slv3@york.ac.uk`
`maths.york.ac.uk/www/slv3`

References

[1] J. An. Badziahin-Pollington-Velani's theorem and Schmidt's game. *Bull. London Math. Soc.* **45** (2013), no. 4, 721–733.

[2] J. An. Two dimensional badly approximable vectors and Schmidt's game. *Duke Math. J.* (2015), doi:10.1215/00127094-3165862

[3] J. An, V. Beresnevich and S. L. Velani. Badly approximable points on planar curves and winning. Pre-print: arXiv:1409.0064.

[4] D. Badziahin. On multiplicatively badly approximable numbers. *Mathematika* **59** (2013), no. 1, 31–55.

[5] D. Badziahin and J. Levesley. A note on simultaneous and multiplicative Diophantine approximation on planar curves. *Glasgow Mathematical Journal* **49** (2007), no. 2, 367–375.

[6] D. Badziahin and S. Harrap. Cantor-winning sets and their applications. Preprint: arXiv:1503.04738.

[7] D. Badziahin, A. Pollington and S. L. Velani. On a problem in simultaneous diophantine approximation: Schmidt's conjecture. *Ann. of Math.* (2) **174** (2011), no. 3, 1837–1883.

[8] D. Badziahin and S. L. Velani. Multiplicatively badly approximable numbers and generalised Cantor sets. *Adv. Math.* **225** (2011), 2766–2796.

[9] D. Badziahin and S. L. Velani. Badly approximable points on planar curves and a problem of Davenport. *Mathematische Annalen* **359** (2014), no. 3, 969–1023.

[10] A. Baker and W. M. Schmidt. Diophantine approximation and Hausdorff dimension. *Proc. Lond. Math. Soc.* **21** (1970), 1–11.

[11] V. Beresnevich. Rational points near manifolds and metric Diophantine approximation. *Ann. of Math.* (2) **175** (2012), no. 1, 187–235.

[12] V. Beresnevich. Badly approximable points on manifolds. *Inventiones Mathematicae* **202** (2015), no. 3, 1199–1240.

[13] V. Beresnevich, V. Bernik, M. Dodson and S. L. Velani. *Classical Metric Diophantine Approximation Revisited*, Roth Festschrift – essays in honour of Klaus Roth on the occasion of his 80th birthday. Editors: W. Chen, T. Gowers, H. Halberstam, W. M. Schmidt and R. C. Vaughan. Cambridge University Press. 2009, pp. 38–61.

[14] V. Beresnevich, D. Dickinson and S. L. Velani. Measure theoretic laws for limsup sets. *Mem. Amer. Math. Soc.* **179** (2006), no. 846, 1–91.

[15] V. Beresnevich, D. Dickinson and S. L. Velani. Diophantine approximation on planar curves and the distribution of rational points. *Ann. of Math.* (2) **166** (2007), pp. 367–426. With an Appendix II by R. C. Vaughan.

[16] V. Beresnevich, A. Haynes and S. L. Velani. Multiplicative zero-one laws and metric number theory. *Acta Arith.* **160** (2013), no. 2, 101–114.

[17] V. Beresnevich, A. Haynes and S. L. Velani. Sums of reciprocals of fractional parts and multiplicative Diophantine approximation. Pre-print: arXiv:1511.06862.

[18] V. Beresnevich, R. Vaughan and S. L. Velani. Inhomogeneous Diophantine approximation on planar curves. *Math. Ann.* **349** (2011), 929–942.

[19] V. Beresnevich, R. Vaughan, S. L. Velani and E. Zorin. Diophantine approximation on manifolds and the distribution of rational points: contributions to the convergence theory. Pre-print: arXiv:1506.09049

[20] V. Beresnevich, R. Vaughan, S. L. Velani and E. Zorin. Diophantine approximation on manifolds and the distribution of rational points: contributions to the divergence theory. In preparation.

[21] V. Beresnevich and S. L. Velani. A mass transference principle and the Duffin-Schaeffer conjecture for Hausdorff measures. *Ann. Math.* **164** (2006), 971–992.

[22] V. Beresnevich and S. L. Velani. Schmidt's theorem, Hausdorff measures and slicing. *IMRN* 2006, Article ID 48794, 24 pages.

[23] V. Beresnevich and S. L. Velani. A note on simultaneous Diophantine approximation on planar curves. *Math. Ann.* **337** (2007), no. 4, 769–796.

[24] V. Beresnevich and S. L. Velani. An inhomogeneous transference principle and Diophantine approximation. *Proc. Lond. Math. Soc.* (3) **101** (2010), no. 3, 821–851.

[25] V. Beresnevich and S. L. Velani. Ubiquity and a general logarithm law for geodesics, Dynamical systems and Diophantine approximation. *Séminaires et Congrès* 20 (2009), 21–36. Editors: Yann Bugeaud, Francoise Dal'Bo and Cornelia Drutu.

[26] V. Beresnevich and S. L. Velani. Simultaneous inhomogeneous diophantine approximation on manifolds, Diophantine and analytical problems in number theory. Conference Proceedings in honour of A. O. Gelfond: Moscow State University, *Fundam. Prikl. Mat.*, 16:5 (2010), 3–17.

[27] V. Beresnevich and S. L. Velani. Classical metric Diophantine approximation revisited: the Khintchine–Groshev theorem. *International Math. Research Notes*, (2010), 69–86.

[28] V. Beresnevich and S. L. Velani. A note on three problems in metric Diophantine approximation. *Dynamical system and Ergodic Theory Conference Proceedings in honour of S. G. Dani on the occasion of his 65th birthday*: University of Baroda, India, 26–29 Dec 2012. Contemporary Mathematics (2015), Volume 631. Editors: S. Bhattacharya, T. Das, A. Ghosh and R. Shah.

[29] V. Beresnevich and S. L. Velani. Divergence Borel–Cantelli Lemma revisited. In preparation.

[30] V. Beresnevich and E. Zorin. Explicit bounds for rational points near planar curves and metric Diophantine approximation. *Adv. Math.* **225** (2010), no. 6, 3064–3087.

[31] V. Bernik and M. Dodson. *Metric Diophantine Approximation on Manifolds*. Cambridge Tracts in Mathematics, 137. Cambridge University Press, 1999. xii+172 pp.

[32] A. S. Besicovitch. Sets of fractional dimensions (IV): on rational approximation to real numbers. *J. Lond. Math. Soc.* **9** (1934), 126–131.

[33] E. Borel. Sur un problème de probabilités aux fractions continues. *Math. Ann.* **72** (1912), 578–584.

[34] S. Brewer. Continued fractions of algebraic numbers. Preprint.

[35] Y. Bugeaud. A note on inhomogeneous Diophantine approximation. *Glasgow Math. J.* **45** (2003), 105–110.

[36] Y. Bugeaud and N. Chevallier. On simultaneous inhomogeneous Diophantine approximation. *Acta Arith.* **123** (2006), 97–123.

[37] J. W. S. Cassels. *An Introduction to Diophantine Approximation*. Cambridge University Press, 1957.

[38] J. W. S. Cassels. Some metrical theorems in Diophantine approximation. *I, Proc. Cambridge Philos. Soc.*, **46** (1950), 209–218.

[39] J. W. S. Cassels and H. P. F. Swinnerton-Dyer. On the product of three homogeneous linear forms and indefinite ternary quadratic forms. *Philos. Trans. Roy. Soc. London.* Ser. A, **248** (1955), 73–96.

[40] P. A. Catlin. Two problems in metric Diophantine approximation I. *J. Number Th.* **8** (1976), 282–288.

[41] H. Davenport. A note on Diophantine approximation. In *Studies in Mathematical Analysis and Related Topics*. Stanford University Press, 1962, pp. 77–82.

[42] H. Davenport. A note on Diophantine approximation II. *Mathematika* **11** (1964), 50–58
[43] H. Davenport and W. M. Schmidt. Dirichlet's theorem on Diophantine approximation, *Inst. Alt. Mat. Symp. Math.* **4** (1970), 113–132.
[44] L. Guan and J. Yu. Badly approximable vectors in higher dimension. Preprint: arXiv:1509.08050.
[45] H. Dickinson and S. L. Velani. Hausdorff measure and linear forms. *J. reine angew. Math.* **490** (1997), 1–36.
[46] M. Dodson, B. Rynne and J. Vickers. Diophantine approximation and a lower bound for Hausdorff dimension. *Mathematika*, **37** (1990), 59–73.
[47] M. Dodson, B. Rynne and J. Vickers. Khintchine type theorems on manifolds. *Acta Arithmetica* **57** (1991), no. 2, 115–130.
[48] R. J. Duffin and A. C. Schaeffer. Khintchine's problem in metric Diophantine approximation, *Duke Math. J.* **8** (1941), 243–255.
[49] M. Einsiedler, A. Katok and E. Lindenstrauss. Invariant measures and the set of exceptions to Littlewood's conjecture. *Ann. of Math.* (2) **164** (2006), no. 2, 513–560.
[50] K. Falconer. *The Geometry of Fractal Sets*, Cambridge Tracts in Mathematics, No. 85, Cambridge University Press, 1985.
[51] L. Fishman, D. Simmons and M. Urbański. Diophantine approximation and the geometry of limit sets in Gromov hyperbolic metric spaces. Preprint: arXiv:1301.5630.
[52] P. X. Gallagher. Approximation by reduced fractions, *J. Math. Soc. Japan* **13** (1961), 342–345.
[53] P. X. Gallagher. Metric simultaneous Diophantine approximation. *Jou. L.M.S.*, **37** (1962), 387–390.
[54] P. X. Gallagher. Metric simultaneous Diophantine approximation II. *Mathematika* **12** (1965), 123–127.
[55] A. Gorodnik and P. Vishe. Inhomogeneous multiplicative Littlewood conjecture and log savings. arXiv:1601.03525
[56] A. Gorodnik, F. Ramírez and P. Vishe. Inhomogeneous multiplicative approximation and Dirichlet non-improvability of typical points on planar curves. In preparation.
[57] G. H. Hardy and E. M. Wright. *An Introduction to the Theory of Numbers*. Sixth edition, Oxford University Press, Oxford, 2008.
[58] G. Harman. *Metric Number Theory*, LMS Monographs New Series, vol. 18, Clarendon Press, 1998.
[59] V. Jarník. Zur metrischen Theorie der diophantischen Appoximationen. *Prace Mat.-Fiz.* **36** (1928–29), 91–106.
[60] V. Jarník. Diophantischen Approximationen und Hausdorffsches Mass. *Mat. Sbornik* **36** (1929), 371–382.
[61] V. Jarník. Über die simultanen diophantischen Approximationen. *Math. Z.* **33** (1931), 505–543.
[62] L. K. Hua. *Introduction to Number Theory*. Springer-Verlag, 1982.
[63] Jing-Jing Huang. Rational points near planar curves and Diophantine approximation. *Advances in Mathematics* **274** (2015), 490–515.

[64] A. I. Khintchine. Einige Sätze über Kettenbruche, mit Anwendungen auf die Theorie der Diophantischen Approximationen. *Math. Ann.* **92** (1924), 115–125.
[65] A. I. Khintchine. Über die angenäherte Auflösung linearer Gleichungen in ganzen Zahlen. *Rec. math. Soc. Moscou Bd.* **32** (1925), 203–218.
[66] A. Y. Khintchine. *Continued Fractions*. Dover Publications, 1997.
[67] D. Kleinbock and G. Margulis. Flows on homogeneous spaces and Diophantine approximation on manifolds. *Ann. Math.* **148** (1998), 339–360.
[68] D. Kleinbock and B. Weiss. Modified Schmidt games and Diophantine approximation with weights. *Adv. Math.* **223** (2010), no. 4, 1276–1298.
[69] S. Kristensen, R. Thorn and S. L. Velani. Diophantine approximation and badly approximable sets. *Adv. Math.* **203** (2006), no. 1, 132–169.
[70] J. Kurzweil. On the metric theory of inhomogeneous Diophantine approximations. *Studia mathematica* **XV** (1955), 84–112.
[71] S. Lang and H. Trotter. Continued fractions for some algebraic numbers. In *Introduction to Diophantine Approximations*, p. 93, 1972.
[72] P. Mattila. *Geometry of Sets and Measures in Euclidean Space*. Cambridge University Press, 1995.
[73] L. G. Peck. Simultaneous rational approximations to algebraic numbers. *Bull. A.M.S.* **67** (1961), 197–201.
[74] O. Perron. Über diophantische Approximationen. *Math. Ann.* **83** (1921), 77–84.
[75] A. D. Pollington and R. C. Vaughan. The k-dimensional Duffin and Schaeffer conjecture. *Mathematika* **37** (1990), 190–200.
[76] A. D. Pollington and S. L. Velani. On a problem in simultaneously Diophantine approximation: Littlewood's conjecture. *Acta Math.* **66** (2000), 29–40.
[77] A. Pollington and S. L. Velani. On simultaneously badly approximable numbers. *J. London Math. Soc. (2)* **66** (2002), no.1, 29–40.
[78] F. Ramírez. Khintchine types of translated coordinate hyperplanes. *Acta Arith.* 170 (2015), no. 3, 243–273.
[79] F. Ramírez, D. Simmons and F. Süess. Rational approximation of affine coordinate subspaces of Euclidean spaces. Pre-print: arXiv:1510.05012.
[80] A. Rockett and P. Szüsz. *Continued Fractions*. World Scientific, Singapore, 1992.
[81] K. F. Roth. On irregularities of distribution. *Mathematika* **7** (1954), 73–79.
[82] K. F. Roth. Rational approximation to algebraic numbers. *Mathematika* **2** (1955), 1–20, with a corrigendum on p. 168.
[83] J. Schmeling and S. Troubetzkoy. Inhomogeneous Diophantine Approximation and Angular Recurrence for Polygonal Billards. *Math. Sbornik* **194** (2003), 295–309.
[84] W. M. Schmidt. Metrical theorems on fractional parts of sequences. *Trans. Amer. Math. Soc.* **110** (1964), 493–518.
[85] W. M. Schmidt. On badly approximable numbers and certain games. *Trans. Amer. Math. Soc.* **123** (1966), 178–199.
[86] W. M. Schmidt. Irregularities of Distribution. VII. *Acta. Arith.* **21** (1972), 45–50.
[87] W. M. Schmidt. *Diophantine Approximation*. Lecture Notes in Maths. 785, Springer-Verlag, 1975.
[88] N. A. Shah. Equidistribution of expanding translates of curves and Dirichlet's theorem on Diophantine approximation. *Invent. Math.* **177** (2009), no. 3, 509–532.

[89] U. Shapira. A solution to a problem of Cassels and Diophantine properties of cubic numbers. *Ann. of Math.* (2) **173** (2011), no. 1, 543–557.
[90] V. G. Sprindžuk. *Metric Theory of Diophantine Approximations*, translated by R. A. Silverman. John Wiley, 1979.
[91] R. C. Vaughan and S. L. Velani. Diophantine approximation on planar curves: the convergence theory. *Invent. Math.* **166** (2006), 103–124.
[92] M. Waldschmidt. Open Diophantine Problems. *Moscow Mathematical Journal* **4** (2004), 245–305.

2

Exponents of Diophantine Approximation

Yann Bugeaud

Abstract

We survey classical and recent results on exponents of Diophantine approximation. We give only a few proofs and highlight several open problems.

2.1 Introduction and Generalities

Although the present survey is essentially self-contained, the reader is assumed to have some basic knowledge of Diophantine approximation and of the theory of continued fractions. Classical references include Cassels' book [29] and Schmidt's monograph [73].

Let ξ be an irrational real number. It follows from the theory of continued fractions that there are infinitely many rational numbers p/q with $q \geq 1$ and such that

$$|q\xi - p| \leq q^{-1}.$$

Expressed differently, for arbitrarily large integers Q, there exist integers p and q with $1 \leq q \leq Q$ and $|q\xi - p| \leq Q^{-1}$. The classical Dirichlet theorem asserts much more, namely that, for *every* integer $Q \geq 1$, there exist integers p and q with $1 \leq q \leq Q$ and $|q\xi - p| \leq Q^{-1}$.

A question arises then naturally: is there some specific irrational real number ξ for which it is possible to improve the above statements, that is, to get the above inequalities with Q^{-1} replaced by Q^{-w} for some real number $w > 1$? Obviously, the quality of approximation strongly depends upon whether we are interested in a uniform statement (that is, a statement valid for every Q, or for every sufficiently large Q) or in a statement valid only for some arbitrarily large values of Q. This leads to the introduction of the exponents of approximation w_1 and $\hat{w}_1$.

Definition 2.1.1 *Let ξ be a real number. We denote by $w_1(\xi)$ the supremum of the real numbers w for which there exist arbitrarily large integers Q and integers p and q with $1 \le q \le Q$ and*

$$|q\xi - p| \le Q^{-w}.$$

We denote by $\hat{w}_1(\xi)$ the supremum of the real numbers $\hat{w}$ such that, for every sufficiently large integer Q, there are integers p and q with $1 \le q \le Q$ and

$$|q\xi - p| \le Q^{-\hat{w}}.$$

As observed by Khintchine [44], every irrational real number ξ satisfies $\hat{w}_1(\xi) = 1$, since there are arbitrarily large integers Q such that the inequality $|q\xi - p| \le 1/(2Q)$ has no solutions in integers p, q with $1 \le q \le Q$; see the proof of Proposition 2.4. However, for any $w > 1$, there exist irrational real numbers ξ such that, for *arbitrarily large* integers Q, the equation

$$|q\xi - p| \le Q^{-w}$$

has a solution in integers p and q with $1 \le q \le Q$. It suffices, for example, to take $\xi = \sum_{j \ge 1} 2^{-\lfloor (w+1)^j \rfloor}$, where $\lfloor \cdot \rfloor$ denotes the integer part.

The general framework is the following. For any (column) vector $\underline{\theta}$ in $\mathbb{R}^n$, we denote by $|\underline{\theta}|$ the maximum of the absolute values of its coordinates and by

$$\|\underline{\theta}\| = \min_{\underline{y} \in \mathbb{Z}^n} |\underline{\theta} - \underline{y}|$$

the maximum of the distances of its coordinates to the rational integers.

Definition 2.1.2 *Let n and m be positive integers and A a real matrix with n rows and m columns. Let $\underline{\theta}$ be an n-tuple of real numbers. We denote by $w(A, \underline{\theta})$ the supremum of the real numbers w for which, for arbitrarily large real numbers X, the inequalities*

$$\|A\underline{x} - \underline{\theta}\| \le X^{-w} \quad \text{and} \quad |\underline{x}| \le X \tag{2.1}$$

have a solution $\underline{x}$ in $\mathbb{Z}^m$. We denote by $\hat{w}(A, \underline{\theta})$ the supremum of the real numbers w for which, for all sufficiently large positive real numbers X, the inequalities (2.1) *have an integer solution $\underline{x}$ in $\mathbb{Z}^m$. The exponent $\hat{w}$ is called a uniform exponent.*

In the sequel, we consistently use the symbol $\hat{}$ to indicate that we require a uniform existence of solutions.

The lower bounds

$$w(A, \underline{\theta}) \ge \hat{w}(A, \underline{\theta}) \ge 0$$

are then obvious. We define furthermore two homogeneous exponents $w(A)$ and $\hat{w}(A)$ as in (2.1) with $\underline{\theta} = {}^t(0, \ldots, 0)$, requiring moreover that the integer solution $\underline{x}$ should be non-zero. In the case where A is the 1×1 matrix (ξ), we then have $w(A) = w_1(\xi)$ and $\hat{w}(A) = \hat{w}_1(\xi)$. The uniform exponent $\hat{w}$ was first introduced and studied by Jarník in the 1930s but with different notation.

The transposed matrix of a matrix A is denoted by ${}^tA$. Furthermore, $1/+\infty$ is understood to be 0. The following result, established in [29, 23], shows that the usual (respectively uniform) inhomogeneous exponents are strongly related to the uniform (respectively usual) homogeneous exponents.

Theorem 2.1.3 *Let n and m be positive integers and A a real matrix with n rows and m columns. For any n-tuple $\underline{\theta}$ of real numbers, we have the lower bounds*

$$w(A, \underline{\theta}) \geq \frac{1}{\hat{w}({}^tA)} \quad \text{and} \quad \hat{w}(A, \underline{\theta}) \geq \frac{1}{w({}^tA)}, \tag{2.2}$$

with equality in (2.2) *for almost all $\underline{\theta}$ with respect to the Lebesgue measure on $\mathbb{R}^n$.*

If the subgroup $G = {}^tA\mathbb{Z}^n + \mathbb{Z}^m$ of $\mathbb{R}^m$ generated by the n rows of the matrix A together with $\mathbb{Z}^m$ has maximal rank $m+n$, then Kronecker's theorem asserts that the dual subgroup $\Gamma = A\mathbb{Z}^m + \mathbb{Z}^n$ of $\mathbb{R}^n$ generated by the m columns of A and by $\mathbb{Z}^n$ is dense in $\mathbb{R}^n$. In this respect, Theorem 2.1.3 may be viewed as a measure of the density of Γ. In the case where the rank of G is smaller than $m + n$, we clearly have

$$\hat{w}({}^tA) = w({}^tA) = +\infty \quad \text{and} \quad \hat{w}(A, \underline{\theta}) = w(A, \underline{\theta}) = 0,$$

for every n-tuple $\underline{\theta}$ located outside a discrete family of parallel hyperplanes in $\mathbb{R}^n$. The assertion of Theorem 2.1.3 is then obvious when the rank of G is smaller than $m + n$.

Cassels' book [29] remains an invaluable reference for these and related questions.

In the sequel of the text, we restrict our attention to the cases where A is either a row or a column matrix:

$$A = (\xi_1, \ldots, \xi_n) \quad \text{or} \quad A = {}^t(\xi_1, \ldots, \xi_n).$$

This amounts to considering small values of the linear form

$$|x_0 + x_1\xi_1 + \ldots + x_n\xi_n|, \quad \text{where } x_0, x_1, \ldots, x_n \in \mathbb{Z},$$

or simultaneous approximation to $\xi_1, \ldots, \xi_n$ by rational numbers with the same denominator, that is, small values of the quantity

$$\max_{1 \le j \le n} |x_0 \xi_j - x_j|, \quad \text{where } x_0, x_1, \ldots, x_n \in \mathbb{Z}.$$

Furthermore, among the elements $\underline{\xi} = (\xi_1, \ldots, \xi_n)$ in $\mathbb{R}^n$, we mainly focus on the points

$$(\xi, \xi^2, \ldots, \xi^n)$$

whose coordinates are the n first successive powers of a real number ξ. However, some of the results stated below hold for a general n-tuple $\underline{\xi}$, as will be indicated in due course. This is in particular the case in Sections 2.6 and 2.7.

The present paper is organised as follows. In Section 2.2, we define six exponents of Diophantine approximation attached to real numbers and give their first properties. We discuss in Section 2.3 how these exponents are interrelated and study the sets of values taken by these exponents in Sections 2.4 and 2.5. Intermediate exponents are introduced in Section 2.6. Parametric geometry of numbers, a deep and powerful new theory introduced by Schmidt and Summerer [75, 76] and developed by Roy [65], is briefly described in Section 2.7. Recent results on the existence of real numbers which are badly approximable by algebraic numbers of bounded degree are discussed in Section 2.8. The final section gathers several open problems and suggestion for further research.

The notation $a \gg_d b$ means that a exceeds b times a constant depending only on d. When $\gg$ is written without any subscript, it means that the constant is absolute. We write $a \asymp b$ if both $a \gg b$ and $a \ll b$ hold.

2.2 Further Definitions and First Results

Here and below, the height $H(P)$ of a complex polynomial $P(X)$ is the maximum of the moduli of its coefficients and the height $H(\alpha)$ of an algebraic number α is the height of its minimal polynomial over $\mathbb{Z}$. We recall a very useful result of Gel'fond [37] (see also Lemma A.3 of [15]) which asserts that, if $P_1(X)$ and $P_2(X)$ are non-zero complex polynomials of degree n_1 and n_2, respectively, then we have

$$2^{-n_1-n_2} H(P_1) H(P_2) \le H(P_1 P_2) \le 2^{n_1+n_2} H(P_1) H(P_2). \tag{2.3}$$

Mahler [52] and Koksma [46] introduced in the 1930s two classifications of real numbers in terms of their properties of approximation by algebraic numbers.

Definition 2.2.1 *Let n be a positive integer and ξ be a real number. We denote by $w_n(\xi)$ the supremum of the real numbers w for which the inequality*

$$0 < |P(\xi)| \le H(P)^{-w} \tag{2.4}$$

is satisfied for infinitely many polynomials $P(X)$ with integer coefficients and degree at most n. We denote by $w_n^*(\xi)$ the supremum of the real numbers w^* for which the inequality

$$0 < |\xi - \alpha| \le H(\alpha)^{-w^*-1} \tag{2.5}$$

is satisfied for infinitely many algebraic numbers α of degree at most n.

Because of the -1 in the exponent of the right-hand side of (2.5), we get immediately that the exponents w_1 and w_1^* coincide. Let us briefly discuss the relation between w_n and w_n^*.

Let $n \ge 2$ be an integer, ξ be a real number, and α be an algebraic number of degree at most n satisfying (2.5) for some real number w^*. Let $P(X)$ denote the minimal defining polynomial of α over $\mathbb{Z}$. We may assume that ξ is not a root of $P(X)$. It follows from (2.3) that the height of the polynomial $P(X)/(X-\alpha)$ is bounded from above by the height of $P(X)$ times a positive constant depending only on n. This implies that

$$0 < |P(\xi)| \ll_{|\xi|,n} |\xi - \alpha| \cdot H(P) = |\xi - \alpha| \cdot H(\alpha), \tag{2.6}$$

which, combined with (2.5), gives

$$0 < |P(\xi)| \ll_{|\xi|,n} H(\alpha)^{-w^*} = H(P)^{-w^*}.$$

We deduce that

$$w_n(\xi) \ge w_n^*(\xi).$$

Let us now discuss the reverse inequality. Let $P(X)$ be an irreducible, integer polynomial of degree $n \ge 2$ and ξ be a real number which is not algebraic of degree at most n. Observe that $P(\xi)P'(\xi) \ne 0$ and

$$\frac{P'(\xi)}{P(\xi)} = \sum_{\alpha : P(\alpha)=0} \frac{1}{\xi - \alpha}.$$

Consequently, if α is the root of $P(X)$ which is closest to ξ, then we have

$$|\xi - \alpha| \le n \left| \frac{P(\xi)}{P'(\xi)} \right|.$$

Since $|P'(\xi)|$ is often roughly equal to $H(P)$ (this is the case unless $P(X)$ has two roots close to ξ), we expect that the estimation

$$|\xi - \alpha| \ll_{|\xi|,n} |P(\xi)| \cdot H(P)^{-1} = |P(\xi)| \cdot H(\alpha)^{-1} \tag{2.7}$$

holds. If $|P(\xi)| \le H(P)^{-w} = H(\alpha)^{-w}$, it would then give

$$|\xi - \alpha| \ll_{|\xi|,n} H(\alpha)^{-w-1}.$$

This shows that $w_n^*(\xi) \ge w_n(\xi)$ holds when there are integer polynomials of arbitrarily large height which satisfy (2.4) and have only one root close to ξ. The upper bound (2.7) does not hold when $|P'(\xi)|$ is small, that is, when $P(X)$ has two or more roots close to ξ.

The behaviour of the sequences $(w_n(\xi))_{n\ge1}$ and $(w_n^*(\xi))_{n\ge1}$ determines the localisation of ξ in Mahler's and Koksma's classifications, respectively (see Chapter 3 of [15]); however, the exact determination of $w_n(\xi)$ and $w_n^*(\xi)$ for a specific real number ξ is usually an extremely difficult problem.

We introduced in [22] four further exponents of Diophantine approximation. They implicitly appeared previously in articles of Jarník, Davenport and Schmidt, among others.

Definition 2.2.2 *Let n be a positive integer and ξ be a real number. We denote by $\lambda_n(\xi)$ the supremum of the real numbers λ such that the inequality*

$$\max_{1\le j\le n} |x_0\xi^j - x_j| \le |x_0|^{-\lambda}$$

has infinitely many solutions in integers $x_0, \ldots, x_n$ with $x_0 \ne 0$.

The three exponents w_1, w_1^* and λ_1 coincide. The three exponents w_n, w_n^* and λ_n have the common feature to be defined by the existence of infinitely many solutions for some set of Diophantine inequalities. We attach to them three exponents defined by a condition of uniform existence of solutions.

Definition 2.2.3 *Let n be a positive integer and ξ be a real number. We denote by $\hat{w}_n(\xi)$ the supremum of the real numbers $\hat{w}$ such that, for any sufficiently large real number X, the inequalities*

$$0 < |x_n\xi^n + \ldots + x_1\xi + x_0| \le X^{-\hat{w}}, \qquad \max_{0\le j\le n} |x_j| \le X,$$

have a solution in integers $x_0, \ldots, x_n$. We denote by $\hat{w}_n^*(\xi)$ the supremum of the real numbers $\hat{w}^*$ such that, for any sufficiently large real number X, there exists an algebraic real number α with degree at most n satisfying

$$0 < |\xi - \alpha| \le H(\alpha)^{-1} X^{-\hat{w}^*} \quad \text{and} \quad H(\alpha) \le X.$$

We denote by $\hat{\lambda}_n(\xi)$ the supremum of the real numbers $\hat{\lambda}$ such that, for any sufficiently large real number X, the inequalities

$$0 < |x_0| \le X, \qquad \max_{1\le j\le n} |x_0\xi^j - x_j| \le X^{-\hat{\lambda}},$$

have a solution in integers $x_0, \ldots, x_n$.

The three exponents $\hat{w}_1$, $\hat{w}_1^*$ and $\hat{\lambda}_1$ coincide.

This survey is mainly devoted to an overview of general results on the six exponents of approximation w_n, w_n^*, λ_n, $\hat{w}_n$, $\hat{w}_n^*$ and $\hat{\lambda}_n$, whose values are connected through various inequalities. The exact determination of the upper bounds is an important problem towards the Wirsing Conjecture (see Section 4) or related questions, such as the approximation of transcendental real numbers by algebraic integers. We begin with some easy properties.

All the exponents w_n, w_n^*, λ_n, $\hat{w}_n$, $\hat{w}_n^*$, $\hat{\lambda}_n$ can a priori take infinite values. However, we will see that the exponents 'hat' are bounded in terms of n. A first result in this direction is the following statement, which goes back to Khintchine [44].

Proposition 2.2.4 *For any irrational real number ξ, we have*

$$\hat{w}_1(\xi) = \hat{w}_1^*(\xi) = \hat{\lambda}_1(\xi) = 1.$$

It immediately follows from Proposition 2.2.4 that $\hat{\lambda}_n(\xi) \leq 1$ for every real number ξ and every integer $n \geq 1$.

Proof Let ξ be an irrational real number and $(p_\ell/q_\ell)_{\ell \geq 1}$ denote the sequence of its convergents. Let $\ell \geq 4$ and q be integers with $1 \leq q \leq q_\ell - 1$. Observe that $q_\ell - q_{\ell-1} \geq q_{\ell-2} \geq q_2 \geq 2$. Then, it follows from the theory of continued fractions that

$$\|q\xi\| \geq \|q_{\ell-1}\xi\| > \frac{1}{q_\ell + q_{\ell-1}} \geq \frac{1}{2(q_\ell - 1)}.$$

This shows that, setting $Q = q_\ell - 1$, the inequality $|q\xi - p| \leq 1/(2Q)$ has no solutions in integers p, q with $1 \leq q \leq q_\ell - 1$. Consequently, $\hat{w}_1(\xi)$ is at most equal to 1. The fact that $\hat{w}_1(\xi)$ is at least equal to 1 follows from the classical Dirichlet theorem. □

We gather in the next theorem several easy results on our six classical exponents of approximation.

Theorem 2.2.5 *For any positive integer n and any real number ξ which is not algebraic of degree $\leq n$, we have*

$$n \leq \hat{w}_n(\xi) \leq w_n(\xi), \qquad \frac{1}{n} \leq \hat{\lambda}_n(\xi) \leq \min\{1, \lambda_n(\xi)\},$$

and

$$1 \leq \hat{w}_n^*(\xi) \leq \min\{w_n^*(\xi), \hat{w}_n(\xi)\} \leq \max\{w_n^*(\xi), \hat{w}_n(\xi)\} \leq w_n(\xi). \tag{2.8}$$

As will be seen in the sequel, all the inequalities in (2.8) are sharp.

Proof Let n and ξ be as in the statement of the theorem. We have shown immediately after Definition 2.2.1 that $w_n^*(\xi) \le w_n(\xi)$. The same argumentation gives also that $\hat{w}_n^*(\xi) \le \hat{w}_n(\xi)$. The upper bounds $\hat{w}_n(\xi) \le w_n(\xi)$, $\hat{w}_n^*(\xi) \le w_n^*(\xi)$ and $\hat{\lambda}_n(\xi) \le \lambda_n(\xi)$ are consequences of the definitions, while the lower bounds $\hat{w}_n(\xi) \ge n$ and $\hat{\lambda}_n(\xi) \ge 1/n$ follow from Dirichlet's box principle (or from Minkowski's theorem). Moreover, we obviously have $\hat{w}_n^*(\xi) \ge \hat{w}_1^*(\xi) = 1$ and $\hat{\lambda}_n(\xi) \le \hat{\lambda}_1(\xi) = 1$. □

The next theorem was pointed out in [22].

Theorem 2.2.6 *For any positive integer n and any real number ξ not algebraic of degree at most n, we have*

$$\hat{w}_n^*(\xi) \ge \frac{w_n(\xi)}{w_n(\xi) - n + 1} \tag{2.9}$$

and

$$w_n^*(\xi) \ge \frac{\hat{w}_n(\xi)}{\hat{w}_n(\xi) - n + 1}. \tag{2.10}$$

Wirsing [82] proved a weaker version of (2.9) in which the left-hand side is replaced by the quantity $w_n^*(\xi)$. His result is also weaker than (2.10), since $\hat{w}_n(\xi) \le w_n(\xi)$.

Proof We follow an argumentation of Wirsing [82]. Let $n \ge 2$ and ξ be as in the statement of the theorem. We first establish (2.9). If $w_n(\xi)$ is infinite, then (2.9) reduces to $\hat{w}_n^*(\xi) \ge 1$, a statement established in Theorem 2.2.5. Assume that $w_n(\xi)$ is finite. Let $\varepsilon > 0$ and set $w = w_n(\xi)(1+\varepsilon)^2$. Minkowski's theorem implies that there exist a constant c and, for any positive real number H, a non-zero integer polynomial $P(X)$ of degree at most n such that

$$|P(\xi)| \le H^{-w}, \qquad \max_{1 \le j \le n-1} |P(j)| \le H \quad \text{and} \quad |P(n)| \le cH^{w-n+1}. \tag{2.11}$$

The definition of $w_n(\xi)$ and the first inequality of (2.11) show that $H(P) > H^{1+\varepsilon}$ if H is large enough. Consequently, $P(X)$ has some (necessarily real) root in the neighbourhood of each of the points $\xi, 1, \ldots, n-1$. Denoting by α its closest root to ξ and recalling that $H(\alpha) \ll_n H(P)$ (see for example Lemma A.3 of [15]), we get

$$|\xi - \alpha| \ll_n \frac{|P(\xi)|}{H(P)} \ll_n H(\alpha)^{-1} \left(H^{w-n+1}\right)^{-w/(w-n+1)}$$

and, by (2.11),

$$H(\alpha) \ll_n H(P) \ll_n H^{w-n+1}.$$

Since all of this is true for every sufficiently large H, we get $\hat{w}_n^*(\xi) \geq w/(w - n + 1)$. Selecting now ε arbitrarily close to 0, we obtain (2.9).

In order to establish (2.10), we may assume that $\hat{w}_n(\xi)$ is finite and set $w = \hat{w}_n(\xi)(1 + \varepsilon)^2$. We follow the same argument as in the proof of (2.9). The definition of $\hat{w}_n(\xi)$ and the first inequality of (2.11) then show that there exist arbitrarily large values of H for which the polynomial $P(X)$ satisfies $H(P) > H^{1+\varepsilon}$. We conclude that there exist algebraic numbers α of arbitrarily large height with

$$|\xi - \alpha| \ll_n H(\alpha)^{-1-w/(w-n+1)}.$$

Thus, we get $w_n^*(\xi) \geq w/(w - n + 1)$ and, selecting ε arbitrarily close to 0, we obtain (2.10). □

The next result shows that if $w_n(\xi) = n$ holds, then the values of the five other exponents at the point ξ are known.

Corollary 2.2.7 *Let n be a positive integer and ξ a real number such that $w_n(\xi) = n$. Then we have*

$$w_n(\xi) = w_n^*(\xi) = \hat{w}_n(\xi) = \hat{w}_n^*(\xi) = n \tag{2.12}$$

and

$$\lambda_n(\xi) = \hat{\lambda}_n(\xi) = \frac{1}{n}. \tag{2.13}$$

Proof Equalities (2.12) follow from Theorems 2.2.5 and 2.2.6. Khintchine's transference theorem (Theorem 2.3.2) shows that $w_n(\xi) = n$ is equivalent to $\lambda_n(\xi) = 1/n$. Combined with Theorem 2.2.5, this gives (2.13). □

Since $w_n(\xi)$ is equal to n for almost all real numbers ξ, with respect to the Lebesgue measure (this was proved by Sprindžuk [78]), the next result is an immediate consequence of Corollary 2.2.7.

Theorem 2.2.8 *For almost all (with respect to Lebesgue measure) real numbers ξ and every positive integer n, we have*

$$w_n(\xi) = w_n^*(\xi) = \hat{w}_n(\xi) = \hat{w}_n^*(\xi) = n$$

and

$$\lambda_n(\xi) = \hat{\lambda}_n(\xi) = \frac{1}{n}.$$

Before discussing the values taken by our exponents at algebraic points, we recall Liouville's inequality (see Theorem A.1 and Corollary A.2 in [15]).

Theorem 2.2.9 (Liouville's Inequality) *Let α and β be distinct real algebraic numbers of degree n and m, respectively. Then,*

$$|\alpha - \beta| \geq (n+1)^{-m}(m+1)^{-n} H(\alpha)^{-m} H(\beta)^{-n}.$$

Furthermore, if $P(X)$ is an integer polynomial of degree n which does not vanish at β, then

$$|P(\beta)| \geq (n+1)^{-m}(m+1)^{-n} H(P)^{-m+1} H(\beta)^{-n}.$$

Let ξ be a real algebraic number of degree $d \geq 1$. It follows from Theorem 2.2.9 that $w_n^*(\xi) \leq w_n(\xi) \leq d-1$ holds for $n \geq 1$. Roth's theorem, which asserts that $w_1(\xi) = 1$ if ξ is irrational, has been considerably extended by Schmidt [73], who showed that $w_n(\xi) = w_n^*(\xi) = n$ holds for $n \leq d-1$. Furthermore, we deduce from Definitions 2.2.2 and 2.2.3 that $\lambda_n(\xi) = \lambda_{d-1}(\xi)$ and $\hat{\lambda}_n(\xi) = \hat{\lambda}_{d-1}(\xi)$ for $n \geq d$. All this and Corollary 2.2.7 enable us to get the values of our six exponents at real algebraic numbers.

Theorem 2.2.10 *Let ξ be a real algebraic number of degree $d \geq 1$ and let n be a positive integer. We have*

$$w_n(\xi) = w_n^*(\xi) = \hat{w}_n(\xi) = \hat{w}_n^*(\xi) = \min\{n, d-1\}$$

and

$$\lambda_n(\xi) = \hat{\lambda}_n(\xi) = \frac{1}{\min\{n, d-1\}}.$$

Theorem 2.2.10 shows that real algebraic numbers of degree greater than n do behave like almost all real numbers, as far as approximation by algebraic numbers of degree less than n is concerned. We may as well consider approximation to complex (non-real) numbers. Quite surprisingly, complex non-real numbers of degree greater than n do not always behave like almost all complex numbers; see [21].

Theorem 2.2.10 shows that we can focus on the values taken by our exponents at transcendental, real numbers. This motivates the following definition.

Definition 2.2.11 *The spectrum of an exponent of approximation is the set of values taken by this exponent at transcendental real numbers.*

We point out an important problem, which will be discussed in the next sections.

Problem 2.2.12 *To determine the spectra of the exponents w_n, w_n^*, λ_n, $\hat{w}_n$, $\hat{w}_n^*$, $\hat{\lambda}_n$.*

Results towards Problem 2.2.12 are gathered in Section 2.5.

2.3 Overview of Known Relations Between Exponents

We begin this section with an easy result on the difference between the exponents w_n and w_n^*.

Theorem 2.3.1 *For any positive integer n and any transcendental real number ξ, we have*

$$w_n(\xi) - n + 1 \le w_n^*(\xi) \le w_n(\xi) \tag{2.14}$$

and

$$\hat{w}_n(\xi) - n + 1 \le \hat{w}_n^*(\xi) \le \hat{w}_n(\xi). \tag{2.15}$$

Proof The right-hand side inequalities of (2.14) and (2.15) have been already stated in Theorem (2.2.5). The left-hand side of inequality (2.14) is inequality (3.11) in [15], whose proof also gives the left-hand side of inequality (2.15). □

It is interesting to note that the left-hand side inequality of (2.15) is sharp since there exist real numbers ξ with $\hat{w}_n(\xi) = n$ and $\hat{w}_n^*(\xi) = 1$; see Corollary 2.5.4. We do not know if the left-hand side inequality of (2.14) is sharp for $n \ge 4$; see Theorem 2.5.7.

We indicate now some transference results linking together the rational simultaneous approximation to $\xi, \dots, \xi^n$ and small values of the linear form with coefficients $\xi, \dots, \xi^n$.

Theorem 2.3.2 *For every integer $n \ge 2$ and every real number ξ which is not algebraic of degree at most n, we have*

$$\frac{1}{n} \le \frac{w_n(\xi)}{(n-1)w_n(\xi) + n} \le \lambda_n(\xi) \le \frac{w_n(\xi) - n + 1}{n} \tag{2.16}$$

and

$$\frac{1}{n} \leq \frac{\hat{w}_n(\xi) - 1}{(n-1)\hat{w}_n(\xi)} \leq \hat{\lambda}_n(\xi) \leq \frac{\hat{w}_n(\xi) - n + 1}{\hat{w}_n(\xi)}. \tag{2.17}$$

Proof The inequalities (2.16) follow directly from Khintchine's transference principle (cf. Theorem B.5 from [15]), whose proof shows that the same inequalities hold for the uniform exponents; see [40]. The latter result is weaker than inequalities (2.17), which have been recently proved by German [38]. □

Observe that (2.17) with $n = 2$ reduces to the following result established by Jarník [40] in 1938; see also [45] for an alternative proof.

Theorem 2.3.3 *For every transcendental real number ξ we have*

$$\hat{\lambda}_2(\xi) = 1 - \frac{1}{\hat{w}_2(\xi)}.$$

Inequalities (2.16) have been recently refined in [49, 25] by means of the introduction of uniform exponents.

Theorem 2.3.4 *For every integer $n \geq 2$ and every real number ξ which is not algebraic of degree at most n, we have*

$$\lambda_n(\xi) \geq \frac{(\hat{w}_n(\xi) - 1)w_n(\xi)}{((n-2)\hat{w}_n(\xi) + 1)w_n(\xi) + (n-1)\hat{w}_n(\xi)}$$

and

$$\lambda_n(\xi) \leq \frac{(1 - \hat{\lambda}_n(\xi))w_n(\xi) - n + 2 - \hat{\lambda}_n(\xi)}{n-1}.$$

Since $\hat{\lambda}_n(\xi) \geq 1/n$ and $\hat{w}_n(\xi) \geq n$, one easily checks that Theorem 2.3.4 implies (2.16).

The first inequality in the next theorem was established by Davenport and Schmidt [33], while investigating the approximation to a real number by algebraic integers. The second one is a recent result of Schleischitz [69].

Theorem 2.3.5 *For any positive integer n and any transcendental real number ξ, we have*

$$w_n^*(\xi) \geq \frac{1}{\hat{\lambda}_n(\xi)} \quad \text{and} \quad \hat{w}_n^*(\xi) \geq \frac{1}{\lambda_n(\xi)}.$$

It is interesting to note that the combination of (2.17) with the first inequality of Theorem 2.3.5 gives (2.10).

Theorem 2.3.5 relates approximation to ξ by algebraic numbers of degree at most n with uniform simultaneous rational approximation to $\xi, \ldots, \xi^n$. Additional explanations can be found in [28] and in Section 3.6 of [15].

Note that Theorem 2.3.5 can be compared with the next result, extracted from [23, 29], which is a particular case of Theorem 2.1.3. If A is the matrix $(\xi, \xi^2, \ldots, \xi^n)$ and $\underline{\theta}$ is a real n-tuple, we simply write $w_n(\xi, \underline{\theta})$ for $w(A, \underline{\theta})$ and $\hat{w}_n(\xi, \underline{\theta})$ for $\hat{w}(A, \underline{\theta})$.

Theorem 2.3.6 *For any integer $n \geq 1$, any transcendental real number ξ and any real n-tuple $\underline{\theta}$, we have*

$$w_n(\xi, \underline{\theta}) \geq \frac{1}{\hat{\lambda}_n(\xi)} \quad \textit{and} \quad \hat{w}_n(\xi, \underline{\theta}) \geq \frac{1}{\lambda_n(\xi)}.$$

The next results were proved by Schmidt and Summerer [76, 77]; see also Moshchevitin [54, 55].

Theorem 2.3.7 *For any transcendental real number ξ we have*

$$w_n(\xi) \geq \hat{w}_n(\xi) \frac{(n-1)(\hat{w}_n(\xi) - 1)}{1 + (n-2)\hat{w}_n(\xi)},$$

for $n \geq 2$, and

$$w_3(\xi) \geq \hat{w}_3(\xi) \frac{\sqrt{4\hat{w}_3(\xi) - 3} - 1}{2}.$$

The case $n = 2$ of Theorem 2.3.7 was proved by Jarník [42].

Theorem 2.3.8 *For any transcendental real number ξ we have*

$$\lambda_n(\xi) \geq \hat{\lambda}_n(\xi) \frac{\hat{\lambda}_n(\xi) + n - 2}{(n-1)(1 - \hat{\lambda}_n(\xi))},$$

for $n \geq 2$, and

$$\lambda_3(\xi) \geq \hat{\lambda}_3(\xi) \frac{\hat{\lambda}_3(\xi) + \sqrt{\hat{\lambda}_3(\xi)(4 - 3\hat{\lambda}_3(\xi))}}{2(1 - \hat{\lambda}_3(\xi))}.$$

Actually, Theorems 2.3.2 to 2.3.4 and 2.3.6 to 2.3.8 are valid not only for tuples of the shape $(\xi, \xi^2, \ldots, \xi^n)$, but also for general tuples $\underline{\xi}$, whose coordinates are, together with 1, linearly independent over $\mathbb{Z}$.

2.4 Bounds for the Exponents of Approximation

Let n be a positive integer. As we have seen in Section 2.2, the Dirichlet *Schubfachprinzip* (or, if one prefers, Minkowski's theorem) readily implies that $w_n(\xi)$ is at least equal to n for any positive integer n and any real number ξ not algebraic of degree $\le n$. It is a longstanding problem, which was first formulated by Wirsing [82], to decide whether the same result remains true for the quantity $w_n^*(\xi)$.

Conjecture 2.4.1 (Wirsing) *For any positive integer n and any transcendental real number ξ we have $w_n^*(\xi) \ge n$.*

The seminal paper of Wirsing [82] and the study of his conjecture, which has been up to now confirmed only for $n = 1$ (this follows from the theory of continued fractions) and $n = 2$ (by Davenport and Schmidt [31]), have motivated many works.

Theorem 2.4.2 *Let ξ be a real number which is neither rational, nor quadratic. Then, for any real number c greater than $160/9$, there exist infinitely many rational or quadratic real numbers α satisfying*

$$|\xi - \alpha| \le c \max\{1, |\xi|^2\} H(\alpha)^{-3}.$$

In particular, we have $w_2^*(\xi) \ge 2$.

Theorem 2.4.2 was proved in [31]; see also [73]. It has been extended by Davenport and Schmidt [32] (up to the value of the numerical constant) as follows.

Theorem 2.4.3 *Let $n \ge 2$ be an integer and ξ be a real number which is not algebraic of degree at most n. Then there exist an effectively computable constant c, depending only on ξ and on n, an integer d with $1 \le d \le n-1$, and infinitely many integer polynomials $P(X)$ of degree n whose roots $\alpha_1, \dots, \alpha_n$ can be numbered in such a way that*

$$|(\xi - \alpha_1) \dots (\xi - \alpha_d)| \le c\, H(P)^{-n-1}.$$

Theorem 2.4.2 has recently been improved by Moshchevitin [56] as follows.

Theorem 2.4.4 *For any real number ξ which is neither rational, nor a quadratic irrationality, we have*

$$w_2^*(\xi) \ge \hat{w}_2(\xi)(\hat{w}_2(\xi) - 1) \ge 2.$$

The proof of Theorem 2.4.4 combines ideas from [31] with an argument used by Jarník [41, 42] in his proof of the case $n = 2$ of Theorem 2.3.7.

The first statement of Theorem 2.4.5 was proved by Wirsing [82] and the second one by Bernik and Tishchenko [12].

Theorem 2.4.5 *Let n be a positive integer and ξ be a transcendental real number. Then we have*

$$w_n^*(\xi) \geq \frac{w_n(\xi)+1}{2} \geq \frac{n+1}{2}$$

and

$$w_n^*(\xi) \geq \frac{n}{4} + \frac{\sqrt{n^2+16n-8}}{4}. \tag{2.18}$$

For proofs of Theorem 2.4.5 and related results, the reader may consult Chapter 3 of [15]. Slight improvements on (2.18) have been subsequently obtained by Tishchenko [79, 80], with very technical proofs. See also [19] for an intermediate result between Wirsing's lower bound $w_n^*(\xi) \geq (n+1)/2$ and Theorem 2.4.3.

The next result, extracted from [27], provides, under a suitable assumption, an upper bound for $\hat{w}_n^*$ in terms of w_m, when m is at most equal to n.

Theorem 2.4.6 *Let m, n be positive integers with $1 \leq m \leq n$ and ξ a real number. If $w_m(\xi) \geq m+n-1$, then we have*

$$\hat{w}_n^*(\xi) \leq \frac{m w_m(\xi)}{w_m(\xi)-n+1}.$$

Proof It is inspired by the proof of Proposition 2.1 of [22]. Let m, n and ξ be as in the statement of the theorem. Assume for convenience that $|\xi| \leq 1$. Let ε be a real number with $0 < \varepsilon \leq 1/2$. Set $w := w_m(\xi)$. Let $P(X)$ be an integer polynomial of degree at most m and height $H := H(P)$ large such that

$$H^{-w-\varepsilon} < |P(\xi)| < H^{-w+\varepsilon}. \tag{2.19}$$

By using (2.3), we may assume without any loss of generality that $P(X)$ is irreducible and primitive. Let v be a positive real number and set $M = H^{w/(v(1+\varepsilon))}$. Let α be the root of $P(X)$ which is the closest to ξ. If $|\xi - \alpha| \leq H(\alpha)^{-1} H^{-w(1+2\varepsilon)/(1+\varepsilon)}$, then, by (2.6), we have

$$|P(\xi)| \ll_n |\xi - \alpha| \cdot H(P) \ll_n H^{-w(1+2\varepsilon/3)} \ll_n H^{-w-4\varepsilon/3},$$

using that $\varepsilon \le 1/2$ and $w \ge 2$. This gives a contradiction to (2.19) if H is large enough. Consequently, we have

$$|\xi - \alpha| > H(\alpha)^{-1} M^{-v(1+2\varepsilon)}, \tag{2.20}$$

provided that H is large enough.

Assume that there exists an algebraic number β of height at most M and degree at most n which satisfies

$$|\xi - \beta| \le H(\beta)^{-1} M^{-v(1+2\varepsilon)}. \tag{2.21}$$

It follows from (2.20) that $\beta \neq \alpha$. Liouville's inequality (Theorem 2.2.9) then gives

$$|P(\beta)| \gg_n H^{-n+1} H(\beta)^{-m}. \tag{2.22}$$

By Rolle's theorem and the fact that $|\xi| \le 1$, we have

$$|P(\beta)| \le |\xi - \beta| \cdot \max_{t:|t|\le 2} |P'(t)| + |P(\xi)| \le m2^m |\xi - \beta| H + H^{-w+\varepsilon}. \tag{2.23}$$

If $H^{-w+\varepsilon} \ge |\xi - \beta| \cdot H$, then (2.22) and (2.23) imply

$$H(\beta)^{-m} \ll_n H^{n-1-w+\varepsilon}$$

and, since $H(\beta) \le H^{w/(v(1+\varepsilon))}$, we get

$$v \le \frac{mw}{(w+1-n-2\varepsilon)(1+\varepsilon)}, \tag{2.24}$$

provided that H is large enough.

If $H^{-w+\varepsilon} \le |\xi - \beta| \cdot H$, then, by (2.21), (2.22) and (2.23), we get

$$H(\beta)^{-m+1} H^{-n} \ll_n H^{-w(1+2\varepsilon)/(1+\varepsilon)},$$

hence,

$$H^{(m-1)w/(v(1+\varepsilon))} \gg_n H^{-n+w(1+2\varepsilon)/(1+\varepsilon)}.$$

This implies

$$v \le \frac{(m-1)w}{w(1+2\varepsilon) - n(1+\varepsilon) - \varepsilon}, \tag{2.25}$$

provided that H is large enough. Consequently, no such algebraic number β can exist unless v satisfies (2.24) and (2.25). We deduce that $\hat{w}_n^*(\xi) < v(1+2\varepsilon)$ as soon as v exceeds the left-hand sides of (2.24) and (2.25). Since ε can be taken arbitrarily close to 0, this gives

$$\hat{w}_n^*(\xi) \le \max\left\{\frac{mw}{w+1-n}, \frac{(m-1)w}{w-n}\right\} = \frac{mw}{w+1-n},$$

using that $w \ge m+n-1$. This ends the proof of the theorem. □

Davenport and Schmidt [33] have given uniform upper bounds for the exponents $\hat{\lambda}_n$ and $\hat{w}_n$. Some of their results have been subsequently improved by Laurent [47] and Bugeaud and Schleischitz [27]. For a positive real number x, we denote by $\lceil x \rceil$ the smallest integer greater than or equal to x. The next theorem gathers results obtained in [33, 47, 27].

Theorem 2.4.7 *For any positive integer n and any transcendental real number ξ, we have*

$$\hat{\lambda}_n(\xi) \leq \frac{1}{\lceil n/2 \rceil} \quad \textit{and} \quad \hat{w}_n^*(\xi) \leq \hat{w}_n(\xi) \leq n - \frac{1}{2} + \sqrt{n^2 - 2n + \frac{5}{4}}.$$

The right-hand side of the last inequality can be written $2n - \frac{3}{2} + \varepsilon_n$, where ε_n is positive for $n \geq 3$ and $\lim_{n \to +\infty} \varepsilon_n = 0$.

We refer to the original articles for a proof of Theorem 2.4.7 and content ourselves to establish a weaker result.

Proof Davenport and Schmidt [33] established that $\hat{w}_n(\xi) \leq 2n - 1$, which is weaker than the result given here and established in [27]. We only explain how to get the easy estimate $\hat{w}_n^*(\xi) \leq 2n - 1$. Let $w < \hat{w}_n^*(\xi)$ be a real number. By definition of $\hat{w}_n^*$, there exist arbitrarily large integers H and distinct algebraic numbers α_1, α_2 of degree at most n such that $H(\alpha_1) \leq H(\alpha_2) \leq H$ and

$$|\xi - \alpha_1| < H(\alpha_1)^{-1} H^{-w}, \quad |\xi - \alpha_2| < H(\alpha_2)^{-1} H^{-w}.$$

This implies that $|\alpha_1 - \alpha_2| < 2H(\alpha_1)^{-1} \cdot H^{-w}$, while Theorem 2.2.9 (Liouville's inequality) gives that

$$|\alpha_1 - \alpha_2| \gg_n H(\alpha_1)^{-n} \cdot H(\alpha_2)^{-n} \gg_n H(\alpha_1)^{-1} \cdot H^{-2n+1}.$$

By combining these two inequalities, we deduce that $w \leq 2n - 1$. This proves the upper bound $\hat{w}_n^*(\xi) \leq 2n - 1$. Arguing now with polynomials instead of algebraic numbers, this can be strengthened to $\hat{w}_n(\xi) \leq 2n - 1$, as was shown in [33]. □

Theorem 2.4.7 has been improved for small values of n.

Theorem 2.4.8 *For any transcendental real number ξ, we have*

$$\hat{\lambda}_2(\xi) \leq \frac{\sqrt{5} - 1}{2} \quad \textit{and} \quad \hat{w}_2(\xi) \leq \frac{3 + \sqrt{5}}{2}, \tag{2.26}$$

and both inequalities are best possible. For any transcendental real number ξ, we have

$$\hat{\lambda}_3(\xi) \le \frac{2+\sqrt{5}-\sqrt{7+2\sqrt{5}}}{2} = 0.4245\ldots$$

The bound for $\hat{\lambda}_2(\xi)$ given in (2.26) was proved in [33] and the one for $\hat{w}_2$ in [2], whose authors were at that time unaware of Jarník's result (Theorem 2.3.3).

The last assertion of Theorem 2.4.8 was proved by Roy [63], who improved the estimate $\hat{\lambda}_3(\xi) \le 1/2$ given in Theorem 2.4.7. He further indicated that his upper bound for the exponent is not best possible.

Roy [57, 59] showed that the inequalities (2.26) are sharp. The set of values taken by the exponents $\hat{w}_2$ and $\hat{w}_2^*$ has been studied in [22, 34, 35, 36, 60, 61, 62]. Among other results, we know that the spectrum of $\hat{w}_2$ is dense in $[2, (3+\sqrt{5})/2]$ and that there exists a real number $c < (3+\sqrt{5})/2$ such that the intersection of this spectrum with $[c, (3+\sqrt{5})/2]$ is countable.

We end this section with a recent result of Schleischitz [70].

Theorem 2.4.9 *For any positive integer n and any transcendental real number ξ, we have*

$$\hat{\lambda}_n(\xi) \le \max\left\{\frac{1}{n}, \frac{1}{\lambda_1(\xi)}\right\}.$$

Proof The theorem is clearly true if $n = 1$ or if $\lambda_1(\xi) = 1$. Assume that $n \ge 2$ and that ξ is in $(0, 1)$ with $\lambda_1(\xi) > 1$. Let q be a large positive integer and v be a real number greater than 1 such that $q^{-v} < \min\{\xi, 1-\xi, 1/(2nq)\}$ and

$$\|q\xi\| \le q^{-v}. \tag{2.27}$$

Let p be the integer such that $|q\xi - p| = \|q\xi\|$. Without any loss of generality, we may assume that p and q are coprime. Observe that $0 < p/q < 1$ and that, for $j = 1, \ldots, n$, we have

$$\left|\xi^j - \frac{p^j}{q^j}\right| = \left|\xi - \frac{p}{q}\right| \cdot \left|\xi^{j-1} + \ldots + \left(\frac{p}{q}\right)^{j-1}\right| \le \frac{n}{q^{1+v}} \le \frac{1}{2q^2}. \tag{2.28}$$

Let v' be a real number with $1 < v' < \min\{v, n\}$ and set $X = q^{v'}$. Let x be a positive integer with $x < X$ and express x in base q. There exist integers $b_0, b_1, \ldots, b_{n-1}$ in $\{0, 1, \ldots, q-1\}$ such that

$$x = b_0 + b_1 q + b_2 q^2 + \ldots + b_{n-1} q^{n-1}.$$

Let u in $\{1, 2, \ldots, n\}$ be the smallest index such that b_{u-1} is non-zero. Then,

$$\|xp^u/q^u\| = \|b_{u-1}p^u/q\| \geq 1/q. \tag{2.29}$$

Furthermore, it follows from (2.28) and the fact that $v' < v$ that

$$\left|x\left(\xi^u - \frac{p^u}{q^u}\right)\right| \leq \frac{nq^{v'}}{q^{1+v}} < \frac{1}{2q},$$

if q is sufficiently large. Denoting by y the nearest integer to $x\xi^u$, the triangle inequality and (2.29) then give

$$\begin{aligned}\max_{1\leq j\leq n} \|x\xi^j\| \geq \|x\xi^u\| &= \left|\left(\frac{xp^u}{q^u} - y\right) + x\left(\xi^u - \frac{p^u}{q^u}\right)\right| \\ &\geq \frac{1}{q} - \frac{1}{2q} \geq \frac{1}{2q} = \frac{1}{2X^{1/v'}}.\end{aligned}$$

If (2.27) is satisfied for arbitrarily large integers q, this shows that

$$\hat{\lambda}_n(\xi) \leq \frac{1}{v'}.$$

As v' can be chosen arbitrarily close to $\min\{\lambda_1(\xi), n\}$, we have proved that

$$\hat{\lambda}_n(\xi) \leq \frac{1}{\min\{\lambda_1(\xi), n\}} = \max\left\{\frac{1}{n}, \frac{1}{\lambda_1(\xi)}\right\}.$$

This completes the proof of the theorem. □

We point out an immediate corollary of Theorem 2.4.9.

Corollary 2.4.10 *For any positive integer n, any transcendental real number ξ with $w_1(\xi) \geq n$ satisfies $\hat{w}_k(\xi) = k$ and $\hat{\lambda}_k(\xi) = 1/k$ for $k = 1, \ldots, n$.*

Proof This follows from Theorem 2.4.9 combined with Theorem 2.3.2. □

2.5 Spectra

This section is mainly devoted to the study of the spectra of the six exponents of approximation defined in Section 2.2.

We begin with an auxiliary result, extracted from [16], which confirms the existence of real numbers ξ for which $w_1(\xi) = w_n(\xi)$, for any given integer $n \geq 2$.

Theorem 2.5.1 *Let $n \geq 1$ be an integer. For any real number $w \geq 2n-1$, there exist uncountably many real numbers ξ such that*

$$w_1(\xi) = w_1^*(\xi) = \ldots = w_n(\xi) = w_n^*(\xi) = w.$$

The spectra of w_n and w_n^* include the interval $[2n-1, +\infty]$.

Proof This is clear for $w = n = 1$. Let $w > 1$ be a real number. Let M be a large positive integer and consider the real number ξ_w given by its continued fraction

$$\xi_w = [0; 2, M\lfloor q_1^{w-1} \rfloor, M\lfloor q_2^{w-1} \rfloor, M\lfloor q_3^{w-1} \rfloor, \ldots],$$

where $q_1 = 2$ and q_j is the denominator of the rational number $p_j/q_j = [0; 2, M\lfloor q_1^{w-1} \rfloor, \ldots, M\lfloor q_{j-1}^{w-1} \rfloor]$, for $j \geq 2$. By construction, we have

$$q_{j+1} \asymp M q_j^w \quad \text{and} \quad \left| \xi_w - \frac{p_j}{q_j} \right| \asymp \frac{1}{M q_j^{w+1}},$$

for $j \geq 1$. Consequently, we get

$$w = w_1(\xi_w) \leq \ldots \leq w_n(\xi_w). \tag{2.30}$$

Using triangle inequalities, it is shown in [16] that, if M is sufficiently large and $w \geq 2n-1$, then

$$|P(\xi_w)| \gg_{n,M} H(P)^{-w}$$

holds for every polynomial $P(X)$ of degree at most n and sufficiently large height, hence $w_n(\xi_w) \leq w$ and the inequalities in (2.30) are indeed equalities. An additional argument is needed to show that $w_1^*(\xi) = \ldots = w_n^*(\xi) = w$; see [16] for the complete proof. □

It would be desirable to replace the assumption $w \geq 2n-1$ (which is at the moment the best known) in Theorem 2.5.1 by a weaker one. Actually, the value $2n-1$ comes from Theorem 2.2.9 (Liouville's inequality), which is widely used in the proof of Theorem 2.5.1.

Theorem 2.5.1 is a key tool to get results on the spectra of various exponents of approximation.

The next result, also established in [16], gives a relationship between the exponents λ_n and λ_m when m divides n.

Lemma 2.5.2 *For any positive integers k and n, and any transcendental real number ξ we have*

$$\lambda_{kn}(\xi) \geq \frac{\lambda_k(\xi) - n + 1}{n}.$$

A similar inequality holds between the uniform exponents, but it gives nothing interesting since $\hat{\lambda}_k(\xi) \leq 1$ for every integer $k \geq 1$ and every irrational real number ξ.

Proof Let v be a positive real number and q be a positive integer such that

$$\max_{1 \leq j \leq k} |q\xi^j - p_j| \leq q^{-v},$$

for suitable integers $p_1, \ldots, p_k$. Let h be an integer with $1 \leq h \leq kn$. Write $h = j_1 + \ldots + j_m$ with $m \leq n$ and $1 \leq j_1, \ldots, j_m \leq k$. Then,

$$|q^m \xi^h - p_{j_1} \cdots p_{j_m}| \ll_m q^{m-1} q^{-v}$$

and

$$\|q^n \xi^h\| \ll q^{n-m} \|q^m \xi^h\| \ll_m q^{n-1-v} \ll_m (q^n)^{-(v-n+1)/n},$$

independently of h. This proves the lemma. □

We display an immediate consequence of Lemma 2.5.2.

Corollary 2.5.3 *Let ξ be a real irrational number. Then, $\lambda_n(\xi) = +\infty$ holds for every positive integer n if, and only if, $\lambda_1(\xi) = +\infty$.*

Combined with Theorems 2.2.5 and 2.4.6 and Corollary 2.4.10, Corollary 2.5.3 allows us to determine the values taken at Liouville numbers (recall that a Liouville number is, by definition, a real number ξ satisfying $w_1(\xi) = +\infty$) by our six exponents of approximation.

Corollary 2.5.4 *For any positive integer n and any Liouville number ξ, we have*

$$w_n(\xi) = w_n^*(\xi) = \lambda_n(\xi) = +\infty,$$

$$\hat{w}_n(\xi) = n, \quad \hat{w}_n^*(\xi) = 1, \quad \textit{and} \quad \hat{\lambda}_n(\xi) = \frac{1}{n}.$$

The proof of Theorem 2.5.1 shows how the theory of continued fractions allows us to construct explicitly real numbers ξ having any arbitrarily prescribed value for $\lambda_1(\xi)$ (recall that $\lambda_1(\xi) = w_1(\xi)$). The same question for an exponent λ_n with $n \geq 2$ is not yet solved. At present, the best known result was proved in [16] and is reproduced below.

Theorem 2.5.5 *Let $n \geq 2$ be an integer and $w \geq n$ be a real number. If the real number ξ_w satisfies $w_1(\xi_w) = \ldots = w_n(\xi_w) = w$, then*

$$\lambda_n(\xi_w) = \frac{w - n + 1}{n}, \quad \hat{w}_n^*(\xi_w) = \frac{w}{w - n + 1},$$

and

$$\lambda_j(\xi_w) = \frac{w - j + 1}{j}, \quad \hat{w}_j(\xi_w) = j, \quad j = 1, \ldots, n.$$

Proof Let $m \geq 2$ be an integer and ξ be a transcendental real number. Lemma 2.5.2 with $k = 1$ implies the lower bound

$$\lambda_m(\xi) \geq \frac{w_1(\xi) - m + 1}{m}.$$

On the other hand, Theorem 2.3.2 gives the upper bound

$$\lambda_m(\xi) \leq \frac{w_m(\xi) - m + 1}{m}.$$

Let ξ_w be such that

$$w_1(\xi_w) = \ldots = w_n(\xi_w) = w.$$

Then, the equalities

$$\lambda_m(\xi_w) = \frac{w - m + 1}{m}, \quad m = 1, \ldots, n,$$

hold; in particular,

$$\lambda_n(\xi_w) = \frac{w - n + 1}{n},$$

and this establishes the first statement of the theorem.

Combining Theorem 2.2.6 with the case $m = 1$ of Theorem 2.4.6 gives

$$\frac{w_n(\xi_w)}{w_n(\xi_w) - n + 1} \leq \hat{w}_n^*(\xi_w) \leq \frac{w_1(\xi_w)}{w_1(\xi_w) - n + 1},$$

thereby proving the second statement of the theorem.

Without any loss of generality, assume that $0 < \xi_w < 1$ and $w > n \geq 2$. Let ε be a real number satisfying $0 < \varepsilon < w - n$. Let p/q be a rational number such that $q \geq 1$, $\gcd(p, q) = 1$ and $|\xi_w - p/q| < q^{-1-w+\varepsilon}$. Let $P(X)$ be an integer polynomial of degree j at most n and height at most $q - 1$. Observe that $P(p/q)$ is a non-zero rational number satisfying $|P(p/q)| \geq 1/q^j$.

By Rolle's theorem, there exists t lying between ξ and p/q such that

$$P(\xi_w) = P(p/q) + (\xi_w - p/q)P'(t).$$

Observe that $|P'(t)| \le 2^n n^2 q$ and $|\xi_w - p/q| \cdot |P'(t)| \le 1/(2q^n)$, if q is large enough. We then deduce that $|P(\xi_w)| \ge 1/(2q^j)$ if q is large enough. This shows that $\hat{w}_j(\xi_w) = j$, as asserted. □

Theorem 2.5.5 allows us to get some information on the spectra of the exponents λ_n and $\hat{w}_n^*$.

Theorem 2.5.6 *For any positive integer n, the spectrum of λ_n includes the interval $[1, +\infty]$, the spectrum of $\hat{w}_n^*$ includes the interval $[1, 2 - 1/n]$ and the spectrum of $\hat{w}_n - \hat{w}_n^*$ includes the interval $[n - 2 + 1/n, n - 1]$.*

Proof This has been already proved for $n = 1$. For $n \ge 2$, the statement follows from the combination of Theorem 2.5.1 with Theorem 2.5.5. □

Recall that, by Theorem 2.3.1, the spectra of $w_n - w_n^*$ and of $\hat{w}_n - \hat{w}_n^*$ are included in $[0, n - 1]$, for $n \ge 1$. The first assertion of the next result was proved by Bugeaud and Dujella [20] by means of an explicit construction of families of polynomials with close roots.

Theorem 2.5.7 *For any positive integer n, the spectrum of $w_n - w_n^*$ includes the interval*

$$\left[0, \frac{n}{2} + \frac{n-2}{4(n-1)}\right).$$

Moreover, the spectrum of $w_2 - w_2^*$ is equal to $[0, 1]$ and that of $w_3 - w_3^*$ is equal to $[0, 2]$.

Explicit examples of real numbers ξ for which $w_2(\xi)$ exceeds $w_2^*(\xi)$ can be found in [18].

Very recently, Schleischitz [70] established that, under some extra assumption, the inequality proved in Lemma 2.5.2 is indeed an equality.

Theorem 2.5.8 *Let n be a positive integer and ξ be a real number. If $\lambda_n(\xi) > 1$, then we have*

$$\lambda_1(\xi) = n\lambda_n(\xi) + n - 1 \tag{2.31}$$

and

$$\lambda_j(\xi) = \frac{n\lambda_n(\xi) - j + n}{j}, \quad \hat{\lambda}_j(\xi) = \frac{1}{j}, \quad (j = 1, \ldots, n).$$

Conversely, if $\lambda_1(\xi) > 2n-1$, *then we have*

$$\lambda_n(\xi) = \frac{\lambda_1(\xi) - n + 1}{n} \tag{2.32}$$

and

$$\hat{\lambda}_j(\xi) = \frac{1}{j}, \quad (j = 1, \ldots, n). \tag{2.33}$$

Proof Assume that ξ is in $(0, 1)$ and satisfies $\lambda_n(\xi) > 1$. Let q be a large positive integer and v be a real number greater than 1 such that $q^{-v} < \min\{\xi, 1-\xi, 1/(2nq)\}$ and

$$\max_{1 \le j \le n} \|q\xi^j\| \le q^{-v}. \tag{2.34}$$

Let p be the integer such that $|q\xi - p| = \|q\xi\|$. Then, p and q may not be coprime, but p/q is a convergent to ξ. Let d be the greatest prime factor of p and q and set $p_0 = p/d$ and $q_0 = q/d$. Observe that $0 < p/q < 1$ and, for $j = 1, \ldots, n$, we have

$$\left|\xi^j - \frac{p^j}{q^j}\right| = \left|\xi - \frac{p}{q}\right| \cdot \left|\xi^{j-1} + \ldots + \left(\frac{p}{q}\right)^{j-1}\right| \le \frac{n}{q^{1+v}} \le \frac{1}{2q^2}. \tag{2.35}$$

Assume that $q < q_0^n$ and express q in base q_0. Recalling that q_0 divides q, there exist integers $b_1, \ldots, b_{n-1}$ in $\{0, 1, \ldots, q_0 - 1\}$ such that

$$q = b_1 q_0 + b_2 q_0^2 + \ldots + b_{n-1} q_0^{n-1}.$$

Let u in $\{2, \ldots, n\}$ be the smallest index such that b_{u-1} is non-zero. Then,

$$\|q p_0^u / q_0^u\| = \|b_{u-1} p_0^u / q_0\| \ge 1/q_0. \tag{2.36}$$

Furthermore, it follows from (2.35) that

$$\left|q\left(\xi^u - \frac{p^u}{q^u}\right)\right| = \left|q\left(\xi^u - \frac{p_0^u}{q_0^u}\right)\right| \le \frac{1}{2q}. \tag{2.37}$$

Let y be the integer such that $|q\xi^u - y| = \|q\xi^u\|$ and observe that

$$\|q\xi^u\| \ge \left|y - q\frac{p_0^u}{q_0^u}\right| - \left|q\left(\xi^u - \frac{p_0^u}{q_0^u}\right)\right|,$$

using the triangle inequality. Combined with (2.36) and (2.37), this gives

$$\max_{1 \le j \le n} \|q\xi^j\| \ge \|q\xi^u\| \ge \frac{1}{q_0} - \frac{1}{2q} \ge \frac{1}{2q},$$

a contradiction to (2.34).

Consequently, $b_1 = \ldots = b_{n-1} = 0$ and we have established that $q \geq q_0^n$ (actually, our proof shows that q must be an integer multiple of q_0^n). In particular, we have

$$d \geq q_0^{n-1}. \tag{2.38}$$

Since

$$\|q_0\xi\| = |q_0\xi - p_0| = d^{-1}\|q\xi\|,$$

it follows from (2.34) and (2.38) that

$$\|q_0\xi\| \leq q_0^{-n+1} q_0^{-nv} = q_0^{-nv-n+1}.$$

Since v can be taken arbitrarily close to $\lambda_n(\xi)$, we deduce that

$$\lambda_1(\xi) \geq n\lambda_n(\xi) + n - 1.$$

Combined with Lemma 2.5.2, this proves the first statement of the theorem.

In particular, we get $\lambda_1(\xi) > 2n - 1$, and it follows from Theorem 2.4.9 that $\hat{\lambda}_j(\xi) = 1/j$ for $j = 1, \ldots, n$.

Let j be an integer with $2 \leq j \leq n-1$. Since $\lambda_n(\xi) > 1$, we have $\lambda_j(\xi) > 1$ and $\lambda_1(\xi) = j\lambda_j(\xi) + j - 1$. Combined with (2.31), this gives $j\lambda_j(\xi) = n\lambda_n(\xi) - j + n$, as claimed.

If the real number ξ satisfies $\lambda_1(\xi) > 2n - 1$, then we get by Lemma 2.5.2 that $\lambda_n(\xi) > 1$ and (2.32) and (2.33) follow from the first assertions of the theorem. □

The condition $\lambda_n(\xi) > 1$ in the statement of Theorem 2.5.8 cannot be removed in view of Theorem 4.3 of [16], which confirms the existence of uncountably many real numbers ξ satisfying $\lambda_n(\xi) = 1$ for every $n \geq 1$. Furthermore, Theorem 4.4 of [16] asserts that, for an arbitrary real number λ in $[1, 3]$, there exist uncountably many real numbers ξ satisfying $\lambda_1(\xi) = \lambda$ and $\lambda_2(\xi) = 1$.

We display an immediate consequence of Theorem 2.5.8.

Corollary 2.5.9 *Let n be a positive integer and ξ be a transcendental real number. Then, $\lambda_n(\xi) > 1$ holds if and only if $\lambda_1(\xi) > 2n - 1$ holds.*

The restriction $w \geq 2n - 1$ in the statement of Theorem 2.5.1 prevents us to get the whole spectra of the exponents w_n and λ_n by the method described above. Actually, we need the help of metric number theory to determine the whole spectra of the exponents w_n and λ_2.

Theorem 2.5.10 *For any positive integer n, the spectrum of w_n is equal to the whole interval $[n, +\infty]$ and the spectrum of w_n^* includes the whole interval $[n, +\infty]$.*

The first statement of Theorem 2.5.10 was proved by Bernik [11] and the second one is a result of Baker and Schmidt [5].

We display an immediate consequence of results by Beresnevich, Dickinson, Vaughan and Velani [10, 81].

Theorem 2.5.11 *The spectrum of λ_2 is equal to $[1/2, +\infty]$.*

More is known than the mere statement of Theorems 2.5.10 and 2.5.11. Indeed, for an integer $n \geq 1$ and a real number $w \geq n$, the Hausdorff dimension of the set of real numbers ξ for which $w_n(\xi) = w$ (resp., $w_n^*(\xi) = w$) is equal to $(n+1)/(w+1)$. Furthermore, the Hausdorff dimension of the set of real numbers ξ for which $\lambda_2(\xi) = \lambda$ is equal to $1/(1+\lambda)$ if $\lambda \geq 1$ and to $(2-\lambda)/(1+\lambda)$ if $1/2 \leq \lambda \leq 1$.

The spectra of the exponents $\hat{w}_n$ and $\hat{\lambda}_n$ remain very mysterious for $n \geq 3$, since we cannot even exclude that they are, respectively, reduced to $\{n\}$ and $\{1/n\}$ (recall that, by spectrum, we mean the set of values taken at transcendental points).

For $n = 2$, the situation is slightly better. By Jarník's Theorem 2.3.3, the value of $\hat{w}_2$ determines that of $\hat{\lambda}_2$, thus it is sufficient to determine the range of $\hat{w}_2$; see also Theorem 2.4.8.

As for the exponent $\hat{w}_n^*$, it is likely that its spectrum includes the interval $[1, n]$, but this is not yet proved.

2.6 Intermediate Exponents

Let $n \geq 2$ be an integer and $\underline{\theta}$ be a point in $\mathbb{R}^n$. In [49], Laurent introduced new exponents $\omega_{n,d}(\underline{\theta})$ (simply denoted by $\omega_d(\underline{\theta})$ in [49], since n is fixed throughout that paper) measuring the sharpness of the approximation to $\underline{\theta}$ by linear rational varieties of dimension d. Actually, Schmidt [72] was the first to investigate the properties of these exponents $\omega_{n,d}$, but he did not introduce them explicitly. We briefly recall their definition and we consider new exponents $w_{n,d}$ defined over $\mathbb{R}$ by restricting $\omega_{n,d}$ to the Veronese curve $(x, x^2, \ldots, x^n)$. It is convenient to view $\mathbb{R}^n$ as a subset of $\mathbb{P}^n(\mathbb{R})$ via the usual embedding $(x_1, \ldots, x_n) \mapsto (1, x_1, \ldots, x_n)$. We shall identify $\underline{\theta} = (\theta_1, \ldots, \theta_n)$ with its image in $\mathbb{P}^n(\mathbb{R})$. Denote by d the projective distance on $\mathbb{P}^n(\mathbb{R})$ and, for any real linear subvariety L of $\mathbb{P}^n(\mathbb{R})$, set

$$\mathrm{d}(\underline{\theta}, L) = \min_{P \in L} \mathrm{d}(\underline{\theta}, P)$$

the minimal distance between $\underline{\theta}$ and the real points P of L. When L is rational over $\mathbb{Q}$, we indicate by $H(L)$ its height, that is the Weil height of any system of Plücker coordinates of L. We refer to [49, 25] for precise definitions of the projective distance, heights, etc.

Definition 2.6.1 *Let $n \geq 2$ and d be integers with $0 \leq d \leq n-1$. Let $\underline{\theta}$ be in $\mathbb{R}^n$. We denote by $\omega_{n,d}(\underline{\theta})$ the supremum of the real numbers ω for which there exist infinitely many rational linear subvarieties $L \subset \mathbb{P}^n(\mathbb{R})$ of dimension d such that*

$$H(L) d(\underline{\theta}, L) \leq H(L)^{-\omega}.$$

We denote by $\hat{\omega}_{n,d}(\underline{\theta})$ the supremum of the real numbers $\hat{\omega}$ such that, for every sufficiently large value of H, there exists a rational linear subvariety $L \subset \mathbb{P}^n(\mathbb{R})$ of dimension d with

$$H(L) d(\underline{\theta}, L) \leq H^{-\hat{\omega}}.$$

If there exists ξ such that $\underline{\theta} = (\xi, \xi^2, \ldots, \xi^n)$, then we set $w_{n,d}(\xi) = \omega_{n,d}(\underline{\theta})$ and $\hat{w}_{n,d}(\xi) = \hat{\omega}_{n,d}(\underline{\theta})$.

We observe that the functions λ_n and $w_{n,0}$ (resp. w_n and $w_{n,n-1}$) coincide. The exponents $\hat{\omega}_{n,d}$ were introduced in [24, 66].

The following transference theorem was proved in [72, 49]. It shows how the Khintchine transference principle Theorem 2.3.2 splits into $n-1$ intermediate estimates which relate the exponents $\omega_{n,d}(\underline{\theta})$ for $d = 0, 1, \ldots, n-1$ (see also [25]).

Theorem 2.6.2 *Let n be a positive integer. For any non-zero vector $\underline{\theta}$ in $\mathbb{R}^n$, we have $\omega_{n,0}(\underline{\theta}) \geq 1/n$ and*

$$\frac{j\omega_{n,j}(\underline{\theta})}{\omega_{n,j}(\underline{\theta}) + j + 1} \leq \omega_{n,j-1}(\underline{\theta}) \leq \frac{(n-j)\omega_{n,j}(\underline{\theta}) - 1}{n-j+1}, \quad j = 1, \ldots, n-1,$$

with the convention that the left-hand side is equal to j if $\omega_{n,j}(\underline{\theta})$ is infinite.

Let the spectrum of the function $\omega_{n,d}$ denote the set of values taken by the exponents $\omega_{n,d}(\underline{\theta})$ when $\underline{\theta} = (\theta_1, \ldots, \theta_n)$ ranges over $\mathbb{R}^n$, with $1, \theta_1, \ldots, \theta_n$ linearly independent over the rationals. Using a result of Jarník [39], Laurent [49] established the following statement.

Theorem 2.6.3 *Let d and n be integers with $n \geq 2$ and $0 \leq d \leq n-1$. For every ω in $[(d+1)/(n-d), +\infty]$, there exists $\underline{\theta}$ such that $\omega_{n,d}(\underline{\theta}) = \omega$. Furthermore, $\omega_{n,d}(\underline{\theta}) = (d+1)/(n-d)$ for almost all $\underline{\theta}$ in $\mathbb{R}^n$.*

By means of the numbers ξ_w defined in the proof of Theorem 2.5.1, we get some information on the spectra of the exponents $w_{n,d}$.

Theorem 2.6.4 *For $n \geq 2$ and $0 \leq d \leq n-1$, the spectrum of $w_{n,d}$ contains the whole interval $[(n+d)/(n-d), +\infty]$ and $w_{n,d}(\xi) = (d+1)/(n-d)$ for almost all real numbers ξ.*

Theorem 2.6.4 plainly includes the last assertion of Theorem 2.5.1 and the first assertion of Theorem 2.5.6.

Proof We follow the proof of the Corollary from [49], where it is established that, for any λ with $1/n \leq \lambda \leq +\infty$ and for any point $\underline{\theta}$ in $\mathbb{R}^n$ such that $\omega_{n,0}(\underline{\theta}) = \lambda$ and $\omega_{n,n-1}(\underline{\theta}) = n\lambda + n - 1$, we have

$$\omega_{n,d}(\underline{\theta}) = \frac{n\lambda + d}{n-d}, \quad (d = 0, 1, \ldots, n-1). \tag{2.39}$$

For $w \geq 2n-1$, the numbers ξ_w defined in the proof of Theorem 2.5.1 satisfy

$$n\lambda_n(\xi_w) = w_n(\xi_w) - n + 1 = w - n + 1,$$

that is,

$$\omega_{n,n-1}(\xi_w, \ldots, \xi_w^n) = n\omega_{n,0}(\xi_w, \ldots, \xi_w^n) + n - 1.$$

We then get from (2.39) that

$$w_{n,d}(\xi_w) = \frac{n\lambda_n(\xi_w) + d}{n-d}, \quad (d = 0, 1, \ldots, n-1).$$

The first assertion of the theorem follows since $\lambda_n(\xi_w)$ takes every value between 1 and $+\infty$ as w varies from $2n-1$ to $+\infty$. The second assertion is an immediate consequence of (2.39) and the fact that $n\lambda_n(\xi) = w_n(\xi) - n + 1 = 1$ holds for almost every real number ξ. □

We conclude this section by mentioning that Laurent [48] determined the set of values taken by the quadruple of functions $(\omega_{2,0}, \omega_{2,1}, \hat{\omega}_{2,0}, \hat{\omega}_{2,1})$ at real points.

2.7 Parametric Geometry of Numbers

In 2009, Schmidt and Summerer [75, 76] introduced a new theory, called the parametric geometry of numbers. They studied the joint behaviour of the $n+1$ successive minima of certain one parameter families of convex bodies in $\mathbb{R}^{n+1}$, as a function of the parameter. They further showed how their results allow them to recover classical inequalities relating various exponents of Diophantine approximation attached to points in $\mathbb{R}^n$ and to find new relations. Some aspects of their theory have been simplified and completed by Roy [65], who was then able to derive several spectacular results [66, 67].

Let n be a positive integer and $\underline{\theta}$ be a non-zero vector in $\mathbb{R}^{n+1}$. For each real number $Q \geq 1$, we form the convex body

$$\mathcal{C}_{\underline{\theta}}(Q) = \{\underline{x} \in \mathbb{R}^{n+1} \ ; \ \|\underline{x}\|_2 \leq 1, \ |\underline{x} \cdot \underline{\theta}| \leq Q^{-1}\},$$

where $| \cdot |$ denotes the scalar product and $\| \cdot \|_2$ the Euclidean norm. For $j = 1, \ldots, n+1$, we denote by $\lambda_j\big(\mathcal{C}_{\underline{\theta}}(Q)\big)$ the jth successive minimum of $\mathcal{C}_{\underline{\theta}}(Q)$, namely the smallest positive real number λ such that $\lambda \mathcal{C}_{\underline{\theta}}(Q)$ contains at least j linearly independent points of $\mathbb{Z}^{n+1}$. Schmidt and Summerer [76] defined

$$L_{\underline{\theta},j}(q) = \log \lambda_j(\mathcal{C}_{\underline{\theta}}(\mathrm{e}^q)), \quad q \geq 0, \ 1 \leq j \leq n+1,$$

and considered the map $\mathbf{L}_{\underline{\theta}} \colon [0, \infty) \to \mathbb{R}^{n+1}$ given by

$$\mathbf{L}_{\underline{\theta}}(q) = (L_{\underline{\theta},1}(q), \ldots, L_{\underline{\theta},n+1}(q)), \quad q \geq 0.$$

They established many properties of this map. For instance, each of its components $L_{\underline{\theta},j} : [0, +\infty) \to \mathbb{R}$ is continuous and piecewise linear with slopes 0 and 1. Schmidt and Summerer showed that every function $\mathbf{L}_{\theta}$ can be approximated up to bounded difference by functions from a certain class, and Roy [65] showed that the same property holds within a simpler class.

For $j = 1, \ldots, n+1$, Roy [66] also introduced

$$\underline{\psi}_j(\underline{\theta}) = \liminf_{q \to +\infty} \frac{L_{\underline{\theta},1}(q) + \cdots + L_{\underline{\theta},j}(q)}{q}$$

and

$$\overline{\psi}_j(\underline{\theta}) = \limsup_{q \to +\infty} \frac{L_{\underline{\theta},1}(q) + \cdots + L_{\underline{\theta},j}(q)}{q}.$$

The following result, established in [66], relates these quantities to those introduced in the previous section.

Proposition 2.7.1 *Let n be a positive integer and $\underline{\theta} = (1, \theta_1, \ldots, \theta_n)$ be a vector in $\mathbb{R}^{n+1}$. For $j = 0, \ldots, n-1$, we have*

$$\omega_{n,j}((\theta_1, \ldots, \theta_n)) = \frac{1}{\underline{\psi}_{n-j}(\underline{\theta})} - 1 \text{ and } \hat{\omega}_{n,j}((\theta_1, \ldots, \theta_n)) = \frac{1}{\overline{\psi}_{n-j}(\underline{\theta})} - 1.$$

We quote below the main result of [66] and observe that it implies the first statement of Theorem 2.6.3.

Theorem 2.7.2 *Let n be a positive integer. For any $\omega_0, \ldots, \omega_{n-1} \in [0, +\infty]$ satisfying $\omega_0 \geq 1/n$ and*

$$\frac{j\omega_{n,j}}{\omega_{n,j} + j + 1} \leq \omega_{n,j-1} \leq \frac{(n-j)\omega_{n,j} - 1}{n - j + 1}, \quad 1 \leq j \leq n-1,$$

there exists a point $\underline{\theta} \in \mathbb{R}^n$, whose coordinates are, together with 1, linearly independent over $\mathbb{Q}$, such that

$$\omega_{n,j}(\underline{\theta}) = \omega_j \quad \text{and} \quad \hat{\omega}_{n,j}(\underline{\theta}) = \frac{j+1}{n-j}, \quad 0 \leq j \leq n-1.$$

Furthermore, the point of view of parametric geometry of numbers has led Schmidt and Summerer to introduce the following exponents of approximations.

Definition 2.7.3 *Let $\underline{\theta} = (\theta_1, \ldots, \theta_n)$ be in $\mathbb{R}^n$. For $i = 1, \ldots, n+1$, we denote by $\lambda_{i,n}(\underline{\theta})$ (resp. $\hat{\lambda}_{i,n}(\underline{\theta})$) the supremum of the real numbers λ such that the system of inequalities*

$$0 < |x_0| \leq X, \qquad \max_{1 \leq j \leq n} |x_0\theta_j - x_j| \leq X^{-\lambda}$$

has i linearly independent solutions $(x_0, x_1, \ldots, x_n)$ in $\mathbb{Z}^n$ for arbitrarily large X (resp. for every sufficiently large X).

Schmidt and Summerer [76] observed that $\lambda_{1,n}(\underline{\theta}) = \lambda_n(\underline{\theta})$, $\hat{\lambda}_{1,n}(\underline{\theta}) = \hat{\lambda}_n(\underline{\theta})$, $\lambda_{n+1,n}(\underline{\theta}) = 1/\hat{w}_n(\underline{\theta})$ and $\hat{\lambda}_{n+1,n}(\underline{\theta}) = 1/w_n(\underline{\theta})$, by Mahler's theorem on polar reciprocal bodies [53].

These exponents have been studied by Schleischitz [68, 71].

2.8 Real Numbers Which Are Badly Approximable by Algebraic Numbers

This short section is devoted to Problems 24 and 25 of [15], which were solved by Badziahin and Velani [4] for $n = 2$ and by Beresnevich [8] for $n \geq 3$.

Recall that an irrational real number ξ is called a *badly approximable number* if there exists a positive real number c such that

$$|qx - p| > \frac{c}{|q|}, \quad \text{for every } p, q \text{ in } \mathbb{Z} \text{ with } q \neq 0.$$

This notion can be extended as follows.

Definition 2.8.1 *Let n be a positive integer. A real number ξ is called n-badly approximable if there exists a positive constant $c(\xi, n)$ such that*

$$|P(\xi)| \geq c(\xi, n) H(P)^{-n} \text{ for any integer polynomial } P(X) \text{ of degree} \leq n.$$

Observe that it follows from Liouville's inequality (Theorem 2.2.9) that, for any positive integer n, any real algebraic number of degree $n + 1$ is n-badly approximable.

Davenport [30] asked whether there exist 2-badly approximable transcendental real numbers. His question remained open for nearly fifty years, until it was finally solved by Badziahin and Velani [4], using an intricate construction inspired by their proof [3] of a conjecture of Schmidt. Their result was subsequently extended a few years later by Beresnevich [8], who confirmed the existence of n-badly approximable transcendental real numbers, for every given positive integer n. For $n = 2$, his proof differs greatly from that of [4].

A closely related problem deals with transcendental numbers badly approximable by algebraic numbers of degree at most equal to some integer n. As well, it has been solved by Badziahin and Velani [4], for $n = 2$, and by Beresnevich [8], for $n \geq 3$.

Theorem 2.8.2 *Let n be a positive integer. There exist transcendental real numbers ξ which are n-badly approximable and for which there exist positive real numbers $c_1(\xi, n)$ and $c_2(\xi, n)$ such that*

$$|\xi - \alpha| \geq c_1(\xi, n)\, H(\alpha)^{-n-1}, \quad \text{for any real algebraic number } \alpha \text{ of degree} \leq n,$$

and

$$|\xi - \alpha| \le c_2(\xi, n)\, H(\alpha)^{-n-1},$$
for infinitely many real algebraic numbers α of degree $\le n$.

Moreover, the set of real numbers with this property has full Hausdorff dimension.

Actually, Beresnevich [8] established that any intersection of finitely many of the sets arising in Theorem 2.8.2 has full Hausdorff dimension.

2.9 Open Problems

In Chapter 10 of [15] we listed several open questions. As we have already seen above, some of them have now been solved. We gather below some of the still open problems mentioned in [15], and add a few supplementary ones.

We begin with the conjecture of Wirsing [82] dealing with the approximation of real transcendental numbers by real algebraic numbers of bounded degree. This celebrated open problem has motivated many works in this area.

Problem 2.9.1 (Wirsing's Conjecture) *For any integer $n \ge 1$ and for any real transcendental number ξ, we have $w_n^*(\xi) \ge n$.*

We may even ask for a stronger version of Wirsing's conjecture, namely whether, for any positive integer n and any real transcendental number ξ, there exist a constant $c(\xi, n)$ and infinitely many real algebraic numbers α of degree less than or equal to n such that

$$|\xi - \alpha| \le c(\xi, n)\, H(\alpha)^{-n-1}.$$

Davenport and Schmidt [31] gave a positive answer to this question in the case $n = 2$; see Theorem 2.4.2. Theorem 2.8.2 implies that their result is best possible up to the value of the numerical constant. However, we do not know if we can fix the exact degree of the approximants instead of just an upper bound for it.

Problem 2.9.2 *For any integer $n \ge 2$ and any real transcendental number ξ, there exist a constant $c(\xi, n)$ and infinitely many real algebraic numbers α of degree exactly n such that*

$$|\xi - \alpha| \le c(\xi, n)\, H(\alpha)^{-n-1}.$$

Results of Roy [58, 59] could speak in favour of the existence of transcendental numbers which do not satisfy the conclusion of Problem 2.9.2, even for $n = 2$.

The next problem was called the 'Main Problem' in [15].

Problem 2.9.3 *Let $(w_n)_{n\geq 1}$ and $(w_n^*)_{n\geq 1}$ be two non-decreasing sequences in $[1, +\infty]$ such that*

$$n \leq w_n^* \leq w_n \leq w_n^* + n - 1, \quad \textit{for any } n \geq 1.$$

Then there exists a real transcendental number ξ such that

$$w_n(\xi) = w_n \quad \textit{and} \quad w_n^*(\xi) = w_n^* \quad \textit{for any } n \geq 1.$$

Since Problem 2.9.3 does not take the exponents λ_n into account, we propose a more general formulation.

Problem 2.9.4 *For $n \geq 2$, determine the joint spectrum of the triple of exponents (w_n, w_n^*, λ_n), that is, the set of triples $(w_n(\xi), w_n^*(\xi), \lambda_n(\xi))$, when ξ runs through the set of transcendental real numbers.*

Theorem 2.3.2 shows how the exponents λ_n are related to the exponents w_n by means of a transference theorem.

Problem 2.9.5 *Let $n \geq 2$ be an integer, $\lambda_n \geq 1/n$ and $w_n \geq n$ be real numbers satisfying*

$$\frac{w_n}{(n-1)w_n + n} \leq \lambda_n \leq \frac{w_n - n + 1}{n}.$$

There exist real numbers ξ such that $w_n(\xi) = w_n$ and $\lambda_n(\xi) = \lambda_n$.

Corollary 2.2.7 asserts that $w_n^*(\xi) = n$ holds if $w_n(\xi) = n$, but the converse is an open question.

Problem 2.9.6 *For any positive integer n, we have $w_n(\xi) = n$ if $w_n^*(\xi) = n$.*

We now consider uniform exponents.

Problem 2.9.7 *Does there exist ξ such that $\hat{w}_2(\xi) > 2$ and $\hat{w}_2(\xi) > \hat{w}_2^*(\xi)$?*

Problem 2.9.8 *The spectrum of $\hat{w}_2^*$ includes the interval $[1, 2]$.*

Problem 2.9.9 *For any integer $n \geq 3$ and any real transcendental number ξ, we have $\hat{w}_n(\xi) = n$. At least, obtain a better upper bound than $\hat{w}_n(\xi) \leq n - \frac{1}{2} + \sqrt{n^2 - 2n + \frac{5}{4}}$.*

Approximation by algebraic integers (i.e. by algebraic numbers, whose minimal defining polynomial over $\mathbb{Z}$ is monic) has been first studied by Davenport and Schmidt [33]. The next problem is the analogue of Wirsing's conjecture for approximation by algebraic integers.

Problem 2.9.10 *For any integer $n \geq 4$, any positive real number ε, and any real transcendental number ξ, there exist a constant $c(\xi, n, \varepsilon)$ and infinitely many real algebraic integers α of degree less than or equal to n such that*

$$|\xi - \alpha| \leq c(\xi, n, \varepsilon)\, H(\alpha)^{-n+\varepsilon}.$$

Roy [58] proved the existence of real numbers ξ which are very badly approximable by quadratic integers, in the sense that they satisfy

$$|\xi - \alpha| > cH(\alpha)^{-(1+\sqrt{5})/2},$$

for some positive real number c and every real quadratic number α. His result shows that the conclusion of 2.9.10 does not hold for $n = 3$.

In view of auxiliary results from [33], the answer to Problem 2.9.10 is positive for some integer $n \geq 4$ if one can prove that any real transcendental number ξ satisfies $\hat{w}_{n-1}(\xi) = n - 1$.

Despite the recent, spectacular progress made in [4, 8], the following problem remains open.

Problem 2.9.11 *There exist a real transcendental number ξ and a sequence $(c(\xi, n))_{n \geq 1}$ of positive real numbers such that*

$$|P(\xi)| \geq c(\xi, n)\, H(P)^{-n}$$

for any integer n and any polynomial $P(X)$ of degree $\leq n$.

It is likely that the answer to Problem 2.9.11 is positive and that, moreover, the set of real numbers ξ with this property has full Hausdorff dimension.

We continue with a problem proposed by Schleischitz [70], which corrects and refines a problem posed in [16].

Problem 2.9.12 *Let m, n be integers with $1 \leq n \leq m$. Does the inequality*

$$\lambda_m(\xi) \geq \frac{n\lambda_n(\xi) - m + n}{m}$$

hold for every transcendental real number ξ?

The next problem extends a question posed by Beresnevich, Dickinson, and Velani [9] in the case of (simultaneous) rational approximation.

Problem 2.9.13 *Let n be a positive integer and let $\tau > 1$ be real. Is the set of real numbers ξ for which there exists a positive constant $c(\xi)$ such that*

$$|\xi - \alpha| \le H(\alpha)^{-\tau(n+1)}, \quad \text{for infinitely many algebraic numbers } \alpha \text{ of degree} \le n,$$

and

$$|\xi - \alpha| \ge c(\xi)\, H(\alpha)^{-\tau(n+1)}, \quad \text{for every algebraic number } \alpha \text{ of degree} \le n,$$

non-empty? If yes, determine its Hausdorff dimension.

Problem 2.9.13 has been solved [14] when $n = 1$. One may also replace the approximation functions $x \mapsto x^{-\tau(n+1)}$ by more general non-increasing functions.

Problems 2.9.14 and 2.9.15 deal with metrical results.

Problem 2.9.14 *Let $n \ge 2$ be an integer. Let λ_n be a real number with $\lambda_n \ge 1/n$. Determine the Hausdorff dimension of the set of real numbers ξ such that $\lambda_n(\xi) = \lambda_n$.*

Problem 2.9.15 *Determine the Hausdorff dimension of the set of real numbers ξ such that $\hat{w}_2(\xi) > 2$ (resp. $\hat{w}_2^*(\xi) > 2$).*

Let $m \ge 2$ be an integer. According to LeVeque [50], a real number ξ is a U_m-number if $w_m(\xi)$ is infinite and $w_{m-1}(\xi)$ is finite. Furthermore, the U_1-numbers are precisely the Liouville numbers.

It is proved in [1] (see also Section 7.6 of [15]) that, for any integer $m \ge 2$, there exist uncountably many real U_m-numbers ξ with

$$w_n^*(\xi) \le m + n - 1, \quad \text{for } n = 1, \dots, m - 1. \tag{2.40}$$

Schmidt [74] showed that $w_n^*(\xi)$ can be replaced by $w_n(\xi)$ in (2.40).

Problem 2.9.16 *Let $m \ge 2$ be an integer. There exist real U_m-numbers ξ satisfying $w_n(\xi) = n$, for $n = 1, \dots, m - 1$.*

Corollary 2.5.4 shows that the values taken by our exponents of approximation at U_1-numbers are known.

Problem 2.9.17 *Let m and n be integers with $n \geq m \geq 2$. Study the values taken by the exponents of approximation λ_n, $\hat{w}_n^*$, ... at U_m-numbers.*

Among many questions concerning the exponents $w_{n,d}$ and $\hat{w}_{n,d}$ defined in Section 2.6, let us point out the following three ones.

Problem 2.9.18 *Let d and n be integers with $n \geq 2$ and $0 \leq d \leq n-1$. Find a real number $C_{n,d}$ as small as possible such that every transcendental real number ξ satisfies $\hat{w}_{n,d}(\xi) \leq C_{n,d}$.*

Problem 2.9.19 *Let d and n be integers with $n \geq 2$ and $0 \leq d \leq n-1$. Is the spectrum of the function $w_{n,d}$ equal to $[(d+1)/(n-d), +\infty]$?*

A positive answer of Problem 2.9.19 would (probably) follow from the resolution of the next problem.

Problem 2.9.20 *Let d and n be integers with $n \geq 2$ and $0 \leq d \leq n-1$. Let w be a real number satisfying $w > (d+1)/(n-d)$. Determine the Hausdorff dimension of the sets*

$$\{\underline{\xi} \in \mathbb{R}^n : \omega_{n,d}(\underline{\xi}) \geq w\}$$

and

$$\{\xi \in \mathbb{R} : w_{n,d}(\xi) \geq w\}.$$

Throughout this survey, we focused our attention on approximation to real numbers. However, we may as well consider approximation to complex numbers or to p-adic numbers; see the references given in Chapter 9 of [15] and the works [21, 83, 17, 13, 7, 43, 26].

There are as well several recent papers on uniform Diophantine approximation on curves in $\mathbb{R}^2$; see [51, 64, 6].

Acknowledgements. The author is very grateful to Dmitry Badziahin for many helpful comments which helped him to improve the presentation of the survey.

Yann Bugeaud
Département de mathématiques, Université de Strasbourg
7, rue René Descartes, 67084 STRASBOURG (FRANCE)
email: bugeaud@math.unistra.fr

References

[1] K. Alniaçik, Y. Avci and Y. Bugeaud. On Um numbers with small transcendence measure, *Acta Math. Hungar.* 99 (2003), 271–277.

[2] B. Arbour and D. Roy. A Gel'fond type criterion in degree two, *Acta Arith.* 111 (2004), 97–103.

[3] D. Badziahin, A. Pollington, and S. Velani. On a problem in simultaneous diophantine approximation: Schmidt's conjecture, *Ann. of Math.* 174 (2011), 1837–1883.

[4] D. Badziahin and S. Velani. Badly approximable points on planar curves and a problem of Davenport, *Math. Ann.* 359 (2014), 969–1023.

[5] A. Baker and W. M. Schmidt. Diophantine approximation and Hausdorff dimension, *Proc. London Math. Soc.* 21 (1970), 1–11.

[6] G. Batzaya. On simultaneous approximation to powers of a real number by rational numbers, *J. Number Theory* 147 (2015), 141–155.

[7] P. Bel. Approximation simultanée d'un nombre v-adique et de son carré par des nombres algébriques, *J. Number Theory* 133 (2013), 3362–3380.

[8] V. Beresnevich. Badly approximable points on manifolds, *Invent. Math.* To appear.

[9] V. V. Beresnevich, H. Dickinson and S. L. Velani. Sets of exact 'logarithmic order' in the theory of Diophantine approximation, *Math. Ann.* 321 (2001), 253–273.

[10] V. Beresnevich, D. Dickinson and S. L. Velani. Diophantine approximation on planar curves and the distribution of rational points, with an appendix by R. C. Vaughan: 'Sums of two squares near perfect squares', *Ann. of Math.* 166 (2007), 367–426.

[11] V. I. Bernik. Application of the Hausdorff dimension in the theory of Diophantine approximations, *Acta Arith.* 42 (1983), 219–253 (in Russian). English transl. in *Amer. Math. Soc. Transl.* 140 (1988), 15–44.

[12] V. I. Bernik and K. Tishchenko. Integral polynomials with an overfall of the coefficient values and Wirsing's problem, *Dokl. Akad. Nauk Belarusi* 37 (1993), no. 5, 9–11 (in Russian).

[13] N. Budarina, Y. Bugeaud, D. Dickinson and H. O'Donnell. On simultaneous rational approximation to a p-adic number and its integral powers, *Proc. Edinb. Math. Soc.* 54 (2011), 599–612.

[14] Y. Bugeaud. Sets of exact approximation order by rational numbers, *Math. Ann.* 327 (2003), 171–190.

[15] Y. Bugeaud. *Approximation by Algebraic Numbers*, Cambridge Tracts in Mathematics, Cambridge University Press 2004.

[16] Y. Bugeaud. On simultaneous rational approximation to a real number and its integral powers, *Ann. Inst. Fourier* (Grenoble) 60 (2010), 2165–2182.

[17] Y. Bugeaud. On simultaneous uniform approximation to a p-adic number and its square, *Proc. Amer. Math. Soc.* 138 (2010), 3821–3826.

[18] Y. Bugeaud. Continued fractions with low complexity: Transcendence measures and quadratic approximation, *Compos. Math.* 148 (2012), 718–750.

[19] Y. Bugeaud. On a theorem of Wirsing in Diophantine approximation, *Proc. Amer. Math. Soc.* 144 (2016), 1905–1911.

[20] Y. Bugeaud and A. Dujella. Root separation for irreducible integer polynomials, *Bull. Lond. Math. Soc.* 43 (2011), 1239–1244.

[21] Y. Bugeaud and J.-H. Evertse. Approximation of complex algebraic numbers by algebraic numbers of bounded degree, *Ann. Scuola Normale Superiore di Pisa* 8 (2009), 333–368.

[22] Y. Bugeaud and M. Laurent. Exponents of Diophantine approximation and Sturmian continued fractions, *Ann. Inst. Fourier* (Grenoble) 55 (2005), 773–804.

[23] Y. Bugeaud and M. Laurent. Exponents of homogeneous and inhomogeneous Diophantine approximation, *Moscow Math. J.* 5 (2005), 747–766.

[24] Y. Bugeaud and M. Laurent. Exponents of Diophantine approximation. In: *Diophantine Geometry Proceedings*, Scuola Normale Superiore Pisa, Ser. CRM, vol. 4, 2007, 101–121.

[25] Y. Bugeaud and M. Laurent. On transfer inequalities in Diophantine approximation, II, *Math. Z.* 265 (2010), 249–262.

[26] Y. Bugeaud and T. Pejkovic. Quadratic approximation in $\mathbb{Q}_p$, *Intern. J. Number Theory* 11 (2015), 193–209.

[27] Y. Bugeaud and J. Schleischitz. On uniform approximation to real numbers. *Acta Arith.* To appear.

[28] Y. Bugeaud et O. Teulié. Approximation d'un nombre réel par des nombres algébriques de degré donné, *Acta Arith.* 93 (2000), 77–86.

[29] J. W. S. Cassels. An Introduction to Diophantine Approximation, Cambridge Tracts in Math. and Math. Phys., vol. 99, Cambridge University Press, 1957.

[30] H. Davenport. A note on Diophantine approximation. *II, Mathematika* 11 (1964), 50–58.

[31] H. Davenport and W. M. Schmidt. Approximation to real numbers by quadratic irrationals, *Acta Arith.* 13 (1967), 169–176.

[32] H. Davenport and W. M. Schmidt. A theorem on linear forms, *Acta Arith.* 14 (1967/1968), 209–223.

[33] H. Davenport and W. M. Schmidt. Approximation to real numbers by algebraic integers, *Acta Arith.* 15 (1969), 393–416.

[34] S. Fischler. Spectres pour l'approximation d'un nombre réel et de son carré, *C. R. Acad. Sci. Paris* 339 (2004), 679–682.

[35] S. Fischler. Palindromic prefixes and episturmian words, *J. Combin. Theory Ser. A* 113 (2006), 1281–1304.

[36] S. Fischler. Palindromic prefixes and Diophantine approximation, *Monatsh. Math.* 151 (2007), 11–37.

[37] A. O. Gel'fond. *Transcendental and Algebraic Numbers*. Translated from the first Russian edition by L. F. Boron. Dover Publications, Inc., New York, 1960.

[38] O. N. German. On Diophantine exponents and Khintchine's transference principle, *Mosc. J. Comb. Number Theory* 2 (2012), 22–51.

[39] V. Jarník. Über einen Satz von A. Khintchine, 2. Mitteilung, *Acta Arith.* 2 (1936), 1–22.

[40] V. Jarník. Zum Khintchineschen "Übertragungssatz", *Trav. Inst. Math. Tbilissi* 3 (1938), 193–212.

[41] V. Jarník. Une remarque sur les approximations diophantiennes linéaires, *Acta Sci. Math. Szeged* 12 (1950), 82–86.

[42] V. Jarník. Contribution á la théorie des approximations diophantiennes linéaires et homogénes, *Czechoslovak Math. J.* 4 (1954), 330–353 (in Russian, French summary).

[43] G. Kekec. On Mahler's p-adic Um-numbers, *Bull. Aust. Math. Soc.* 88 (2013), 44–50.

[44] A. Ya. Khintchine. Über eine Klasse linearer diophantischer Approximationen, *Rendiconti Circ. Mat. Palermo* 50 (1926), 170–195.

[45] A. Ya. Khintchine. On some applications of the method of the additional variable, *Uspehi Matem. Nauk* 3, (1948), 188–200 (in Russian). English translation: *Amer. Math. Soc. Translation* (1950), no. 18, 14 pp.

[46] J. F. Koksma. Über die Mahlersche Klasseneinteilung der transzendenten Zahlen und die Approximation komplexer Zahlen durch algebraische Zahlen, *Monatsh. Math. Phys.* 48 (1939), 176–189.

[47] M. Laurent. Simultaneous rational approximation to the successive powers of a real number, *Indag. Math.* 11 (2003), 45–53.

[48] M. Laurent. Exponents of Diophantine approximation in dimension two, *Canad. J. Math.* 61 (2009), 165–189.

[49] M. Laurent. On transfer inequalities in Diophantine approximation. In: *Analytic Number Theory*, 306–314, Cambridge University Press, 2009.

[50] W. J. LeVeque. On Mahler's U-numbers, *J. London Math. Soc.* 28 (1953), 220–229.

[51] S. Lozier and D. Roy. Simultaneous approximation to a real number and to its cube by rational numbers, *Acta Arith.* 156 (2012), 39–73.

[52] K. Mahler. Zur Approximation der Exponentialfunktionen und des Logarithmus. I, II, *J. reine angew. Math.* 166 (1932), 118–150.

[53] K. Mahler. Ein Übertreibungsprinzip für konvexe Körper, *Casopis Pěst.Mat. Fyz.* 68 (1939), 93–102.

[54] N. G. Moshchevitin. Singular Diophantine systems of A. Ya. Khinchin and their application, *Uspekhi Mat. Nauk* 65 (2010), 43–126; English translation in *Russian Math. Surveys* 65 (2010), 433–511.

[55] N. Moshchevitin. Exponents for three-dimensional simultaneous Diophantine approximations, *Czechoslovak Math. J.* 62 (2012), 127–137.

[56] N. Moshchevitin. A note on two linear forms, *Acta Arith.* 162 (2014), 43–50.

[57] D. Roy. Approximation simultanée d'un nombre et son carré, *C. R. Acad. Sci. Paris* 336 (2003), 1–6.

[58] D. Roy. Approximation to real numbers by cubic algebraic numbers, II, *Ann. of Math.* 158 (2003), 1081–1087.

[59] D. Roy. Approximation to real numbers by cubic algebraic numbers, I, *Proc. London Math. Soc.* 88 (2004), 42–62.

[60] D. Roy. Diophantine approximation in small degree, In: *Number Theory*, 269–285, CRM Proc. Lecture Notes 36, Amer. Math. Soc., Providence, RI, 2004.

[61] D. Roy. On two exponents of approximation related to a real number and its square, *Canad. J. Math.* 59 (2007), 211–224.

[62] D. Roy. On the continued fraction expansion of a class of numbers. In: *Diophantine Approximation, Festschrift for Wolfgang Schmidt*, Developments in Math. vol. 16, Eds: H. P. Schlickewei, K. Schmidt and R. Tichy, Springer-Verlag, 2008, 347–361.

[63] D. Roy. On simultaneous rational approximations to a real number, its square, and its cube, *Acta Arith.* 133 (2008), 185–197.

[64] D. Roy. Rational approximation to real points on conics, *Ann. Inst. Fourier (Grenoble)* 63 (2013), 2331–2348.

[65] D. Roy. On Schmidt and Summerer parametric geometry of numbers, *Ann. of Math.* 182 (2015), 739–786.

[66] D. Roy. Spectrum of the exponents of best rational approximation. *Math. Z.* To appear.

[67] D. Roy. Construction of points realizing the regular systems of Wolfgang Schmidt and Leonard Summerer, *J. Théor. Nombres Bordeaux* 27 (2015), 591–603.

[68] J. Schleischitz. Diophantine approximation and special Liouville numbers, *Commun. Math.* 21 (2013), 39–76.

[69] J. Schleischitz. Two estimates concerning classical Diophantine approximation constants, *Publ. Math.* Debrecen 84 (2014), 415–437.

[70] J. Schleischitz. On the spectrum of Diophantine approximation constants. Preprint. arXiv:1409.1472.

[71] J. Schleischitz. On approximation constants for Liouville numbers, *Glas. Mat.* To appear.

[72] W. M. Schmidt. On heights of algebraic subspaces and diophantine approximations, *Ann. of Math.* 85 (1967), 430–472.

[73] W. M. Schmidt. *Diophantine Approximation*. Lecture Notes in Math. 785, Springer, Berlin, 1980.

[74] W. M. Schmidt. Mahler and Koksma classification of points in $\mathbb{R}$n and $\mathbb{C}$n, *Funct. Approx. Comment. Math.* 35 (2006), 307–319.

[75] W. M. Schmidt and L. Summerer. Parametric geometry of numbers and applications, *Acta Arith.* 140 (2009), 67–91.

[76] W. M. Schmidt and L. Summerer. Diophantine approximation and parametric geometry of numbers, *Monatsh. Math.* 169 (2013), 51–104.

[77] W. M. Schmidt and L. Summerer. Simultaneous approximation to three numbers, *Moscow J. Comb. Number Th.* 3 (2013), 84–107.

[78] V. G. Sprindžuk. Mahler's problem in metric number theory. Izdat. "Nauka i Tehnika", Minsk, 1967 (in Russian). English translation by B. Volkmann, Translations of Mathematical Monographs, Vol. 25, American Mathematical Society, Providence, RI, 1969.

[79] K. I. Tishchenko. On approximation of real numbers by algebraic numbers, *Acta Arith.* 94 (2000), 1–24.

[80] K. I. Tsishchanka, On approximation of real numbers by algebraic numbers of bounded degree, *J. Number Theory* 123 (2007), 290–314.

[81] R. C. Vaughan and S. Velani. Diophantine approximation on planar curves: the convergence theory, *Invent. Math.* 166 (2006), 103–124.

[82] E. Wirsing. Approximation mit algebraischen Zahlen beschránkten Grades, *J. reine angew. Math.* 206 (1961), 67–77.

[83] D. Zelo. Simultaneous approximation to real and p-adic numbers. PhD thesis. Univ. Ottawa. 2009. `arXiv:0903.0086`.

3

Effective Equidistribution of Nilflows and Bounds on Weyl Sums

Giovanni Forni

Abstract

In these lectures we will present an approach developed in collaboration with L. Flaminio in [FF2], [FF3], [FF4], to effective equidistribution for nilflows, based on representation theory and renormalization or scaling. Bounds on Weyl sums can be derived from equidistribution bounds on nilflows.

Our bounds on Weyl sums are comparable with the best available bounds to date, obtained by T. D. Wooley [Wo2], [Wo3], [Wo4], [Wo5], at least as far as behavior of the exponent of the power function for large degree is concerned, but are significantly weaker than Wooley's as they only hold *almost everywhere* (that is, for almost all coefficients of lower degree for polynomials with a given Diophantine leading coefficient). However, our effective distribution results on nilflows are more general and our approach can in principle be generalized to other homogeneous flows and higher-rank actions (see, for instance, [BF], [CF], [FF1], [FFT]). Moreover, uniform bounds with respect to coefficients of lower degree of the polynomial are reduced to uniform estimates on the *average rate of close returns* for nilflows.

We were so far unable to prove the conjectural uniform bounds on close returns (except for step-2 and step-3 nilflows) which would immediately imply corresponding uniform bounds on Weyl sums (under Diophantine conditions on the leading coefficient). Our proof therefore relies in general on a Borel–Cantelli argument based on the maximal ergodic theorem, which can only prove almost everywhere bounds or bounds on L^p mean for all $p \in [1, 2)$. In the special case of 2-step and 3-step filiform nilflows we are able to prove uniform bounds, although in this case the classical Weyl bounds (respectively for polynomials of degree 2 and 3) are still the best available.

3.1 Introduction

We summarize known results on the effective equidistribution of nilflows and on bounds on Weyl sums and state our main results.

Bounds on Weyl Sums

The problem of bounds on Weyl sums has a long history going back at least a century to the seminal work of Hardy and Littlewood [HL] on the circle method followed by the work of H. Weyl [We] on the exponential sums which now bear his name.

Weyl sums are exponential sums for polynomials sequences. Let

$$P_k(\mathbf{a}, X) := a_k X^k + \cdots + a_1 X + a_0$$

be a polynomial of degree $k \geq 2$ with real coefficients $\mathbf{a} := (a_k, \ldots, a_0) \in \mathbb{R}^{k+1}$. The Weyl sums for the polynomial sequence $\{P_k(\mathbf{a}, n) | n \in \mathbb{N}\}$ are defined, for all $N \in \mathbb{N}$, as the exponential sums

$$W_N(a_k, \ldots, a_0) := \sum_{n=0}^{N-1} e^{2\pi i P_k(\mathbf{a}, n)}.$$

Weyl sums for quadratic polynomials, often called Gaussian sums or theta sums, are very well understood. In particular it has been proved that they display square root cancellations for Roth-type irrational coefficients [HL], [FJK], [FK]. More recently several results have been proved about limit their distributions [Ma], [GM], [CM]. These results are based on the self-similarity properties of exponential sums of quadratic polynomials [HL], [FK], which are related to the existence of a *renormalization* for 2-step (Heisenberg) nilflows introduced in [FF2] and presented in these lectures.

Weyl sums for higher-degree polynomials are much harder to estimate and, despite recent advances, no definitive results are known. In this case, no self-similarity of the exponential sums is known. The absence of self-similarity of exponential sums for higher-degree polynomials is related to the absence of renormalization for higher-step nilflows. The main goal of these lectures is to outline a generalization of the renormalization method from the renormalizable case of 2-step nilflows to the non-renormalizable case of higher-step nilflows. This effort is motivated by questions of effective equidistribution for general unipotent flows for which no renormalization is known (with the exception of linear toral flows, horocycle flows, Heisenberg nilflows).

Results in the analytic number theory literature are often formulated as follows. Let

$$W_N^{(k)}(a_k) := \max_{a_{k-1},\dots,a_0} |W_N(a_k,\dots,a_0)|.$$

The Weyl exponent w_k is the supremum of all $w > 0$ such that if there exists p/q with $(p,q)=1$, $|a_k - p/q| < 1/q^2$ and $N \le q \le N^{k-1}$, then

$$W_N^{(k)}(a_k) = O(N^{1-w}).$$

It is interesting to derive corresponding bounds on Weyl sums under classical Diophantine conditions. We recall that $a_k \in DC_\nu$ for $\nu \ge 1$ if there exists a constant $C := C(a_k) > 0$ such that

$$|qa_k - p| \ge \frac{C}{q^\nu}.$$

An elementary argument based on the theory of continued fractions gives that for all $\epsilon > 0$ and for all $a_k \in DC_\nu$ with $\nu \le k-1+\epsilon$ we have

$$W_N^{(k)}(a_k) = O_\epsilon(N^{1-w_k+\epsilon}).$$

We recall some landmark results on the Weyl exponent:

- $1/w_k \le 2^{k-1}$ for all $k \ge 2$ (Weyl [We], see also [Va]);
- $1/w_k \le (4+o(1))k^2 \log k$ for large $k >> 2$ (Hua [Hu]);
- $1/w_k \le (\frac{3}{2}+o(1))k^2 \log k$ for large $k >> 2$ (Wooley [Wo1]);
- $1/w_k \le 2(k-1)(k-2)$ for $k \ge 3$ (Wooley [Wo5]).

All improvements on Weyl's bound for large $k >> 2$ were derived from *Vinogradov mean value theorem* [Vi] and its refinements. To the author's best knowledge, for small $k \ge 2$ (say $k \le 5$) the Weyl's bound is still unsurpassed in general. The best available bounds to date follow from the *efficient congruencing* approach to the Vinogradov mean value theorem recently introduce by Wooley [Wo2], [Wo3]. Wooley's best result to date in fact states in particular that for any p/q with $(p,q)=1$, such that $|a_k - p/q| < 1/q^2$, one has ([Wo5], Theorem 7.3)

$$W_N^{(k)}(a_k) = O_\epsilon\left(N^{1+\epsilon}(q^{-1}+N^{-1}+qN^{-k})^{1/2(k-1)(k-2)}\right).$$

The most optimistic conjecture on Weyl sums, inspired by the optimal bounds available for quadratic polynomials ($k=2$), can be stated as follows (see for instance [Bo], [Va]). If there exists p/q with $(p,q)=1$, $|a_k - p/q| < 1/q^2$ and $N \le q \le N^{k-1}$, then for every $\epsilon > 0$ we have the conjectural bound

$$W_N^{(k)}(a_k) = O_\epsilon\left(N^{1+\epsilon}(q^{-1}+qN^{-k})^{1/k}\right).$$

The conjecture easily implies $1/w_k \leq k$, a result much stronger than the best bounds available mentioned above. It is also remarkable that square root cancellations would follow whenever a_k is a Roth-type irrational, that is, whenever $a_k \in DC_\nu$ for all $\nu > 1$. In fact, in that case it easy to derive that we would have

$$W_N^{(k)}(a_k) = O_\epsilon(N^{\frac{1}{2}+\epsilon}).$$

It is unclear whether the most optimist conjecture holds. However, square root cancellations indeed occur for a full measure set of polynomials. In fact, a general result in theory of uniform distribution (modulo 1) of sequences (Erdös–Gál–Koksma) implies such that the following bound holds. For any sequence $\mathbf{x} := \{x_n\}$ of real numbers and for all $N \in \mathbb{N}$ the Nth *discrepancy* (modulo 1) of $\mathbf{x}$ is the number (see [DT], Def. 1.5)

$$D_N(\mathbf{x}) = D_N(x_0, \ldots, x_{N-1}) := \sup_{I \subset \mathbb{T}} |\frac{1}{N} \sum_{n=0}^{N-1} \chi_I(\{x_n\}) - \text{Leb}(I)|.$$

(In the above formula χ_I denotes the characteristic function of the interval $I \subset \mathbb{T}$ and $\{x\} \in [0, 1)$ denotes the fractional part of $x \in \mathbb{R}$).

For every fixed vector $(a_{k-1}, \ldots, a_0)$ there exists a full measure set of a_k such that for the vector of coefficients $\mathbf{a} = (a_k, \ldots, a_0)$ the sequence $D_N(\mathbf{a})$ of discrepancies of the polynomial sequence $\{P(\mathbf{a}, n) | n \in \mathbb{N}\}$ satisfies the following bound (see [DT], Theorem 1.158): for all $\epsilon > 0$,

$$D_N(\mathbf{a}) = O_\epsilon(N^{-1/2}(\log N)^{5/2+\epsilon}), \quad \text{for all } N \in \mathbb{N}.$$

It is unclear whether the full measure set in the above result can be fixed independently of the vector $(a_{k-1}, \ldots, a_0)$ and whether it can be described in terms of classical Diophantine conditions.

Effective Equidistribution of Nilflows

The problem of effective equidistribution of nilflows has a much more recent history. The relation between Weyl sums and skew-shifts over rotations (which are return maps of nilflows) already appears in the work of H. Furstenberg [Fu], who derived basic qualitative results on Weyl sums from a proof of unique ergodicity of skew-shifts. This relation is described in detail in these lectures (see Section 3.2). A general theory on the effective equidistribution of polynomial sequences on nilmanifolds has been developed by B. Green and T. Tao [GT1] in their work on the *Möbius and Nilsequences conjecture* [GT2]. For nilflows it implies the following result. Let $\{\phi_t^X\}$ denote a nilflow on a nilmanifold $M := \Gamma \backslash N$, that is, a flow generated by a smooth vector field X on M induced by a left-invariant vector field on the nilpotent group N. If the projected linear toral flow has a Diophantine frequency, then there exist a constant $C > 0$ and an exponent $\delta \in (0, 1)$ such that, for all Lipschitz function f on M and for all $(x, T) \in M \times \mathbb{R}^+$ we have

$$|\frac{1}{T}\int_0^T f \circ \phi_t^X(x)dt - \int_M f\, d\text{vol}| \leq C\|f\|_{Lip}T^{-\delta}.$$

The approach of the Green and Tao is a far-reaching extension of Weyl's method to estimate Weyl sums, based on the van der Corput inequality (see [GT1], Lemma 4.1). It is therefore based on the special structure of nilflows as successive Abelian extensions of linear flows on tori. It is unclear what is the best exponent accessible by this method in the general case as it is not explicitly derived in the paper. It is reasonable to assume that it would be similar to Weyl's bound on Weyl sum stated above, hence it would decay exponentially with respect to $k \geq 2$ for k-step nilmanifolds.

In these lectures we shall be concerned with similar effective ergodicity results with emphasis on optimal exponents and on methods which can be in principle be generalized to other classes of homogeneous flows.

Statement of Results

We state below our results for Heisenberg nilflows. This is the main 2-step case and it is related to Weyl sums for quadratic polynomials. Our results hold for general functions on a nilmanifold M which are sufficiently smooth in the Sobolev sense. For every $\sigma \in \mathbb{R}$, let $W^\sigma(M)$ denote the usual *Sobolev space* [Ad], [He] of functions on the compact manifold M.

Theorem 3.1.1 *There exists a set $\mathcal{B}$ of full Hausdorff dimension (and zero measure) of vector fields on an Heisenberg nilmanifold M (the set $\mathcal{B}$ consists of all flows which project onto a linear toral vector field with slope an irrational number of bounded type) such that the following holds. For every $\sigma > 5/2$ and for every $X \in \mathcal{B}$, there exists a constant $C_\sigma(X) > 0$ such that, for all function $f \in W^\sigma(M)$ and all $(x, T) \in M \times \mathbb{R}^+$, we have*

$$|\frac{1}{T}\int_0^T f(\phi_t^X(x))dt - \int_M f dvol| \leq C_\sigma(X)\|f\|_\sigma T^{-1/2}.$$

Theorem 3.1.2 *Let $\beta : [1, +\infty) \to \mathbb{R}^+$ be any non-decreasing function satisfying the integral condition*

$$\int_1^{+\infty} \frac{1}{T\beta^4(T)} dT < +\infty.$$

There exists a full measure set $\mathcal{F}_\beta$ of vector fields on an Heisenberg nilmanifold M such that the following holds. For every $\sigma > 5/2$ and for every $X \in \mathcal{F}_\beta$, there exists a constant $C_\sigma(X) > 0$ such that, for all function $f \in W^\sigma(M)$ and for all $(x, T) \in M \times \mathbb{R}^+$, we have

$$|\frac{1}{T}\int_0^T f(\phi_t^X(x))dt - \int_M f dvol| \leq C_\sigma(X)\|f\|_\sigma T^{-1/2}\beta(T).$$

For quadratic Weyl sums (theta-series) the above result was first established by H. Fiedler, W. Jurkat, and O. Körner [FJK] who also proved that it is optimal, in the sense that if the function β does not satisfy the above integral condition, then the set of flows for which the conclusion of the theorem holds has measure zero. We remark that in the logarithmic scale we can have

$$\beta(T) = (\log T)^{\frac{1}{4}+\delta}, \qquad \text{for any } \delta > 0,$$

while the original result of Hardy and Littlewood [HL] established a weaker bound for

$$\beta(T) = (\log T)^{\frac{1}{2}+\delta}, \qquad \text{for any } \delta > 0.$$

For any higher-step filiform nilflow we have the following weaker result.

Theorem 3.1.3 *Let $\{\phi_t^X\}$ be a nilflow on a k-step filiform nilmanifold M which projects to a linear toral flow on $\mathbb{T}^2$ with Diophantine frequency of exponent $\nu \in [1, k/2]$. Let $\sigma > k^2$. For every $\epsilon > 0$ there exists a full measure set $\mathcal{G}_\epsilon \subset M$ of good points and a measurable function $K_{\sigma,\epsilon} : \mathcal{G}_\epsilon \to \mathbb{R}^+$, with $K_{\sigma,\epsilon} \in L^p(M)$ for every $p \in [1, 2[$, such that for every function $f \in W^\sigma(M)$ and for all $(x, T) \in \mathcal{G}_\epsilon \times \mathbb{R}^+$ we have*

$$\left|\frac{1}{T}\int_0^T f \circ \phi_s^X(x)\, ds - \int_M f vol\right| \leq K_{\sigma,\epsilon}(x) T^{-\frac{2}{3(k-1)k}+\epsilon} \|f\|_\sigma .$$

A slight refinement of the above theorem, derived from the remark that the set of good points is invariant under the action on M of the center $Z := Z(G)$ of the nilpotent group G and under the action on M/Z of the center $Z(G/Z)$ of the quotient group G/Z, implies the following bound on Weyl sums:

Corollary 3.1.4 *Let $a_k \in \mathbb{R} \setminus \mathbb{Q}$ be a Diophantine number of exponent $\nu \leq k/2$. For every $\epsilon > 0$, there exists a measurable positive function $K_\epsilon \in L^p(\mathbb{T}^{k-2})$, for all $p \in [1, 2)$, such that the following bound holds. For all $a_0, a_1 \in \mathbb{R}^2$, for almost all $(a_2, \ldots, a_{k-1}) \in \mathbb{R}^{k-2}$ and for every $N \geq 1$,*

$$|W_N(a_k, \ldots, a_0)| \leq K_\epsilon(a_2, \ldots, a_{k-1}) N^{1 - \frac{2}{3k(k-1)} + \epsilon}.$$

It is quite remarkable that the best exponent accessible by our approach, which is determined by the optimal scaling of invariant distributions, essentially coincides with the exponent derived in the work of Wooley [Wo2] by entirely different methods. Our argument in fact proves bounds on ergodic averages under a quantitative condition on close returns of orbits of filiform nilflows. It is reasonable to conjecture that such a condition holds for all points, hence the bound is uniform (as in Wooley's theorem). However, so far we have only been able to prove by a Borel–Cantelli argument that the set of *good points* has full measure (with the exception of the case of 3-step filiform nilflows, for which we prove a uniform bound in these lectures). In a recent paper, Wooley [Wo6] has been able to derive, from his mean value theorem by a Borel–Cantelli argument, a much stronger bound on Weyl sums, with nearly square root cancellation for large degree, for a full measure set of coefficients. It would be interesting to derive corresponding results for (filiform) nilflows.

In the case of 3-step filiform nilflows we are able to prove by our approach the following uniform bounds:

Theorem 3.1.5 *Let $\{\phi_t^X\}$ be a nilflow on a 3-step filiform nilmanifold M which projects to a toral linear flow on $\mathbb{T}^2$ with Diophantine frequency of Roth type. For every $\sigma > 6$ and for every $\epsilon > 0$ there exists a constant $C_{\sigma,\epsilon}(X) > 0$ such that the following holds: for all function $f \in W^\sigma(M)$ and all $(x, T) \in M \times \mathbb{R}^+$, we have*

$$|\frac{1}{T} \int_0^T f(\phi_t^X(x))dt - \int_M f dvol| \leq C_{\sigma,\epsilon}(X) \|f\|_\sigma T^{-\frac{1}{6}+\epsilon}.$$

3.2 Nilflows and Weyl Sums

We recall the classical theory of nilflows, which states in particular that all nilflows with minimal toral projection are minimal and uniquely ergodic. In these lectures we will focus mainly on the special case of *filiform* nilflows, although our results in [FF4] are proved for a wider class of nilflows on nilmanifolds which we call *quasi-Abelian*. We then recall the well-known relation of nilflows to skew-shifts on tori and to Weyl sums. Filiform nilflows are enough for applications to classical Weyl sums.

Nilflows

Let N be a connected, simply connected nilpotent Lie group and $\Gamma \subset N$ be a co-compact lattice. The quotient space $M = \Gamma\backslash N$ is called a *nilmanifold.* A *nilflow* on M is the flow generated by a left-invariant vector field corresponding to an element X of Lie algebra $\mathfrak{n}$ of N, that is, it is the flow $\{\phi_t^X\}_{t\in\mathbb{R}}$ defined as

$$\phi_t^X(\Gamma x) = \Gamma x \exp(tX)\,, \qquad \text{for all } (x, t) \in N \times \mathbb{R}.$$

Any nilflow on M is volume-preserving, namely, it preserves the (normalized) volume measure vol on M induced by the Haar measure on the nilpotent group N. Let $\bar{N} := N/[N, N]$ denote the Abelianized group and $\bar{\Gamma} := \Gamma/[\Gamma, \Gamma]$ denote the Abelianized lattice. The quotient manifold $\bar{M}$ is therefore a torus. There exists a canonical projection $\pi : M \to \bar{M}$ which maps the nilflow $\{\phi_t^X\}_{t\in\mathbb{R}}$ onto a linear toral flow $\{\psi_t^{\bar{X}}\}_{t\in\mathbb{R}}$, generated by the projection $\bar{X} := \pi_*(X)$ of the vector field X onto the Abelian Lie algebra $\bar{\mathfrak{n}}$ of the Abelian Lie group $\bar{N}$.

The topological dynamics and ergodic theory of nilmanifolds are reduced to the corresponding theories for linear toral flows (hence circle rotations) by the following now classical theorem.

Theorem 3.2.1 ([Gr], [AGH]) *The following conditions are equivalent:*

1. *The nilflow* $(\{\phi_t^X\}_{t\in\mathbb{R}}, vol)$ *is ergodic.*
2. *The nilflow* $\{\phi_t^X\}_{t\in\mathbb{R}}$ *is uniquely ergodic.*
3. *The nilflow* $\{\phi_t^X\}_{t\in\mathbb{R}}$ *is minimal.*
4. *The projected flow* $\{\psi_t^{\bar{X}}\}_{t\in\mathbb{R}}$ *is an irrational linear flow on a torus and hence (uniquely) ergodic and minimal.*

In fact, more is true: the projected flow $\{\psi_t^{\bar{X}}\}_{t\in\mathbb{R}}$ is the isometric factor of $\{\phi_t^X\}_{t\in\mathbb{R}}$ and the latter flow is relatively weak mixing. In fact, the flow $(\{\phi_t^X\}_{t\in\mathbb{R}}, \text{vol})$ has countable Lebesgue spectrum on the orthogonal complement of $\pi^*(L^2(\bar{M}))$.

Examples

1. Heisenberg nilmanifolds and nilflows. The three-dimensional *Heisenberg group* is the 2-step nilpotent Lie group $H_3(\mathbb{R})$ consisting of the matrices

$$[p, q, r] := \begin{pmatrix} 1 & p & r \\ 0 & 1 & q \\ 0 & 0 & 1 \end{pmatrix}, \qquad p, q, r \in \mathbb{R}. \tag{3.1}$$

The matrices $\{[0, 0, r] \mid r \in \mathbb{R}\}$ form the center $Z(H_3(\mathbb{R}))$ and the commutator subgroup of $H_3(\mathbb{R})$; the maps

$$t \mapsto [0, 0, t] \qquad \text{and} \qquad [p, q, r] \mapsto (p, q). \tag{3.2}$$

define a (non-split!) exact sequence

$$0 \to \mathbb{R} \to H_3(\mathbb{R}) \to \mathbb{R}^2 \to 0\,, \tag{3.3}$$

which exhibits $H_3(\mathbb{R})$ as a line bundle over $\mathbb{R}^2$.

The maps

$$(p, t) \in \mathbb{R}^2 \mapsto [p, 0, t] \in H_3(\mathbb{R}) \qquad \text{and} \qquad [p, q, r] \mapsto q \in \mathbb{R} \tag{3.4}$$

yield a split exact sequence

$$0 \to \mathbb{R}^2 \to H_3(\mathbb{R}) \to \mathbb{R} \to 0. \tag{3.5}$$

We take as a basis of the Lie algebra $\mathfrak{h}_3$ of $H_3(\mathbb{R})$ the three vectors X_0, Y_0, Z_0 corresponding to the matrices

$$X_0 = \begin{pmatrix} 0 & 1 & 0 \\ 0 & 0 & 0 \\ 0 & 0 & 0 \end{pmatrix}, \quad Y_0 = \begin{pmatrix} 0 & 0 & 0 \\ 0 & 0 & 1 \\ 0 & 0 & 0 \end{pmatrix}, \quad Z_0 = \begin{pmatrix} 0 & 0 & 1 \\ 0 & 0 & 0 \\ 0 & 0 & 0 \end{pmatrix}. \tag{3.6}$$

The triple (X_0, Y_0, Z_0) satisfies the well-known Heisenberg commutation relations, which for a triple (X, Y, Z) are

$$[X, Y] = Z \qquad \text{and} \qquad [X, Z] = [Y, Z] = 0. \tag{3.7}$$

Let Γ be a lattice subgroup of $H_3(\mathbb{R})$. It is well-known that $\Gamma\backslash H_3(\mathbb{R})$ is compact and that there exists a positive integer E such that, up to an automorphism of $H_3(\mathbb{R})$, we have

$$\Gamma := \left\{ \begin{pmatrix} 1 & p & r/E \\ 0 & 1 & q \\ 0 & 0 & 1 \end{pmatrix} \mid p, q, r \in \mathbb{Z} \right\}.$$

The compact manifold $M := \Gamma\backslash H_3(\mathbb{R})$ is called a *Heisenberg nilmanifold*.

The homomorphism $H_3(\mathbb{R}) \to \mathbb{R}^2$ defined in (3.2) induces a Seifert fibration $\pi : M \to \mathbb{T}^2 \approx \mathbb{Z}^2\backslash\mathbb{R}^2$, i.e. M is a circle bundle over the 2-torus $\mathbb{T}^2$ with fibers given by the orbits of flow by right translation by the central one-parameter subgroup $Z(H_3(\mathbb{R})) = (\exp tZ)_{t\in\mathbb{R}}$. The left invariant fields X_0, Y_0 define a connection whose total curvature (the Euler characteristic of the fibration) is exactly E.

Considering instead the split sequence (3.5) we can also see that M is a $\mathbb{T}^2$-torus bundle over S^1; the holonomy, obtained via the splitting map, is given by the unipotent matrix $\begin{pmatrix} 1 & E \\ 0 & 1 \end{pmatrix}$.

A presentation of Γ it is given by

$$\langle x, y, z \mid [x, z] = [y, z] = 1,\ [x, y] = z^E \rangle.$$

Homogeneous flows on Heisenberg nilmanifolds, that is, flows generated by a projection to M of a left-invariant vector field on $H_3(\mathbb{R})$, are called *Heisenberg nilflows*. In other terms, for any element $X \in \mathfrak{h}$ of the Heisenberg Lie algebra, the Heisenberg nilflow $\{\phi_t^X\}_{t\in\mathbb{R}}$ is the flow defined as follows:

$$\phi_t^X(x) = x \exp(tX)\,, \qquad \text{for } x = \Gamma y \text{ with } y \in H_3(\mathbb{R}) \text{ and } t \in \mathbb{R}.$$

2. Filiform nilmanifolds and nilflows. A *filiform Lie algebra* is a nilpotent Lie algebra $\mathfrak{f}_k$ whose descending central sequence has length $k = \dim \mathfrak{f}_k - 1$. In these lectures we shall consider those special k-step nilpotent filiform Lie algebras that have a basis $\{\xi, \eta_1, \ldots, \eta_k\}$ satisfying the commutation relations

$$\begin{aligned} &[\xi, \eta_i] = \eta_{i+1}, \quad \text{for all } i = 1, \ldots, k-1, \text{ and } \quad [\xi, \eta_k] = 0 \\ &[\eta_i, \eta_j] = 0\,, \quad \text{for all } i, j = 1, \ldots, k. \end{aligned} \tag{3.8}$$

A basis $(\xi, \eta) := (\xi, \eta_1, \ldots, \eta_k)$ satisfying the above commutation relations will be called a *filiform basis*.

The filiform Lie algebras defined above are the simplest higher-step generalization of the Heisenberg Lie algebra (on two generators). In these lectures a *(k-step) filiform Lie group* is a simply connected, connected group F_k whose Lie algebra is $\mathfrak{f}_k$ and a *filiform nilflow* is a homogeneous flow on a compact filiform nilmanifold.

As we shall explain below the classical Weyl sums (that is, exponential sums) for polynomials of degree $k \geq 2$ can be interpreted as ergodic integrals for filiform nilflows on k-step filiform nilmanifolds for special smooth functions.

The groups F_k have also another description. Let $h : \mathbb{R} \to \mathrm{Aut}(\mathbb{R}^k)$ be the unique continuous homomorphism such that

$$h(1)(\mathbf{s}) = (s_1, s_2 + s_1, \ldots, s_i + s_{i-1}, \ldots, s_k + s_{k-1}), \tag{3.9}$$

where $\mathbf{s} := (s_1, s_2, \ldots, s_k)$. Let G_k be the twisted product $\mathbb{R} \ltimes_h \mathbb{R}^k$. We can view G_k as an algebraic subgroup of the real algebraic group $GL_k(\mathbb{R}) \ltimes \mathbb{R}^k$. Since $h(\mathbb{Z}) \subset \mathrm{Aut}(\mathbb{Z}^k)$, the twisted product $\Gamma_k = \mathbb{Z} \ltimes_{h|\mathbb{Z}} \mathbb{Z}^k$ is well-defined and it is a Zariski dense discrete subgroup of G_k, hence a lattice of G_k, generated by elements $x, y_1, \ldots, y_k$ satisfying the commutation relations

$$x y_i x^{-1} = \begin{cases} y_i\, y_{i+1}\,, & \text{for } 1 \le i < k \\ y_k\,, & \text{for } i = k \end{cases} \qquad y_i y_j = y_j y_i, \quad 1 \le i, j \le k. \tag{3.10}$$

(We have taken for x the element $(1, (0, \ldots, 0)) \in \Gamma_k$ which acts by conjugation on $\mathbb{Z}^k$ by the automorphism $h(1)$ defined in (3.9), and for elements $y_1, \ldots, y_k$ the elements of the standard basis $(0, (1, 0, \ldots, 0))$, …, $(0, (0, 0, \ldots, 1))$ of $\{0\} \times \mathbb{Z}^k$.)

Let g be the Lie algebra of G and let $\log : G \to g$ the inverse of the exponential map. The elements

$$\xi = \log x \qquad \tilde\eta_i = \log y_i, \quad i = 1, \ldots, k \tag{3.11}$$

form a basis of g and satisfy the commutation relations

$$[\xi, \tilde\eta_j] = \sum_{i=j+1}^{k} \frac{(-1)^{i-j-1}}{i-j}\, \tilde\eta_i = \tilde\eta_{j+1} - \tfrac{1}{2}\,\tilde\eta_{j+2} + \tfrac{1}{3}\,\tilde\eta_{j+3} + \ldots, \qquad 1 \le j < k\,, \tag{3.12}$$

all other commutators being equal to zero. We obtain a filiform basis defined by induction

$$\eta_1 = \tilde\eta_1, \quad \eta_{j+1} = [\xi, \eta_j]\,, \quad j = 1, \ldots, k-1.$$

Thus $\mathfrak{g}$ is isomorphic to $\mathfrak{f}_k$ and G to F_k. Clearly there exists strictly upper triangular rational matrices $R, S \in M_k(\mathbb{Q})$ such that

$$\eta_j = \tilde\eta_j + \sum_{i=j+1}^{k} R_{ij} \tilde\eta_j \tag{3.13}$$

$$\tilde\eta_j = \eta_j + \sum_{i=j+1}^{k} S_{ij} \eta_j \tag{3.14}$$

for all $j = 1, \ldots, k-1$. Thus, by the formulas (3.14) and by taking exponentials, we can associate to each filiform basis $(\xi, \eta_1, \ldots, \eta_k)$ of $\mathfrak{f}_k$ a lattice of F_k.

Since Γ_k is discrete and Zariski dense in F_k the quotient $\Gamma_k \backslash F_k$ is a compact nilmanifold M_k.

Definition 3.2.2 *The quotient $M_k = \Gamma_k \backslash F_k$ will be called a k-step* filiform nilmanifold.

Observe that for any filiform basis $(\xi, \eta_1, \ldots, \eta_k)$ the center $z(\mathfrak{f}_k)$ of $\mathfrak{f}_k$ is spanned by the vector η_k and therefore the vectors $(\xi, \eta_1, \ldots, \eta_{k-1})$ project onto a filiform basis of $\mathfrak{f}_k/z(\mathfrak{f}_k) \approx \mathfrak{f}_{k-1}$. At the group level, the center $Z(F_k)$ of F_k, that is the group $(\exp t\eta_k)_{t\in\mathbb{R}} = (y_k^t)_{t\in\mathbb{R}}$, meets Γ_k into the subgroup generated by y_k, which is the center $Z(\Gamma_k)$ of Γ_k. Hence $F_k/Z(F_k) \approx F_{k-1}$ and the elements $x, \ldots, y_{k-1})$ project onto the generators of Γ_{k-1} in F_{k-1}. This shows that a k-step filiform Riemannian nilmanifold M_k has a structure of circle bundle over a $(k-1)$-step filiform nilmanifold M_{k-1} the fibers of this fibration being the orbits of the right action of the center $Z(F_k)$ on M_k:

$$0 \to \mathbb{S}^1 \to M_k \to M_{k-1} \to 0. \tag{3.15}$$

We shall call this fibration *the central fibration of M_k*.

Composing the central fibrations $M_k \to M_{k-1} \to M_{k-2} \to \cdots \to M_1$ we have also another fibration of

$$pr : M_k \to M_1 \approx \mathbb{T}^2. \tag{3.16}$$

This fibration can be obtained directly considering that the Abelian group $\overline{F_k} = F_k/[F_k, F_k]$ is isomorphic to $\mathbb{R}^2$ and the subgroup $\overline{\Gamma}_k = \Gamma_k/[\Gamma_k, \Gamma_k]$ is a co-compact lattice of $\overline{F_k}$.

Finally there also a fibration

$$0 \to \mathbb{T}^k \to M_k \to M_0 = \mathbb{T}^1 \to 0 \tag{3.17}$$

which is induced by the homomorphism $\mathfrak{f}_k \to \mathfrak{f}_k/\langle \eta_1, \ldots, \eta_k \rangle$.

All the fibrations (3.15), (3.16), and (3.17) can be seen in a unified way in the following way. Let E_j be the Abelian normal subgroup of F_k generated by the Abelian ideal $\langle \eta_j, \eta_{j+1}, \ldots, \eta_k \rangle$ of $\mathfrak{f}_k$. The subgroup E_j meets the lattice Γ_k into a lattice Γ_k^j. Hence the orbits of the right action of E_j on $M_k = \Gamma_k \backslash F_k$ are closed and they are the fibers of the fibration $M_k \to M_{j-1}$.

For any element $X \in \mathfrak{f}_k$ we denote $\{\phi_t^X\}_{t\in\mathbb{R}}$ the *filiform nilflow* on M_k generated by X, that is the flow given by right multiplication by the one-parameter subgroup $(\exp(tX))_{t\in\mathbb{R}}$ of F_k:

$$\phi_t^X(x) = x \exp(tX), \qquad \text{for } x = \Gamma_k y \text{ with } y \in F_k \text{ and } t \in \mathbb{R}. \tag{3.18}$$

We denote by $\mathbb{T}_0^k$ the k-dimensional torus

$$\Gamma_k \exp(s_1 \tilde{\eta}_1) \cdots \exp(s_k \tilde{\eta}_k), \quad s_i \in \mathbb{R},$$

which is the orbit of the identity coset under the action of the group E_1. Observe that the map

$$(s_1, \ldots, s_k) \in \mathbb{R}^k/\mathbb{Z}^k \mapsto \Gamma_k \exp(s_1 \tilde{\eta}_1) \cdots \exp(s_k \tilde{\eta}_k) \tag{3.19}$$

is a diffeomorphism of $\mathbb{R}^k/\mathbb{Z}^k$ onto $\mathbb{T}_0^k$.

Henceforth we shall use the coordinate $\mathbf{s} = (s_1, \ldots, s_k)$ to denote the point $\Gamma_k \exp(s_1 \tilde{\eta}_1) \cdots \exp(s_k \tilde{\eta}_k)$ of $\mathbb{T}_0^k$.

It follows from this description on the manifold M_k that a fundamental domain for Γ_k acting on F_k is the set

$$\{\exp(t\xi) \exp(s_1 \tilde{\eta}_1) \cdots \exp(s_k \tilde{\eta}_k) \mid t \in [0, 1], s_i \in [0, 1]\};$$

this implies that the left (and right) invariant volume form on F_k defined by $\omega(\xi, \tilde{\eta}_1, \ldots, \tilde{\eta}_k) = 1$ pushes down to a right-invariant volume form on M_k whose density yields a right-invariant *probability* measure on M_k. Observing that the formulas (3.13) and (3.14) imply that $\xi \wedge \tilde{\eta}_1 \wedge \cdots \wedge \tilde{\eta}_k = \xi \wedge \eta_1 \wedge \cdots \wedge \eta_k$, we conclude that, on M_k, there exist a right-invariant volume form ω_{M_k} and a right-invariant probability measure vol_k such that, for the fixed basis $(\xi, \eta_1, \ldots, \eta_k)$, we have

$$\omega_{M_k}(\xi, \eta_1, \ldots, \eta_k) = 1, \quad \text{and} \quad \mathrm{vol}_k = |\omega_{M_k}|. \tag{3.20}$$

Weyl Sums

Let $P_k := P_k(X) \in \mathbb{R}[X]$ be a polynomial of degree $k \geq 2$:

$$P_k(X) := \sum_{j=0}^{k} a_j X^j.$$

Let $f \in C^\infty(\mathbb{T}^1)$ be a smooth periodic function. A *Weyl sum* of degree $k \geq 2$ is the sum

$$W_N(P_k, f) = \sum_{n=0}^{N-1} f\big(P_k(n)\big), \quad \text{for any } N \in \mathbb{N}. \tag{3.21}$$

Classical (complete) Weyl sums are obtained as a particular case when $f(s) = e(s) = \exp(2\pi i s)$, for any $s \in \mathbb{T}^1$.

Return Maps

Let $\alpha := (\alpha_1, \ldots, \alpha_k) \in \mathbb{R}^k$ and let

$$X_\alpha := \log[\, x^{-1} \exp(\sum_{j=1}^{k} \alpha_j \tilde{\eta}_j)\,]. \tag{3.22}$$

The torus $\mathbb{T}_0^k$ is a global section of the nilflow $\{\phi_t^\alpha\}_{t\in\mathbb{R}} := \{\phi_t^{X_\alpha}\}_{t\in\mathbb{R}}$ on M_k, generated by $X_\alpha \in \mathfrak{f}_k$. The lemma below, which is classical (see [Fu]), makes explicit its return maps, which are skew-shifts on tori.

Lemma 3.2.3 *The flow $\{\phi_t^\alpha\}_{t\in\mathbb{R}}$ on M_k is isomorphic to the suspension of its first return map $\Phi_\alpha : \mathbb{T}_0^k \to \mathbb{T}_0^k$ which, in the coordinates $\mathbf{s} \in \mathbb{R}^k/\mathbb{Z}^k$ of formula (3.19), is written as*

$$\Phi_\alpha(\mathbf{s}) = (s_1 + \alpha_1, \ldots, s_j + s_{j-1} + \alpha_j, \ldots, s_k + s_{k-1} + \alpha_k). \tag{3.23}$$

In particular, all return times are constant integer-valued functions on $\mathbb{T}_0^k$. For any $N \in \mathbb{Z}$, the Nth return map $\Phi_\alpha^N : \mathbb{T}_0^k \to \mathbb{T}_0^k$ has the form

$$\begin{aligned}\Phi_\alpha^N(\mathbf{s}) = \Big(&s_1 + N\alpha_1, s_2 + N(s_1+\alpha_2) + \tbinom{N}{2}\alpha_1, \ldots,\\ &s_k + \sum_{i=1}^{N-1} \tbinom{N}{i}(s_{k-i} + \alpha_{k-i+1}) + \tbinom{N}{k}\alpha_1\Big).\end{aligned} \tag{3.24}$$

Proof Let $\Gamma_k \exp(\sum_{j=1}^k s_j \tilde{\eta}_j) \in \mathbb{T}_0^k$. We have

$$\begin{aligned}\exp(\sum_{j=1}^{k} s_j\tilde{\eta}_j)\exp(X_\alpha) &= \exp(\sum_{j=1}^{k} s_j\tilde{\eta}_j)x^{-1}\exp(\sum_{j=1}^{k}\alpha_j\tilde{\eta}_j)\\ &= x^{-1}\exp\big[(s_1+\alpha_1)\tilde{\eta}_1 + \sum_{j=1}^{k-1}(s_j + s_{j+1} + \alpha_{j+1})\tilde{\eta}_{j+1}\big].\end{aligned} \tag{3.25}$$

In fact, the following identity holds:

$$\begin{aligned}x\exp(\sum_{j=1}^{k} s_j\tilde{\eta}_j)x^{-1} &= \exp\Big[e^{\,\mathrm{ad}_\xi}(\sum_{j=1}^{k} s_j\tilde{\eta}_j)\Big]\\ &= \exp\big[s_1\tilde{\eta}_1 + \sum_{j=1}^{k-1}(s_j + s_{j+1})\tilde{\eta}_{j+1}\big].\end{aligned} \tag{3.26}$$

Since $x \in \Gamma_k$, it follows that

$$\begin{aligned}
&\Gamma_k \exp(\sum_{j=1}^{k} s_j \tilde{\eta}_j) \exp(X_\alpha) \\
&\quad = \Gamma_k \exp\big[(s_1 + \alpha_1)\tilde{\eta}_1 + \sum_{j=1}^{k-1} (s_j + s_{j+1} + \alpha_{j+1})\tilde{\eta}_{j+1}\big].
\end{aligned} \tag{3.27}$$

The above formula implies that $t = 1$ is a return time of the flow $\{\phi_t^\alpha\}_{t\in\mathbb{R}}$ to $\mathbb{T}_0^k$ and the map (3.23) is the corresponding return map. In addition, $t = 1$ is the first return time, since it is the first return time of the projection onto $\mathbb{T}_0^k$ of the flow $\{\phi_t^\alpha\}_{t\in\mathbb{R}}$ to the projection of the transverse section $\mathbb{T}_0^k$.

Finally, the formula (3.24) for the Nth return map follows from formula (3.23) by induction on $N \in \mathbb{N}$. □

Reduction

For any $(\alpha, \mathbf{s}) \in \mathbb{R}^k \times \mathbb{R}^k/\mathbb{Z}^k$, let $P_{\alpha,\mathbf{s}}(X) \in \mathbb{R}[X]$ be the polynomial of degree $k \geq 2$ defined (modulo $\mathbb{Z}$) as follows:

$$P_{\alpha,\mathbf{s}}(X) := \binom{X}{k}\alpha_1 + \sum_{j=1}^{k-1} \binom{X}{j}(s_{k-j} + \alpha_{k-j+1}) + s_k. \tag{3.28}$$

We identify a point $x \in \mathbb{T}_0^k$ with its coordinates $\mathbf{s} \in \mathbb{R}^k/\mathbb{Z}^k$ (see (3.19)). The following result holds:

Lemma 3.2.4 *The map $(\alpha, x) \to P_{\alpha,x}(X)$ sends $\mathbb{R}^k \times \mathbb{T}_0^k$ onto the space $\mathbb{R}_k[X]$ of real polynomials (modulo $\mathbb{Z}$) of degree at most $k \geq 2$. The leading coefficient $a_k \in \mathbb{R}$ of the polynomial $P_{\alpha,x}(X)$ is given by the formula:*

$$a_k = \frac{\alpha_1}{k!}.$$

Let $X_\alpha \in \mathfrak{f}_k$ be the vector field on M_k given by formula (3.22) and let $\mathcal{A}_T^\alpha := \mathcal{A}_T^{X_\alpha}$ be the ergodic averaging operator defined, for all $f \in C^\infty(M)$ and all $(x, T) \in M_k \times \mathbb{R}$, by the formula

$$\mathcal{A}_{x,T}^\alpha(f) = \mathcal{A}_T^\alpha(f)(x) = \frac{1}{T}\int_0^T f \circ \phi_t^\alpha(x)\, dt. \tag{3.29}$$

It is possible to derive bounds for the Weyl sums $\{W_N(P_{\alpha,\mathbf{s}}, f)\}_{N\in\mathbb{N}}$ for any smooth function $f \in C^\infty(\mathbb{T}^1)$ from Sobolev bounds on the ergodic averaging operators $\mathcal{A}_{s,T}^\alpha$ introduced above.

Lemma 3.2.5 *For any $\alpha \in \mathbb{R}^k$, there exists a bounded injective linear operator*

$$F = F_\alpha : L^2(\mathbb{T}^1) \to L^2(M_k)$$

such that the following holds. For any $r \geq 0$, the operator F maps continuously $W^r(\mathbb{T}^1)$ into $W^r(M_k)$; furthermore, for any $r > 1/2$, there exists a constant $C_r > 0$ such that, for any function $f \in W^r(\mathbb{T}^1)$, for all $x \in \mathbb{T}_0^k$ and all $N \in \mathbb{N}$, we have

$$|\sum_{n=0}^{N} f\left(P_{\alpha,x}(n)\right) - N\mathcal{A}^{\alpha}_{x,N}(F(f))| \leq C_r \, \|f\|_{W^r(\mathbb{T}^1)}. \tag{3.30}$$

Proof For any $\varepsilon \in]0, 1/2[$ the map

$$(x,t) \in \mathbb{T}_0^k \times]-\varepsilon, \varepsilon[\;\mapsto \phi_t^\alpha(x) = x \exp(tX_\alpha) \tag{3.31}$$

is an embedding of $\mathbb{T}_0^k \times]-\varepsilon, \varepsilon[$ onto a tubular neighborhood $\mathcal{U}_\varepsilon$ of $\mathbb{T}_0^k$ in M_k.

Let $pr_Z : \mathbb{T}_0^k \to \mathbb{T}^1 \approx Z(F_k)/Z(\Gamma_k)$ the projection on the central coordinate, that is, the map defined as

$$pr_Z\big(\exp(\sum_{i=1}^{k} s_i \tilde{\eta}_i)\big) = \exp(s_k \tilde{\eta}_k)\,, \quad \text{for all } (s_1, \ldots, s_k) \in \mathbb{R}^k. \tag{3.32}$$

For any $f \in L^2(\mathbb{T}^1)$ and $\chi \in C_0^\infty(]-\varepsilon, \varepsilon[)$, let $F(f,\chi) \in L^2(M)$ be the function defined on the open set $\mathcal{U}_\varepsilon$ as

$$F(f,\chi)(y) = \chi(t)\,(f(pr_Z(x)))\,, \tag{3.33}$$

where $(x,t) \in \mathbb{T}_0^k \times]-\varepsilon, \varepsilon[$ and $y = \phi_t^\alpha(x)$. We extend $F(f,\chi)$ as zero on $M \setminus \mathcal{U}_\varepsilon$.

The function $F(f,\chi)$ is well-defined and square-integrable on M, since $\chi \in C_0^\infty(]-\varepsilon, \varepsilon[)$ and the map (3.31) is an embedding. In addition, we have $F(f,\chi) \in C^0(M)$ if $f \in C^0(S^1)$ and, for any $r \geq 0$, $F(f,\chi) \in W^r(M)$ if $f \in W^r(\mathbb{T}^1)$.

Let $f \in C^0(\mathbb{T}^1)$. We claim that, by the definition (3.33) of the function $F(f,\chi)$, for all $(x,N) \in \mathbb{T}_0^k \times \mathbb{N}$, we have

$$\int_{-\varepsilon}^{N+\varepsilon} F(f,\chi) \circ \phi_t^\alpha(x) dt = \left(\int_{-\varepsilon}^{\varepsilon} \chi(s)\,ds\right) \sum_{n=0}^{N} f\left(P_{\alpha,x}(n)\right). \tag{3.34}$$

In fact, let $\Phi_\alpha^n : \mathbb{T}_0^k \to \mathbb{T}_0^k$ be the nth map of the flow $\{\phi_t^\alpha\}_{t\in\mathbb{R}}$. By Lemma 3.2.3 and by definition (3.28), for all $(x,n) \in \mathbb{T}_0^k \times \mathbb{N}$,

$$pr_Z \circ \Phi_\alpha^n(x) = P_{\alpha,x}(n). \tag{3.35}$$

For all $(x, n) \in \mathbb{T}_0^k \times \mathbb{N}$ and for all $f \in C^0(\mathbb{T}^1)$,

$$\int_{n-\varepsilon}^{n+1-\varepsilon} F(f,\chi) \circ \phi_t^\alpha(x)dt = \int_{n-\varepsilon}^{n+\varepsilon} F(f,\chi) \circ \phi_t^\alpha(x)dt$$
$$= \int_{-\varepsilon}^{\varepsilon} F(f,\chi) \circ \phi_\tau^\alpha \left(\Phi_\alpha^n(x)\right) dt = \left(\int_{-\varepsilon}^{\varepsilon} \chi(s)\, ds\right) f\left(P_{\alpha,x}(n)\right). \tag{3.36}$$

The claim is therefore proved.

Let $\chi \in C_0^\infty(]-\varepsilon, \varepsilon[)$ be a given function such that $\int_{\mathbb{R}} \chi(\tau)\, d\tau = 1$ and let $C_1 = \|\chi\|_\infty$. Define the operator $F : L^2(\mathbb{T}^1) \to L^2(M)$ as

$$F(f) := F(f,\chi), \qquad \text{for all } f \in L^2(\mathbb{T}^1). \tag{3.37}$$

It follows by formula (3.34) that

$$\left| \int_0^N F(f) \circ \phi_t^\alpha(x)dt - \sum_{n=0}^{N} f\left(P_{\alpha,x}(n)\right) \right| \le 2\epsilon \|F(f)\|_\infty. \tag{3.38}$$

For any $r > 1/2$, by the Sobolev embedding theorem [Ad], [He] $W^r(\mathbb{T}^1) \subset C^0(\mathbb{T}^1)$ and there exists a constant $c_r > 0$ such that $\|f\|_\infty \le c_r \|f\|_{W^r(\mathbb{T}^1)}$; since $\|F(f)\|_\infty \le \|\chi\|_\infty \|f\|_\infty$, we obtain the inequality (3.30) and the argument is concluded. □

The problem of establishing bounds on Weyl sums is thus reduced to that of bounds for the nilpotent averages (3.29).

3.3 The Cohomological Equation

In this section we derive results on the *cohomological equation* and *invariant distributions* for fililiform nilflows, following [FF3]. The filiform case is special and simpler, but it is enough for the main applications to Weyl sums. Our analysis of the cohomological equation is based on the Kirillov's theory of irreducible unitary representations of nilpotent groups, in the special case of filiform groups. This analysis shows that nilflows always have an infinite-dimensional space of invariant distributions. For Diophantine nilflows, every sufficiently smooth function which belongs to the kernel of all invariant distributions is a smooth *coboundary* (with finite loss of derivatives for the transfer function), hence almost all nilflows are *stable* in the sense of A. Katok (in addition, the lack of stability can only come from the toral factor). The fundamental theorem of calculus implies that ergodic integrals of *coboundaries* are uniformly bounded.

We recall that this analysis is motivated, on the one hand, by the well-known elementary fact that ergodic integrals of coboundaries with bounded transfer function (that is, of all derivatives of bounded functions along the flow) are uniformly bounded, on the other hand, by the heuristic principle that the growth of ergodic integrals (and the corresponding decay of ergodic averages) of smooth zero-average functions is related to the scaling of the invariant distributions under an appropriate renormalization group action.

Irreducible Unitary Representations

Representation Models

Kirillov's theory yields the following complete classification of *irreducible unitary representations* of filiform Lie groups (up to unitary equivalence).

Let $\mathfrak{a}_k^*$ be the space of $\mathbb{R}$-linear forms on the maximal Abelian ideal $\mathfrak{a}_k$ of the filiform Lie algebra $\mathfrak{f}_k$. For any filiform basis $\mathcal{F} = (\xi, \eta_1, \ldots, \eta_k)$, the ideal $\mathfrak{a}_k$ is generated by the system $\{\eta_1, \ldots, \eta_k\}$. For any $\Lambda \in \mathfrak{a}_k^*$ denote by $\exp \imath \Lambda$ the character χ_Λ of the subgroup $\mathcal{A}_k := \exp(\mathfrak{a}_k)$ defined by $\chi_\Lambda(g) := \exp(\imath \Lambda(Y))$, for $g = \exp Y$ with $Y \in \mathfrak{a}_k$.

The infinite-dimensional irreducible representations of the filiform nilpotent Lie group F_k are unitarily equivalent to the representations $\mathrm{Ind}_{\mathcal{A}_k}^{F_k}(\Lambda)$, obtained by inducing from $\mathcal{A}_k$ to F_k a character $\chi = \exp \imath \Lambda$ not vanishing on $[\mathfrak{f}_k, \mathfrak{f}_k]$. In addition, two linear forms Λ and Λ' determine unitarily equivalent representations if and only if they belong to the same co-adjoint orbit.

For any $X \in \mathfrak{f}_k \setminus \mathfrak{a}_k$, restricting the function of $\mathrm{Ind}_{\mathcal{A}_k}^{F_k}(\Lambda)$ to the subgroup $\exp(tX)$, $t \in \mathbb{R}$, yields the following models for the unitary representations $\mathrm{Ind}_{\mathcal{A}_k}^{F_k}(\Lambda)$. For $Y \in \mathfrak{a}_k$ and $\Lambda \in \mathfrak{a}_k^*$, we denote by $P_{\Lambda,Y}$ the polynomial function $x \to \Lambda(\mathrm{Ad}(e^{xX})Y)$. Let π_Λ^X be the unitary representation of the filiform k-step nilpotent Lie group F_k on the Hilbert space $L^2(\mathbb{R})$ uniquely determined by the derived representation $D\pi_\Lambda^X$ of the filiform Lie algebra $\mathfrak{f}_k$ given by the following formulas:

$$D\pi_\Lambda^X : \begin{cases} X \mapsto \frac{d}{dx} \\ Y \mapsto \imath P_{\Lambda,Y} \mathrm{Id}_{L^2(\mathbb{R})}, \qquad \text{for all } Y \in \mathfrak{a}_k. \end{cases} \tag{3.39}$$

For each $\Lambda \in \mathfrak{a}_k^*$, not vanishing on $[\mathfrak{f}_k, \mathfrak{f}_k]$, the unitary representation π_Λ^X is irreducible and, by Kirillov's theory, each irreducible unitary representation of the filiform k-step nilpotent Lie group F_k, which does not factor through a unitary representation of the Abelian quotient $F_k/[F_k, F_k]$, is unitarily equivalent to a representation of the form π_Λ^X described above.

Definition 3.3.1 *An* adapted basis *of the Lie algebra $\mathfrak{f}_k$ is an ordered basis $(X, Y) := (X, Y_1, \ldots, Y_k)$ of $\mathfrak{f}_k$ such that $X \notin \mathfrak{a}_k$ and $Y := (Y_1, \ldots, Y_k)$ is* *a basis of the Abelian ideal* $\mathfrak{a}_k \subset \mathfrak{f}_k$. *An adapted basis is normalized if the multivector* $X \wedge Y_1 \wedge \cdots \wedge Y_k$ *has volume* 1, *that is,*

$$\omega(X, Y_1, \ldots, Y_k) = 1.$$

Definition 3.3.2 *For any* $Y \in \mathfrak{a}_k$ *we define its* degree $d_Y \in \mathbb{N}$ *with respect to the representation* π^X_Λ *to be the degree of the polynomial* $P_{\Lambda,Y}$. *For any basis* $\mathcal{F} = (X, Y_1, \ldots, Y_k)$ *of the Lie algebra* $\mathfrak{f}_k$ *let* $(d_1, \ldots, d_k) \in \mathbb{N}^k$ *denote the degrees of the elements* $(Y_1, \ldots, Y_k)$. *The degree of the representation* π_Λ *is then defined as the maximum of the degrees of the elements* $(Y_1, \ldots, Y_k)$ *as* $\mathcal{F} = (X, Y_1, \ldots, Y_k)$ *varies over all adapted bases of* $\mathfrak{f}_k$.

Let

$$\mathfrak{a}^*_{k,\Lambda} = \{\Lambda \in \mathfrak{a}^*_k | \Lambda([\mathfrak{f}_k, \mathfrak{f}_k]) \neq 0\}.$$

Observe that the condition $\Lambda \in \mathfrak{a}^*_{k,\Lambda}$ is equivalent to $(d_1, \ldots, d_k) \neq 0$.

Let $\mathrm{ad}(X)$ denote the adjoint operator on $\mathfrak{f}_k$ induced by an element $X \in \mathfrak{f}_k$, that is:

$$\mathrm{ad}(X)(V) = [X, V], \quad \text{for all } V \in \mathfrak{f}_k.$$

For all $i = 1, \ldots, k$ and $j = 1, \ldots, d_i$, we let

$$\Lambda^{(j)}_i(\mathcal{F}) = (\Lambda \circ \mathrm{ad}^j(X))(Y_i). \tag{3.40}$$

Then the representation π^X_Λ can be written as follows:

$$D\pi^X_\Lambda : \begin{cases} X \mapsto \frac{d}{dx} \\ Y_i \mapsto \imath \left(\sum_{j=0}^{d_i} \frac{\Lambda^{(j)}_i(\mathcal{F})}{j!} x^j \right) \mathrm{Id}_{L^2(\mathbb{R})}. \end{cases} \tag{3.41}$$

For any linear form $\Lambda \in \mathfrak{a}^*_{k,\Lambda}$, let $\mathfrak{I}_\Lambda \subset \mathfrak{a}_k$ be the subset defined as follows:

$$\mathfrak{I}_\Lambda := \bigcap_{i=0}^{k-1} \ker(\Lambda \circ \mathrm{ad}^i(X)). \tag{3.42}$$

Since $\mathfrak{f}_k$ is filiform, the set $\mathfrak{I}_\Lambda \subset \mathfrak{f}_k$ is an ideal of the Lie algebra $\mathfrak{f}_k$. Let $F^\Lambda_k \subset F_k$ the normal subgroup defined by exponentiation of the ideal $\mathfrak{I}_\Lambda$. It is clear from the above definition that the ideal $\mathfrak{I}_\Lambda$, hence the subgroup F^Λ_k, depends only on the co-adjoint orbit of the form $\Lambda \in \mathfrak{a}^*_k$.

Sobolev Norms

We introduce below Sobolev norms *transverse* to orbits of ergodic nilflows on filiform nilmanifolds.

Let $\mathcal{F} := (X, Y) = (X, Y_1, \ldots, Y_k)$ be a normalized adapted basis. Let us introduce the associated transverse Laplacian

$$\Delta_Y = -\sum_{j=1}^{k} Y_j^2. \tag{3.43}$$

For any $\sigma \in \mathbb{R}$, we define the *transverse Sobolev norms* associated to the adapted basis $\mathcal{F} := (X, Y_1, \ldots, Y_k)$ as

$$\|f\|_{Y,\sigma} := \|f\|_{\Delta_Y,\sigma} = \|(I+\Delta_Y)^{\frac{\sigma}{2}} f\|_{L^2(M)}, \qquad \text{for all } f \in C^\infty(M). \tag{3.44}$$

The completion of $C^\infty(M)$ with respect to the norm $|\cdot|_{\mathcal{F},\sigma}$ is denoted by $W^\sigma(M, \mathcal{F})$. Endowed with this norm, $W^\sigma(M, \mathcal{F})$ is a Hilbert space.

The transverse Sobolev norms introduced above in formula (3.44) can be written in representation as follows. For the irreducible unitary representation $(\pi_\Lambda^X, H_\Lambda^X)$, described in formulas (3.39) and (3.41), the transverse Laplace operator Δ_Y, introduced in formula (3.43), is represented as the multiplication operator by the non-negative polynomial function

$$\Delta_{\Lambda,\mathcal{F}}(x) := \sum_{i=1}^{k} |P_{\Lambda,\mathcal{F},i}(x)|^2. \tag{3.45}$$

Thus, the transverse Sobolev norms can be written as follows: for every $\sigma \geq 0$ and for every $f \in C^\infty(\pi_\Lambda^X)$,

$$|f|_{\mathcal{F},\sigma} := \left(\int_{\mathbb{R}} [1 + \Delta_{\Lambda,\mathcal{F}}(x)]^{\frac{\sigma}{2}} |f(x)|^2 \, dx \right)^{1/2}. \tag{3.46}$$

For all $\sigma \in \mathbb{R}$, let $W_{\mathcal{F}}^\sigma(\pi_\Lambda^X) \subset C^\infty(\pi_\Lambda^X)$ denote the space of Sobolev functions for the representation π_Λ^X endowed with the norm defined above.

A Priori Estimates

The unique distributional obstruction to the existence of solutions of the cohomological equation

$$Xu = f \tag{3.47}$$

in a given irreducible unitary representation $(\pi_\Lambda^X, H_\Lambda^X)$ is the normalized X-invariant distribution $\mathcal{D}_\Lambda^X \in \mathcal{D}'(\pi_\Lambda^X)$ which can be written as

$$\mathcal{D}_\Lambda^X(f) := \int_{\mathbb{R}} f(x)\, dx\,, \qquad \text{for all } f \in C^\infty(\pi_\Lambda^X). \tag{3.48}$$

The formal *Green operator* G^X_Λ for the cohomological equation (3.47) is given by the formula

$$G^X_\Lambda(f)(x) := \int_{-\infty}^{x} f(y)\,dy\,, \quad \text{for all } f \in C^\infty(\pi^X_\Lambda). \tag{3.49}$$

It is not difficult to prove that the Green operator is well-defined on the kernel $\mathcal{K}^\infty(\pi^X_\Lambda)$ of the distribution $\mathcal{D}^X_\Lambda$ on $C^\infty(\pi^X_\Lambda)$: for all $f \in \mathcal{K}^\infty(\pi^X_\Lambda)$, the function $G^X_\Lambda(f) \in C^\infty(\pi_\Lambda)$ and the following identities hold:

$$G^X_\Lambda(f)(x) = \int_{-\infty}^{x} f(y)\,dy = -\int_{x}^{+\infty} f(y)\,dy. \tag{3.50}$$

For all $\sigma > 1/2$ let $\mathcal{K}^\sigma(\pi^X_\Lambda)$ denote the kernel of the invariant distribution $\mathcal{D}^X_\Lambda$ in the Sobolev space $W^\sigma_{\mathcal{F}}(\pi^X_\Lambda)$. The subspace $\mathcal{K}^\sigma(\pi^X_\Lambda)$ coincides with the closure of $\mathcal{K}^\infty(\pi^X_\Lambda)$ in $W^\sigma_{\mathcal{F}}(\pi^X_\Lambda)$.

We prove below bounds on the transverse Sobolev norms $\|G^X_\Lambda(f)\|_{\mathcal{F},\tau}$ for all functions $f \in \mathcal{K}^\sigma(\pi^X_\Lambda)$ for $\sigma >> \tau$.

For any $\sigma, \tau \in \mathbb{R}_+$ let

$$\begin{aligned} I_\sigma(\Lambda, \mathcal{F}) &:= \left(\int_{\mathbb{R}} \frac{dx}{[1 + \Delta_{\Lambda,\mathcal{F}}(x)]^\sigma} \right)^{1/2}; \\ J^\tau_\sigma(\Lambda, \mathcal{F}) &:= \left(\int\int_{|y|\ge|x|} \frac{[1 + \Delta_{\Lambda,\mathcal{F}}(x)]^\tau}{[1 + \Delta_{\Lambda,\mathcal{F}}(y)]^\sigma}\,dxdy \right)^{1/2}. \end{aligned} \tag{3.51}$$

Lemma 3.3.3 *Let $\mathcal{D}^X_\Lambda \in \mathcal{D}'(\pi^X_\Lambda)$ be the distribution defined in formula* (3.48). *For any $\sigma \in \mathbb{R}_+$, the following holds:*

$$|\mathcal{D}^X_\Lambda|_{\mathcal{F},-\sigma} := \sup_{f\neq 0} \frac{|\mathcal{D}^X_\Lambda(f)|}{|f|_{\mathcal{F},\sigma}} = I_\sigma(\Lambda, \mathcal{F}). \tag{3.52}$$

Proof It follows from the definitions by Hölder inequality. In fact,

$$\mathcal{D}^X_\Lambda(f) = \langle (1 + \Delta_{\Lambda,\mathcal{F}})^{-\frac{\sigma}{2}}, (1 + \Delta_{\Lambda,\mathcal{F}})^{\frac{\sigma}{2}} f \rangle_{L^2(\mathbb{R})}. \tag{3.53}$$

Since $|f|_{\mathcal{F},\sigma} = |(1 + \Delta_{\Lambda,\mathcal{F}})^{\frac{\sigma}{2}} f|_{\mathcal{F},0} = \|(1 + \Delta_{\Lambda,\mathcal{F}})^{\frac{\sigma}{2}} f\|_0$, it follows that

$$\sup_{f\neq 0} \frac{|\mathcal{D}^X_\Lambda(f)|}{|f|_{\mathcal{F},\sigma}} = \|(1 + \Delta_{\Lambda,\mathcal{F}})^{-\frac{\sigma}{2}}\|_{L^2(\mathbb{R})} = I_\sigma(\Lambda, \mathcal{F}).$$

The identity (3.52) is thus proved. □

Lemma 3.3.4 *For any $\sigma \geq \tau$ and for all $f \in \mathcal{K}^\sigma(\pi^X_\Lambda)$,*

$$|G^X_\Lambda(f)|_{\mathcal{F},\tau} \leq J^\tau_\sigma(\Lambda, \mathcal{F})]\,|f|_{\mathcal{F},\sigma}. \tag{3.54}$$

Proof It follows by Hölder inequality from formula (3.50) for the Green operator. In fact, for all $x \in \mathbb{R}$, by Hölder inequality we have

$$|G^X_\Lambda(f)(x)|^2 \le \left(\int_{|y|\ge|x|} \frac{dy}{(1+\Delta_{\Lambda,\mathcal{F}}(y))^\sigma}\right) |f|^2_{\mathcal{F},\sigma}.$$

Another application of Hölder inequality yields the result. □

We have thus reduced Sobolev bounds on the Green operator for the cohomological equation and on the ergodic averages operator (in each irreducible representation) to bounds on the integrals defined in formula (3.51).

Let $\mathcal{F} = (X, Y_1, \ldots, Y_k)$ be any adapted basis. Let $(d_1, \ldots, d_k) \in \mathbb{N}^k$ be the degrees of the elements $(Y_1, \ldots, Y_k)$, respectively; for all $\Lambda \in \mathfrak{a}^*_{k,\Lambda}$, let $\Lambda_i^{(j)}(\mathcal{F}) = \Lambda(\mathrm{ad}(X)^j Y_i)$ be the coefficients appearing in formula (3.41) and set

$$|\Lambda(\mathcal{F})| := \sup_{\{(i,j):\,1\le i\le k,\,0\le j\le d_i\}} \left|\frac{1}{j!}\Lambda_i^{(j)}(\mathcal{F})\right|. \tag{3.55}$$

We introduce on $\mathfrak{a}^*_{k,\Lambda}$ the following weight. For all $\Lambda \in \mathfrak{a}^*_{k,\Lambda}$, let

$$w_{\mathcal{F}}(\Lambda) := \min_{\{i\,:\,d_i\neq 0\}} \left|\frac{\Lambda_i^{(d_i)}(\mathcal{F})}{d_i!}\right|^{-\frac{1}{d_i}}. \tag{3.56}$$

We will prove below estimates for the integrals $I_\sigma(\Lambda, \mathcal{F})$ and $J^\tau_\sigma(\Lambda, \mathcal{F})$ of formula (3.51) in terms of the above weight.

For all $i = 1, \ldots, k$ and $j = 1, \ldots, d_i$ we define the rescaled coefficients

$$\hat\Lambda_i^{(j)}(\mathcal{F}) := \Lambda_i^{(j)}(\mathcal{F})\,(w_{\mathcal{F}}(\Lambda))^j, \tag{3.57}$$

and set, in analogy with (3.55),

$$|\hat\Lambda(\mathcal{F})| := \sup_{\{(i,j):\,1\le i\le k,\,0\le j\le d_i\}} \left|\frac{1}{j!}\hat\Lambda_i^{(j)}(\mathcal{F})\right|. \tag{3.58}$$

Lemma 3.3.5 *For all $\sigma > 1/2$, there exists a constant $C_{k,\sigma} > 0$ such that, for all $\Lambda \in \mathfrak{a}^*_{k,\Lambda}$, the following bounds hold:*

$$\frac{C^{-1}_{k,\sigma}}{(1+|\hat\Lambda(\mathcal{F})|)^\sigma} \le \frac{I_\sigma(\Lambda,\mathcal{F})}{w^{1/2}_{\mathcal{F}}(\Lambda)} \le C_{k,\sigma}(1+|\hat\Lambda(\mathcal{F})|). \tag{3.59}$$

For all $\sigma > \tau(k-1)+1$, there exists a constant $C_{k,\sigma,\tau} > 0$ such that, for all $\Lambda \in \mathfrak{a}^*_{k,\Lambda}$, the following bounds hold:

$$\frac{C^{-1}_{k,\sigma,\tau}}{(1+|\hat\Lambda(\mathcal{F})|)^\sigma} \le \frac{J^\tau_\sigma(\Lambda,\mathcal{F})}{w_{\mathcal{F}}(\Lambda)} \le C_{k,\sigma,\tau}(1+|\hat\Lambda(\mathcal{F})|)^{\tau k+1}. \tag{3.60}$$

Proof By change of variables, for any $w > 0$,

$$I_\sigma(\Lambda, \mathcal{F}) = w^{1/2} \left(\int_{\mathbb{R}} \frac{\mathrm{d}x}{[1 + \Delta_{\Lambda,\mathcal{F}}(wx)]^\sigma} \right)^{1/2} ;$$
$$J_\sigma^\tau(\Lambda, \mathcal{F}) = w \left(\int \int_{|y| \geq |x|} \frac{[1 + \Delta_{\Lambda,\mathcal{F}}(wx)]^\tau}{[1 + \Delta_{\Lambda,\mathcal{F}}(wy)]^\sigma} \mathrm{d}x\mathrm{d}y \right)^{1/2} . \tag{3.61}$$

Let $w := w_{\mathcal{F}}(\Lambda) > 0$. By definitions (3.56), (3.57) and (3.45), for all $i = 1, \dots, k$, the coefficients of the polynomial map $P_{\Lambda,Y_i}(wx)$ are the numbers $\hat{\Lambda}_i^{(j)}(\mathcal{F})/j!$. Thus all these coefficients are bounded by $|\hat{\Lambda}(\mathcal{F})|$ and there exists $i_0 \in \{1, \dots, k\}$ such that the polynomial $P_{\Lambda,Y_{i_0}}(wx)$ is monic. The following inequalities therefore hold: for all $x \in \mathbb{R}$,

$$1 + P^2_{\Lambda,Y_{i_0}}(wx) \leq 1 + \Delta_{\Lambda,\mathcal{F}}(wx) \leq (1 + |\hat{\Lambda}(\mathcal{F})|)^2(1 + x^{2(k-1)}). \tag{3.62}$$

Let $P(x)$ be any non-constant monic polynomial of degree $d \geq 1$ and let $\|P\|$ denote the maximum modulus of its coefficients. We claim that, if $d\sigma > 1/2$, there exists a constant $C_{d,\sigma} > 0$ such that

$$\int_{\mathbb{R}} \frac{\mathrm{d}x}{(1 + P^2(x))^\sigma} \leq C_{d,\sigma}(1 + \|P\|) , \tag{3.63}$$

and, if $d\sigma > (k-1)\tau + 1/2$, there exists a constant $C_{k,d,\sigma,\tau} > 0$ such that

$$\int \int_{|y| \geq |x|} \frac{(1 + x^{2(k-1)})^\tau}{(1 + P^2(y))^\sigma} \mathrm{d}x\mathrm{d}y \leq C_{k,d,\sigma,\tau}(1 + \|P\|)^{2+2\tau(k-1)}. \tag{3.64}$$

In fact, since P is monic, there exists $s \in [1, (1 + \|P\|)]$ such that the polynomial $P_s(x) := s^{-d}P(sx)$ is monic and has all coefficients in the unit ball. It follows that, if $d\sigma > 1/2$, there exists a constant $C_{d,\sigma} > 0$ such that

$$\int_{\mathbb{R}} \frac{\mathrm{d}x}{(1 + P^2(x))^\sigma} = \int_{\mathbb{R}} \frac{s\,\mathrm{d}x}{(1 + P^2(sx))^\sigma} \leq \int_{\mathbb{R}} \frac{s\,\mathrm{d}x}{(1 + P_s^2(x))^\sigma} \leq C_{d,\sigma}s ,$$

hence the bound in formula (3.63) is proved. Similarly, if $\sigma > \tau(k-1) + 1$, there exists a constant $C_{k,d,\sigma,\tau} > 0$ such that

$$\int \int_{|y| \geq |x|} \frac{(1 + x^{2(k-1)})^\tau}{(1 + P^2(y))^\sigma} \mathrm{d}x\mathrm{d}y = \int \int_{|y| \geq |x|} s^2 \frac{(1 + (sx)^{2(k-1)})^\tau}{(1 + P^2(sy))^\sigma} \mathrm{d}x\mathrm{d}y$$
$$\leq \int \int_{|y| \geq |x|} s^2 \frac{(1 + (sx)^{2(k-1)})^\tau}{(1 + P_s^2(sy))^\sigma} \mathrm{d}x\mathrm{d}y \leq C_{k,d,\sigma,\tau}s^{2+2\tau(k-1)} ,$$

hence the bound in formula (3.64) is proved as well.

Finally, applying bounds in (3.63) and (3.64) to the polynomial $P_{\Lambda,Y_i}(wx)$ and taking into account the formulas (3.61) and the estimates (3.62) we obtain the upper bounds (3.59) and (3.60).

The lower bounds are an immediate consequence of the upper bound in formula (3.62), hence the argument is complete. □

3.4 The Heisenberg Case

In this section we discuss the main 2-step case, that is, the case of Heisenberg nilflows, following [FF2]. Effective equidistribution results for this case have been obtained by several authors (J. Marklof [Ma], A. Fedotov and F. Klopp [FK] and others) and are related to bounds on Weyl sums for quadratic polynomials which go back a century to the work of Hardy and Littlewood [HL]. This case is special since Heisenberg nilflows are *renormalizable*, in the sense that the almost all nilflows are approximately self-similar under the action of a hyperbolic dynamical systems on a moduli space of Heisenberg nilflows. We present our point of view on renormalization of Heisenberg nilflows and derive effective equidistribution results. The argument is based on renormalization as well as on the analysis of the cohomological equation and invariant distributions presented in the previous section.

The Renormalization Flow

We introduce below a renormalization flow on a moduli space of Heisenberg structures on a Heisenberg nilmanifold.

Definition 3.4.1 *A* Heisenberg frame (X, Y, Z_0) *on M is a frame of the tangent bundle of M induced by left invariant vector fields on the Heisenberg group with the property that the following commutation relations hold:*

$$[X, Y] = Z_0 \quad \text{and} \qquad [X, Z_0] = [Y, Z_0] = 0.$$

The deformation space $\mathcal{T}$ *of Heisenberg structures on a Heisenberg nilmanifold M is the space of all Heisenberg frames on M.*

A Heisenberg frame determines a basis of the Heisenberg Lie algebra which will be called a *Heisenberg basis*. The deformation space can therefore be viewed as the space of all Heisenberg bases of the Heisenberg Lie algebra.

Definition 3.4.2 *The* moduli space $\mathcal{M}$ *of Heisenberg structures on the Heisenberg manifold M is the quotient of the deformation space $\mathcal{T}$ under the* *action of the group diffeomorphism of M induced by automorphisms of the Lie algebra.*

The deformation space can be identified with the subgroup $A < \mathrm{Aut}(\mathfrak{h})$ of automorphism of the Heisenberg Lie algebra $\mathfrak{h}$ acting as the identity on its center $Z(\mathfrak{h}) = \mathbb{R}Z_0$. In fact, let $\{X_0, Y_0, Z_0\}$ be a fixed (rational) Heisenberg basis of the Heisenberg Lie algebra. There exists a bijective correspondence between the deformation space $\mathcal{T}$ and the group A given by the map defined as

$$a \to (a_*(X_0), a_*(Y_0), Z_0), \qquad \text{for all } a \in A.$$

The subgroup of diffeomorphisms of M induced by automorphisms of the Lie algebra is isomorphic to the subgroup $A_\Gamma < A$ of elements mapping the lattice Γ onto itself. The subgroup A_Γ is a lattice in A. It follows that the moduli space $\mathcal{M}$ of Heisenberg structures on M is given by the isomorphism

$$\mathcal{M} \approx A_\Gamma \backslash A.$$

It can be proved that the group A is isomorphic to the semi-direct product of $SL(2,\mathbb{R})$ acting on $\mathbb{R}^2$ by the contragredient linear action of $SL(2,\mathbb{R})$ on the vector space of linear forms, that is,

$$A \approx SL(2,\mathbb{R}) \ltimes \mathbb{R}^2 ,$$

and A_Γ is a semi-direct product of a finite index subgroup Λ_Γ of $SL(2,\mathbb{Z})$ (either equal to $SL(2,\mathbb{Z})$ or to the congruence lattice $\Gamma(2)$ depending on the Euler constant) with a lattice isomorphic to $\mathbb{Z}^2$. It follows that the moduli space $\mathcal{M}$ is a non-compact finite-volume homogeneous manifold of dimension five which fibers over the non-compact finite-volume quotient $\Lambda_\Gamma \backslash SL(2,\mathbb{R})$ with compact two-dimensional toral fiber.

There is a natural action by multiplication on the right of the group A on the moduli space $A_\Gamma \backslash A$. Let $\{a_t\} < A$ be the one-parameter subgroup defined by the formula

$$a_t(X_0, Y_0, Z_0) = (e^t X_0, e^{-t} Y_0, Z_0), \qquad \text{for all } t \in \mathbb{R}.$$

Definition 3.4.3 *The* renormalization flow $\{\rho_t\}$ *is the flow on* $\mathcal{M}$ *defined by the one-parameter subgroup* $\{\rho_t\} < A$ *by multiplication on the right, that is,*

$$\rho_t(A_\Gamma a) = A_\Gamma a a_t \qquad \textit{for all } (a,t) \in A \times \mathbb{R}.$$

It is not hard to prove that the renormalization flow is a homogeneous *Anosov* flow which project onto a diagonal (geodesic) flow on the finite volume quotient $\Lambda_\Gamma \backslash SL(2,\mathbb{R})$.

In terms of Heisenberg bases, for any $a \in A$ the renormalization flow maps the Heisenberg basis $(X_a, Y_a, Z_0) = (a_*(X_0), a_*(Y_0), Z_0)$ into the rescaled basis $(e^t X_a, e^{-t} Y_a, Z_0)$, in fact

$$\begin{aligned}(\rho_t)_*(X_a, Y_a, Z_0) &= ((aa_t)_*(X_0), (aa_t)_*(Y_0), Z_0)\\ &= (e^t a_*(X_0), e^{-t} a_*(Y_0), Z_0) = (e^t X_a, e^{-t} Y_a, Z_0).\end{aligned}$$

A crucial element of our equidistribution estimates will involve certain *bounds on the geometry* of the rescaled Heisenberg structure along orbits the renormalization flow. For every Heisenberg frame $\mathcal{F} = (X, Y, Z_0)$, let $R_{\mathcal{F}}$ denote the Riemann metric on the Heisenberg nilmanifold M uniquely defined by the condition that the Heisenberg frame (X, Y, Z) is orthonormal. Effective equidistribution results for a Heisenberg nilflow $\{\phi_s^X\}$ on M are based on (exact) square root scaling of the norm of invariant distributions and on lower bounds on the injectivity radius of the metric $R_{\mathcal{F}(t)}$ along the forward orbit $\{\mathcal{F}(t) | t \geq 0\}$ of the frame $\mathcal{F}$ under the renormalization flow on the moduli space $\mathcal{M}$.

Scaling of Invariant Distributions and of the Green Operator

In this section we derive bounds on the scaling of invariant distributions and on the Green operator for Heisenberg nilflows under the renormalization flow. Irreducible unitary representations of the Heisenberg Lie group are determined up to unitary equivalences by the corresponding derived representations of the Heisenberg Lie algebra. By the Stone–Von Neumann theorem, such derived representations can be written as irreducible skew-adjoint representations on the Hilbert space $L^2(\mathbb{R}, dx)$ of the following form:

$$D\pi_\Lambda^X : \begin{cases} X \mapsto \frac{d}{dx} \\ Y \mapsto \imath(\Lambda(Z_0)x + \Lambda(Y))\mathrm{Id}_{L^2(\mathbb{R},dx)} \\ Z_0 \mapsto \imath(\Lambda(Z_0))\mathrm{Id}_{L^2(\mathbb{R},dx)}. \end{cases} \tag{3.65}$$

Representations such that $\Lambda(Z_0) = 0$ appear in the component of the space $L^2(M)$ of square-integrable functions on the nilmanifold M given by all functions which factor through the base torus. We therefore restrict our attention to irreducible unitary representations such that $\Lambda(Z_0) \neq 0$. It is also not restrictive to assume that $\Lambda(Y) = 0$. In fact, for every $\Lambda \in \mathfrak{h}^*$ such that $\Lambda(Z_0) \neq 0$, its co-adjoint orbit contains Λ_0 such that $\Lambda_0(Y) = 0$.

Let $\mathcal{F} = (X, Y, Z_0)$ be any Heisenberg basis and, for any $t \in \mathbb{R}$, let $\mathcal{F}(t)$ denote the rescaled basis defined as follows:

$$\mathcal{F}(t) := (X(t), Y(t), Z_0) = (e^t X, e^{-t} Y, Z_0).$$

In order to compare for all $t \in \mathbb{R}$ and $h \in \mathbb{R}$ the representations $\pi_\Lambda^{X(t+h)}$ and $\pi_\Lambda^{X(t)}$ we introduce the *intertwining operator*

$$U_h(f)(x) = e^{h/2} f(e^h x), \qquad \text{for all } f \in L^2(\mathbb{R}).$$

It follows by the definitions that, for all $V \in \mathfrak{h}$ we have

$$\pi_\Lambda^{X(t)}(V) \circ U_h = U_h \circ \pi_\Lambda^{X(t+h)}(V).$$

For all $\sigma \in \mathbb{R}$ and $h \in \mathbb{R}$, the operator U_h is unitary from the Sobolev spaces $W^\sigma_{\mathcal{F}}(\pi_\Lambda^{X(t+h)})$ for the representation $\pi_\Lambda^{X(t+h)}$ onto the Sobolev spaces $W^\sigma_{\mathcal{F}}(\pi_\Lambda^{X(t)})$ for the representation $\pi_\Lambda^{X(t)}$.

Lemma 3.4.4 *For all $t \in \mathbb{R}$, we have the following identity:*

$$|\mathcal{D}_\Lambda^{X(t+h)}|_{\mathcal{F}(t),-\sigma} = e^{-h/2} |\mathcal{D}_\Lambda^{X(t)}|_{\mathcal{F}(t),-\sigma}.$$

Proof For all $f \in C^\infty(\pi_\Lambda^{X(t+h)}) \subset L^1(\mathbb{R})$, by intertwining and change of variables, for all t and $h \in \mathbb{R}$ we have that

$$\mathcal{D}_\Lambda^{X(t+h)}(f) = \mathcal{D}_\Lambda^{X(t)}(U_h f) = e^{-h/2} \mathcal{D}_\Lambda^{X(t)}(f),$$

which immediately implies the statement. □

Let $G_{X,\Lambda}^{X(t)}$ denote the Green operator for the cohomological equation $X(t)u = f$ in the representation π_Λ^X. We recall that, according to our definitions above, see formula (3.50), $G_\Lambda^{X(t)}$ denotes the Green operator for the same cohomological equation $X(t)u = f$ in the representation $\pi_\Lambda^{X(t)}$.

Lemma 3.4.5 *For all $t \in \mathbb{R}$, the Green operators $G_{X,\Lambda}^{X(t)}$ on $\mathcal{K}^\sigma(\pi_\Lambda^X)$ and $G_\Lambda^{X(t)}$ on $\mathcal{K}^\sigma(\pi_\Lambda^{X(t)})$ are unitarily equivalent as*

$$G_{X,\Lambda}^{X(t)} = U_t \circ G_\Lambda^{X(t)} \circ U_t^{-1}.$$

Consequently, the norm $\|G_{X,\Lambda}^{X(t)}\|_{\sigma,\tau,\Lambda}$ of the operator $G_{X,\Lambda}^{X(t)}$ from $\mathcal{K}^\sigma(\pi_\Lambda^X) \subset W^\sigma_{\mathcal{F}}(\pi_\Lambda^X)$ into $W^\tau_{\mathcal{F}}(\pi_\Lambda^X)$ coincides with the norm $\|G_\Lambda^{X(t)}\|_{\sigma,\tau,\Lambda}$ of operator $G_\Lambda^{X(t)}$ from $\mathcal{K}^\sigma(\pi_\Lambda^{X(t)}) \subset W^\sigma_{\mathcal{F}}(\pi_\Lambda^{X(t)})$ into $W^\tau_{\mathcal{F}}(\pi_\Lambda^{X(t)})$.

Finally, from the general scaling results of Section 3.3 we can easily derive the following bounds:

Lemma 3.4.6 *For all $\sigma > 1/2$, there exists a constant $C_\sigma > 0$ such that, for all $\Lambda \in \mathfrak{h}^*$ such that $\Lambda(Z_0) \neq 0$, for all $t \in \mathbb{R}$, the following bounds hold:*

$$\frac{C_\sigma^{-1}}{|\Lambda(Z_0)|^{1/2}(1+|\Lambda(Z_0)|)^\sigma} \leq |\mathcal{D}_\Lambda^{X(t)}|_{\mathcal{F}(t),-\sigma} \leq C_\sigma \frac{(1+|\Lambda(Z_0)|)}{|\Lambda(Z_0)|^{1/2}}. \quad (3.66)$$

For all $\sigma > \tau + 1$, there exists a constant $G_{\sigma,\tau} > 0$ such that, for all $\Lambda \in \mathfrak{h}^*$ such that $\Lambda(Z_0) \neq 0$, the following bounds hold: for all $f \in \mathcal{K}^\sigma(\pi_\Lambda^X)$, we have

$$|G_\Lambda^{X(t)}(f)|_{\mathcal{F}(t),\tau} \leq G_{\sigma,\tau} \frac{(1+|\Lambda(Z_0)|)^{2\tau+1}}{|\Lambda(Z_0)|} |f|_{\mathcal{F}(t),\sigma}. \quad (3.67)$$

Proof By Lemma 3.3.3 and Lemma 3.3.4, the bounds stated above are reduced to estimates given in Lemma 3.3.5 on the integrals $I_\sigma(\Lambda, \mathcal{F}(t))$ and $J_\sigma(\Lambda, \mathcal{F}(t))$ which were defined in formula (3.51). By Lemma 3.3.5 the above integrals can be estimated in terms of the weights $w_{\mathcal{F}(t)}(\Lambda)$ given in formula (3.56), and $\hat{\Lambda}(\mathcal{F}(t))$, given in formula (3.58). Since by assumption $\Lambda(Z_0) \neq 0$, the vectors $(Y_1, Y_2) := (Y(t), Z_0)$ have degrees $(d_1, d_2) = (1, 0)$ and we have

$$\Lambda_1^{(0)}(\mathcal{F}(t)) = e^{-t}\Lambda(Y) = 0 \quad \text{and} \quad \Lambda_1^{(1)}(\mathcal{F}(t)) = \Lambda_2^{(0)}(\mathcal{F}(t)) = \Lambda(Z_0),$$

which immediately implies that

$$|\hat{\Lambda}(\mathcal{F}(t))| := \max\{1, |\Lambda(Z_0)|\} \quad \text{and} \quad w_{\mathcal{F}(t)}(\Lambda) = |\Lambda(Z_0)|^{-1}.$$

The result then follows from Lemma 3.3.3, Lemma 3.3.4 and Lemma 3.3.5. □

The point of the above result is that the bounds on the norms of the invariant distribution $\mathcal{D}_\Lambda^{X(t)}$ in $W_{\mathcal{F}}^\sigma(\pi_\Lambda^{X(t)})$ and of the Green operator $G_\Lambda^{X(t)}$ from $\mathcal{K}^\sigma(\pi_\Lambda^{X(t)})$ into $W_{\mathcal{F}}^\tau(\pi_\Lambda^{X(t)})$ for the rescaled Heisenberg frame $\mathcal{F}(t) = (X(t), Y(t), Z_0)$ are independent of the scaling.

Orthogonal Decomposition of Ergodic Averages

In this section we decompose ergodic averages, which are given by uniform probability measures along orbit segments, according to the following orthogonal splitting of Sobolev spaces into the kernel of the invariant distribution and its orthogonal complement:

$$W_{\mathcal{F}}^\sigma(\pi_\Lambda^X) = \mathcal{K}^\sigma(\pi_\Lambda^X) \oplus^\perp [\mathcal{K}^\sigma(\pi_\Lambda^X)]^\perp. \quad (3.68)$$

Such a decomposition is possible since by the Sobolev embedding theorem [Ad], [He] all measures are distributions which belong to all Sobolev space of square integrable functions of exponent larger than half the dimension of the space.

In fact, we prove a priori Sobolev bounds which can be derived from the *Sobolev trace theorem* [Ad], [He] and from Sobolev bounds on solutions of the cohomological equations explained in section 3.3. The Sobolev trace theorem is a fundamental result that allows us in particular to estimate the L^1 norm of a function along a submanifold in terms of its Sobolev norms. In our case, we estimate the L^1 norm along orbit segments, which are one-dimensional submanifolds, in terms of Sobolev norms of any exponent larger than half of the dimension of the space minus one. It is in these estimates that the geometry of the nilmanifold appears as the *best Sobolev constant*, that is, the best constant for which the Sobolev trace theorem holds, depends on the injectivity radius of the Riemannian metric associated to a Heisenberg frame.

Let $\mathcal{F} = (X, Y, Z_0)$ be any Heisenberg frame and, for all $T > 0$, let $\mathcal{A}^X(x, T)$ denote the measure defined as

$$\mathcal{A}^X(x, T)(f) := \frac{1}{T}\int_0^T f \circ \phi_s^X(x)ds\,, \qquad \text{for all } f \in C^0(M).$$

By the Sobolev trace theorem $\mathcal{A}^X(x, T) \in W_{\mathcal{F}}^{-\sigma}(M)$ for all $\sigma > 1$. For a fixed $\sigma > 1$, let $\mathcal{A}^X_{\Lambda,\sigma}(x, T)$ denote the orthogonal projection of $\mathcal{A}^X(x, T)$ onto an irreducible component of the dual Sobolev space $W_{\mathcal{F}}^{-\sigma}(M)$ isomorphic to $W_{\mathcal{F}}^{-\sigma}(\pi^X_\Lambda)$. According to the orthogonal splitting in formula (3.68), for any $(x, T) \in M \times \mathbb{R}$ there exist a constant $c^X_{\Lambda,\sigma}(x, T) \in \mathbb{R}$ and a distribution $R^X_{\Lambda,\sigma}(x, T) \in \mathrm{Ann}[\mathcal{K}^\sigma(\pi^X_\Lambda)^\perp] \subset W_{\mathcal{F}}^{-\sigma}(\pi^X_\Lambda)$ such that we have the following *orthogonal decomposition* in $W_{\mathcal{F}}^{-\sigma}(\pi^X_\Lambda)$:

$$\mathcal{A}^X_{\Lambda,\sigma}(x, T) = c^X_{\Lambda,\sigma}(x, T)\mathcal{D}^X_\Lambda + R^X_{\Lambda,\sigma}(x, T). \tag{3.69}$$

For all $\sigma > 1$, let $B_\sigma(\mathcal{F}) > 0$ denote the *best Sobolev constant* of the Riemannian metric $R_{\mathcal{F}}$ associated to the frame $\mathcal{F}$. We recall that the metric $R_{\mathcal{F}}$ is defined as the unique Riemannian metric such that the frame $\mathcal{F}$ is orthonormal. It is convenient to adopt the following definition

$$B_\sigma(\mathcal{F}) = \sup_{f\neq 0} \frac{\|f\|_{C^0(M)}}{|f|_{\mathcal{F},\sigma} + |Xf|_{\mathcal{F},0}}. \tag{3.70}$$

The following a priori bounds are immediate from the definitions:

Lemma 3.4.7 *For any $\sigma > 1$, there exists a constant $C_\sigma > 0$ such that*

$$|c^X_{\Lambda,\sigma}(x, T)||\mathcal{D}^X_\Lambda|_{\mathcal{F},-\sigma} \le |\mathcal{A}^X_{\Lambda,\sigma}(x, T)|_{\mathcal{F},-\sigma} \le C_\sigma B_\sigma(\mathcal{F}).$$

Proof The first inequality (on the left) follows from orthogonality of the splitting in $W_{\mathcal{F}}^{-\sigma}(\pi_{\Lambda}^{X})$. The second inequality follows from the argument below.

For any $t \in \mathbb{R}$ and for all $x \in M$, let

$$f_t(x) := \int_0^t f \circ \phi_s^X(x) ds.$$

By the Heisenberg commutation relations we have

$$Y(f \circ \phi_s^X) = (Y - sZ)(f) \circ \phi_s^X \quad \text{and} \quad Z(f \circ \phi_s^X) = Zf \circ \phi_s^X\,,$$

hence for $\sigma \in \mathbb{N}$ we have

$$X^\sigma f_t(x) = \int_0^t X^\sigma f \circ \phi_s^X(x) ds\,,$$

$$Y^\sigma f_t(x) = \int_0^t (Y - sZ)^\sigma f \circ \phi_s^X(x) ds\,,$$

$$Z^\sigma f_t(x) = \int_0^t Z^\sigma f \circ \phi_s^X(x) ds.$$

It follows by Minkowski integral inequality (for the L^2 norm of an integral) that there exists a constant $C'_\sigma > 0$ such that

$$|f_t|_{\mathcal{F},\sigma} \leq C'_\sigma(1 + |t|^\sigma)|f|_{\mathcal{F},\sigma}.$$

The above estimate can be extended to all $\sigma \in \mathbb{R}^+$ by interpolation, hence it follows that for all $\sigma \geq 1$ we have

$$|f_t|_{\mathcal{F},\sigma} + |Xf_t|_{\mathcal{F},0} \leq C''_\sigma(1 + |t|^\sigma)|f|_{\mathcal{F},\sigma}.$$

From the definition of the best Sobolev constant $B_\sigma(\mathcal{F})$, we can immediately derive that, for all $\sigma > 1$, we have

$$\|f_t\|_{C^0(M)} \leq B_\sigma(\mathcal{F})\{|f_t|_{\mathcal{F},\sigma} + |Xf_t|_{\mathcal{F},0}\} \leq C''_\sigma B_\sigma(\mathcal{F})(1 + |t|^\sigma)|f|_{\mathcal{F},\sigma}.$$

We can split the integral over an orbit segment of length $T \geq 1$ as sum of integrals over at most $[T] + 1$ orbits segments of at most unit length. In fact, for all $x \in M$ we have

$$f_T(x) = \sum_{h=0}^{[T]} f_1 \circ \phi_h^X(x) + f_{\{T\}} \circ \phi_{[T]}^X(x).$$

Since the uniform norm of a function is invariant under composition with a diffeomorphism, we derive the estimate

$$\|f_T\|_{C^0(M)} \leq [T]\|f_1\|_{C^0(M)} + \|f_{\{T\}}\|_{C^0(M)} \leq 2C''_\sigma B_\sigma(\mathcal{F})(1 + T)|f|_{\mathcal{F},\sigma}.$$

By definition $\mathcal{A}^X_{\Lambda,\sigma}(x,T)$ is the orthogonal projection of $\mathcal{A}^X_\sigma(x,T)$ onto an irreducible component of the dual Sobolev space $W^{-\sigma}_{\mathcal{F}}(M)$. For any function $f \in W^\sigma_{\mathcal{F}}(M)$ let $f_{\Lambda,\sigma} \in W^\sigma_{\mathcal{F}}(M)$ denote the orthogonal projection onto the corresponding irreducible component such that

$$\mathcal{A}^X_{\Lambda,\sigma}(x,T)(f) = \mathcal{A}^X_\sigma(x,T)(f_{\Lambda,\sigma}).$$

We can finally conclude that, for all $\sigma > 1$, also by the orthogonality of the decomposition, we have

$$\begin{aligned}|\mathcal{A}^X_{\Lambda,\sigma}(x,T)(f)| = |\mathcal{A}^X_\sigma(x,T)(f_{\Lambda,\sigma})| &\leq \frac{1}{T}\|(f_{\Lambda,\sigma})_T\|_{C^0(M)} \\ &\leq 4C''_\sigma B_\sigma(\mathcal{F})|f_{\Lambda,\sigma}|_{\mathcal{F},\sigma} \leq 4C''_\sigma B_\sigma(\mathcal{F})|f|_{\mathcal{F},\sigma}.\end{aligned}$$

The argument is completed. □

We are finally ready to estimate the dual Sobolev norm of the remainder distribution $R^X_{\Lambda,\sigma}(x,T) \in W^{-\sigma}_{\mathcal{F}}(\pi^X_\Lambda)$.

Lemma 3.4.8 *For all $\sigma > 2$ and $\tau \in (1, \sigma - 1)$ we have*

$$|R^X_{\Lambda,\sigma}(x,T)|_{\mathcal{F},-\sigma} \leq \frac{2}{T} B_\sigma(\mathcal{F})(1 + |G^X_\Lambda|_{\mathcal{F},\sigma,\tau}).$$

Proof By the splitting in formula (3.68), every $f \in W^\sigma_{\mathcal{F}}(\pi^X_\Lambda)$ has an orthogonal decomposition

$$f = f_0 + f_1\,, \quad \text{with } f_0 \in \mathcal{K}^\sigma(\pi^X_\Lambda) \text{ and } f_1 \in \mathcal{K}^\sigma(\pi^X_\Lambda)^\perp.$$

Since $R^X_{\Lambda,\sigma}(x,T) \in \mathrm{Ann}[\mathcal{K}^\sigma(\pi^X_\Lambda)^\perp]$ we have

$$R^X_{\Lambda,\sigma}(x,T)(f) = R^X_{\Lambda,\sigma}(x,T)(f_0) = \mathcal{A}^X(x,T)(f_0).$$

In addition, since $f_0 \in \mathcal{K}^\sigma(\pi^X_\Lambda)$, by Lemma 3.3.4 and Lemma 3.3.5, for any $\tau \in (1, \sigma - 1)$ there exists a solution $u := G^X_\Lambda(f_0)$ of the cohomological equation $Xu = f_0$. It follows that

$$|\mathcal{A}^X(x,T)(f_0)| = \frac{1}{T}|u \circ \phi^X_T(x) - u(x)| \leq \frac{2}{T}\|u\|_{C^0(M)}.$$

Finally, by the definition of the best Sobolev constant and by the bounds on the norm of the Green operator, for any $\sigma > 2$ and any $\tau \in (1, \sigma - 1)$ we have

$$\begin{aligned}\|u\|_{C^0(M)} &\leq B_\tau(\mathcal{F})(|u|_{\mathcal{F},\tau} + |Xu|_{\mathcal{F},0}) \\ &\leq B_\tau(\mathcal{F})(|G^X_\Lambda|_{\mathcal{F},\sigma,\tau}|f_0|_{\mathcal{F},\sigma} + |f|_{\mathcal{F},0}).\end{aligned}$$

By orthogonality, it follows that

$$|R^X_{\Lambda,\sigma}(x,T)(f)| \le \frac{2}{T} B_\tau(\mathcal{F})(|G^X_\Lambda|_{\mathcal{F},\sigma,\tau}|f|_{\mathcal{F},\sigma} + |f|_{\mathcal{F},0})$$

which immediately implies the stated bound. □

Spectral Estimates

In this section we derive Sobolev bounds for the distributions given by ergodic averages. We have proved above that such distributions can be decomposed into an invariant distribution term plus a remainder term. Under renormalization the invariant distribution term scales appropriately as derived above and yields the expected square root decay of ergodic averages with time.

The remainder term appears at first sight to be essentially negligible as it decays as the inverse power of time. However, it contains a multiplicative constant, the best Sobolev constant, which depends on the geometry of the nilmanifold along the renormalization orbit. For almost all renormalization orbits, this geometric term is responsible for the logarithmic or, more generally, subpolynomial terms which appear in equidistribution estimates and bounds on Weyl sums.

From a technical point of view we have to take into account the fact that although the space of invariant distributions is invariant (or, more precisely, equivariant) under the renormalization dynamics, its orthogonal complement is not.

Let $\mathcal{F} := (X, Y, Z_0)$ be any Heisenberg frame. For any $T \ge e$, let $[T]$ denote its integer part and let $h \in [1, 2]$ denote the ratio $\log T/[\log T]$. We define a sequence of Heisenberg frames $\mathcal{F}_j := (X_j, Y_j, Z_0)$ given by the forward orbit of the renormalization flow:

$$(X_j, Y_j, Z_0) := \rho_{jh}(X, Y, Z_0) = (e^{jh}X, e^{-jh}Y, Z_0)\,, \qquad \text{for all } j \in \mathbb{N}.$$

For any $\sigma > 0$, let us define

$$B_\sigma(\mathcal{F}, T) := \sum_{j=1}^{[\log T]} e^{jh/2} B_\sigma(\mathcal{F}_j). \tag{3.71}$$

Lemma 3.4.9 *For all $\sigma > 2$ there exists a constant $C_{\Lambda,\sigma} > 0$ such that, for any Heisenberg frame $\mathcal{F} = (X, Y, Z_0)$ and for all $(x, T) \in M \times \mathbb{R}^+$, we have*

$$|\mathcal{A}^X_{\Lambda,\sigma}(x,T)|_{\mathcal{F},-\sigma} \le C_{\Lambda,\sigma} \frac{B_\sigma(\mathcal{F}, T)}{T}.$$

Proof Let (T_j) be the finite sequence given by the formula

$$T_j := e^{-jh} T, \qquad \text{for all } j \in \mathbb{N}.$$

By change of variable we have the immediate identities

$$\mathcal{A}^X_{\Lambda,\sigma}(x, T) = \mathcal{A}^{X_j}_{\Lambda,\sigma}(x, T_j), \qquad \text{for all } j \in \mathbb{N}.$$

We then write the orthogonal decomposition of the distribution $\mathcal{A}^{X_j}_{\Lambda,\sigma}(x, T_j)$ in the Sobolev space $W^{\sigma}_{\mathcal{F}_j}(\pi^X_\Lambda)$: for all $j \in \mathbb{N}$ we have

$$\mathcal{A}^X_{\Lambda,\sigma}(x, T) = \mathcal{A}^{X_j}_{\Lambda,\sigma}(x, T_j) = c^{X_j}_{\Lambda,\sigma}(x, T_j)\mathcal{D}^{X_j}_\Lambda + R^{X_j}_{\Lambda,\sigma}(x, T_j)$$

which in particular implies the following

$$c^{X_{j-1}}_{\Lambda,\sigma}(x, T_{j-1})\mathcal{D}^{X_{j-1}}_\Lambda + R^{X_{j-1}}_{\Lambda,\sigma}(x, T_{j-1}) = c^{X_j}_{\Lambda,\sigma}(x, T_j)\mathcal{D}^{X_j}_\Lambda + R^{X_j}_{\Lambda,\sigma}(x, T_j).$$

Let now π_j denote the orthogonal projection from $W^{-\sigma}_{\mathcal{F}_j}(\pi^X_\Lambda)$ onto the one-dimensional subspace generated by the invariant distribution $\mathcal{D}^{X_j}_\Lambda$. Since the normalized invariant distributions $\mathcal{D}^{X_j}_\Lambda$ and $\mathcal{D}^{X_{j-1}}_\Lambda$ are proportional, it follows that

$$c^{X_{j-1}}_{\Lambda,\sigma}(x, T_{j-1})\mathcal{D}^{X_{j-1}}_\Lambda = c^{X_j}_{\Lambda,\sigma}(x, T_j)\mathcal{D}^{X_j}_\Lambda + \pi_{j-1} R^{X_j}_{\Lambda,\sigma}(x, T_j),$$

from which we derive the bound

$$\begin{aligned} &|c^{X_{j-1}}_{\Lambda,\sigma}(x, T_{j-1})||\mathcal{D}^{X_{j-1}}_\Lambda|_{\mathcal{F}_{j-1},-\sigma} \\ &\qquad \le |c^{X_j}_{\Lambda,\sigma}(x, T_j)||\mathcal{D}^{X_j}_\Lambda|_{\mathcal{F}_{j-1},-\sigma} + |R^{X_j}_{\Lambda,\sigma}(x, T_j)|_{\mathcal{F}_{j-1},-\sigma}. \end{aligned}$$

On the one hand, by the scaling of invariant distributions (see Lemma 3.4.4 and Lemma 3.4.6) there exists an explicit constant $K_{\Lambda,\sigma} > 0$ (given in Lemma 3.4.6) such that

$$|\mathcal{D}^{X_j}_\Lambda|_{\mathcal{F}_{j-1},-\sigma} = e^{-h/2}|\mathcal{D}^{X_{j-1}}_\Lambda|_{\mathcal{F}_{j-1},-\sigma} \ge K^{-1}_{\Lambda,\sigma} e^{-h/2};$$

on the other hand, by the estimates on solutions of the cohomological equation (see Lemma 3.4.6) there exist constants $K_\sigma > 0$ and $K'_{\Lambda,\sigma} > 0$ such that

$$|R^{X_j}_{\Lambda,\sigma}(x, T_j)|_{\mathcal{F}_{j-1},-\sigma} \le K_\sigma |R^{X_j}_{\Lambda,\sigma}(x, T_j)|_{\mathcal{F}_j,-\sigma} \le K'_{\Lambda,\sigma} \frac{B_\sigma(\mathcal{F}_j)}{T_j}.$$

From the above inequalities we then derive that there exists a constant $K''_{\Lambda,\sigma} > 0$ such that we have the following estimate:

$$|c^{X_{j-1}}_{\Lambda,\sigma}(x, T_{j-1})| \le e^{-h/2}|c^{X_j}_{\Lambda,\sigma}(x, T_j)| + K''_{\Lambda,\sigma} \frac{B_\sigma(\mathcal{F}_j)}{T_j}.$$

By finite (backward) induction over the set $\{0, \ldots, [\log T]\}$ we find

$$|c^X_{\Lambda,\sigma}(x, T)| \leq T^{-1/2}|c^{X_{[\log T]}}_{\Lambda,\sigma}(x, 1)| + \frac{K''_{\Lambda,\sigma}}{T} \sum_{j=1}^{[\log T]} B_\sigma(\mathcal{F}_j)e^{(j+1)h/2}.$$

By the a priori bound in Lemma 3.4.7 we derive that

$$|c^{X_{[\log T]}}_{\Lambda,\sigma}(x, 1)||\mathcal{D}^{X_{[\log T]}}_{\Lambda}|_{\mathcal{F}_{[\log T]},-\sigma} \leq C_\sigma B_\sigma(\mathcal{F}_{[\log T]}) \leq C_\sigma \frac{B_\sigma(\mathcal{F}, T)}{T^{1/2}},$$

hence by Lemma 3.4.6 there exists a constant $K_{\Lambda,\sigma} > 0$ such that

$$|c^{X_{[\log T]}}_{\Lambda,\sigma}(x, 1)| \leq C_\sigma K_{\Lambda,\sigma} \frac{B_\sigma(\mathcal{F}, T)}{T^{1/2}}.$$

By the above inequalities, we have thus proved the following estimate:

$$|c^X_{\Lambda,\sigma}(x, T)| \leq (C_\sigma K_{\Lambda,\sigma} + K''_{\Lambda,\sigma})\frac{B_\sigma(\mathcal{F}, T)}{T}.$$

Finally, the stated estimate follows from the decomposition (3.69) of the distribution $\mathcal{A}^X_{\Lambda,\sigma}(x, T)$, from the above bound and from the bound on the remainder distribution $R^X_{\Lambda,\sigma}(x, T)$ proved in Lemma 3.4.8. □

We have thus proved that estimates on ergodic averages are reduced to bounds on the best Sobolev constant $B_\sigma(\mathcal{F}(t))$ along the forward orbit $\{\mathcal{F}(t)|t \geq 0\}$ of the renormalization flow. For instance, let us assume that there exists a Heisenberg frame $\mathcal{F}$ of *bounded type* in the sense that

$$\sup_{t \geq 0} B_\sigma(\mathcal{F}(t)) < +\infty. \tag{3.72}$$

It follows immediately from Lemma 3.4.9 that for any Heisenberg frame $\mathcal{F}$ of bounded type and for all $(x, T) \in M \times \mathbb{R}^+$ we have

$$|\mathcal{A}^X_{\Lambda,\sigma}(x, T)|_{\mathcal{F},-\sigma} \leq C_{\Lambda,\sigma}T^{-1/2}.$$

For general Heisenberg frames the upper bound on ergodic averages is determined by an upper bound on the rate of growth of the best Sobolev constant, which in turn depends on the degeneration of the geometry of the nilmanifold under the renormalization flow.

Estimates on the Geometry

We have thus reduced the problem of finding bounds on ergodic averages of Heisenberg nilflows to bounds on the best Sobolev constant along orbits of the renormalization flow. Indeed, by its definition, for any $\sigma > 0$, the function $B_\sigma(\mathcal{F})$ is invariant under the action of automorphisms of the nilmanifold M

(that is automorphisms of the Heisenberg group which preserve the center and the lattice), hence it is well-defined (and continuous) on the moduli space $\mathcal{M}$. We recall that $\mathcal{M}$ is not compact, but it projects with compact fibers onto a finite covering of the modular surface. Under this projection, the renormalization flow projects onto the diagonal (geodesic) flow. By a well-known result the set of bounded orbits of the diagonal flow on a finite volume quotient of the group $SL(2,\mathbb{R})$ has full Hausdorff dimension, but zero Lebesgue measure. Hence the set of Heisenberg frames of bounded type is also a set of full Hausdorff dimension and zero measure. A generalization and refinement of this basic result requires the following bound of the best Sobolev constant in terms of the distance of the Heisenberg triple from a compact part of the moduli space.

Let $\delta_{\mathcal{M}} : \mathcal{M} \to \mathbb{R}^+$ denote the distance function on the moduli space $\mathcal{M}$, induced by the hyperbolic distance from a fixed base point (or from the thick part) on the finite volume surface $A_\Gamma\backslash SL(2,\mathbb{R})/SO(2,\mathbb{R})$ endowed with the hyperbolic metric of *curvature* -1.

Lemma 3.4.10 *(see [FF2], Corollary 3.11) For every* $\sigma > 1$, *there exists a constant* $C_\sigma > 0$ *such that*

$$B_\sigma(\mathcal{F}) \leq C_\sigma e^{\delta_{\mathcal{M}}(\mathcal{F})/4}.$$

Proof (Sketch) Let $\mathcal{F} = (X, Y, Z_0)$ be a Heisenberg frame on M at distance $\delta > 0$ from the thick part of the moduli space $\mathcal{M}$. We claim that the base (Abelianized) torus of M, endowed with the projected metric $R_{\bar{\mathcal{F}}}$ induced by the Abelian projected frame $\bar{\mathcal{F}} := (\bar{X}, \bar{Y})$, has shortest loop of length proportional to $e^{-\delta/2}$. In fact, the diagonal flow generated by the one-parameter group

$$g_t := \{\begin{pmatrix} e^t & 0 \\ 0 & e^{-t} \end{pmatrix} | t \in \mathbb{R}\},$$

by right multiplication on $SL(2,\mathbb{R})$, *has speed equal to* 2 with respect to the metric on $SL(2,\mathbb{R})$ induced by the hyperbolic metric of curvature -1 on the hyperbolic plane $SL(2,\mathbb{R})/SO(2,\mathbb{R})$. For any rational frame $\bar{\mathcal{F}} := (\bar{X}, \bar{Y})$, the shortest loop of the torus with respect to the flat metric $R_{\bar{\mathcal{F}}_t}$ corresponding to the Abelian frame $\bar{\mathcal{F}}_t := (\bar{X}_t, \bar{Y}_t) := g_t(\bar{X}, \bar{Y})$ is proportional to e^{-t}, for all $t > 0$. Since any frame $\bar{\mathcal{F}} := (\bar{X}, \bar{Y})$ is at bounded distance from the orbit of any given rational frame, the claim is proved.

It follows that the shortest loop on M with respect to the Riemannian metric $R_{\mathcal{F}}$ associated to the frame $\mathcal{F} := (X, Y, Z_0)$ (respect to which the frame is orthonormal) has length at least proportional to $e^{-\delta/2}$. Finally, it is possible to derive by an elementary scaling argument from the standard Sobolev embedding theorem (or Sobolev trace theorem) that there exists a constant $C_\sigma > 0$ such that the following holds. Let $\mathcal{F} = (X, Y, Z_0)$ be a Heisenberg frame on M such that its shortest loop has length $d \in (0, 1)$ with respect to the flat metric $R_\mathcal{F}$ on M. The best Sobolev constant $B_\sigma(\mathcal{F})$ for the frame $\mathcal{F}$ satisfies the bound

$$B_\sigma(\mathcal{F}) \le C_\sigma d^{-1/2}.$$

The statement then follows immediately. □

Remark It is stated erroneously in the proof of Lemma 5.7 of [FF2] that the renormalization flow on the moduli space has unit speed. As a consequence of this error, the proof of Lemma 5.7 (as well as its statement) is wrong. However, the main results of the paper are correct as we shall see below.

Proof of Theorem 3.1.1 It is a well-known theorem of S. G. Dani [Da] that the set of bounded orbits for the diagonal flow on finite volume quotients of $SL(2, \mathbb{R})$ has full Hausdorff dimension. Since the renormalization flow on the moduli space $\mathcal{M}$ projects onto the diagonal flow on a finite cover of the unit tangent bundle of the modular surface and since the projection has compact fibers in the moduli space, it follows immediately that the set of relatively compact orbits for the renormalization flow has full Hausdorff dimension. By Lemma 3.4.10 it follows that there exists a set of full Hausdorff dimension of *bounded-type* Heisenberg frames defined as in formula (3.72). The statement of Theorem 3.1.1 then follows immediately from Lemma 3.4.9. □

We recall that from the Khintchine–Sullivan theorem (stated for the geodesic flow of speed 2 on the unit tangent bundle of a hyperbolic surface) it follows that for any positive, non-increasing function $\phi : [1, +\infty) \to \mathbb{R}^+$ the set of Heisenberg frames $\mathcal{F}$ such that there exists a constant $C_\mathcal{F}(\phi) > 0$ such that

$$\delta_\mathcal{M}(\mathcal{F}(t)) < \phi(t) + C_\phi(\mathcal{F}), \qquad \text{for all } t \ge 0,$$

has full Lebesgue measure if and only if

$$\int_0^{+\infty} e^{-\phi(t)} dt < +\infty.$$

In particular, in the logarithmic scale we derive that for every $\epsilon > 0$ and for almost all Heisenberg frames $\mathcal{F}$ there exists a constant $C_\epsilon(\mathcal{F}) > 0$ such that

$$\delta_\mathcal{M}(\mathcal{F}(t)) \le (1 + \epsilon) \log t + C_\epsilon(\mathcal{F}).$$

Proof of Theorem 3.1.2 Let β be any positive, non-decreasing function defined on $[1, +\infty)$ such that

$$\int_1^{+\infty} \frac{dT}{T\beta^4(T)} < +\infty.$$

From the Khintchine–Sullivan [Su] theorem it follows that for almost all Heisenberg frames $\mathcal{F}$ there exists a constant $C_{\mathcal{F}} > 0$ such that

$$\delta_{\mathcal{M}}(\mathcal{F}(t)) \leq 4 \log \beta(e^t) + C_\beta(\mathcal{F}),$$

hence by Lemma 3.4.10 we derive the following bound on the best Sobolev constant: for almost all Heisenberg frames $\mathcal{F}$ there exists a constant $C_{\beta,\sigma}(\mathcal{F})$ such that

$$B_\sigma(\mathcal{F}(t)) \leq C_{\beta,\sigma}(\mathcal{F})\beta(e^t), \qquad \text{for all } t \geq 1.$$

From the above estimate, since the function $\beta : [1, +\infty) \to \mathbb{R}^+$ is non-decreasing we derive that there exists a constant $C'_{\beta,\sigma}(\mathcal{F}) > 0$ such that

$$\begin{aligned} B_\sigma(\mathcal{F}, T) &:= \sum_{j=1}^{[\log T]} e^{(j+1)h/2} B_\sigma(\mathcal{F}_j) \\ &\leq C_{\beta,\sigma}(\mathcal{F})\beta(T) \sum_{j=1}^{[\log T]} e^{(j+1)h/2} \leq C'_{\beta,\sigma}(\mathcal{F})\beta(T)T^{1/2}. \end{aligned}$$

From Lemma 3.4.9, for every $\Lambda \in \mathfrak{h}^*$ such that $\Lambda(Z_0) \neq 0$, we then derive for all $\sigma > 2$ the estimate

$$|\mathcal{A}^X_{\Lambda,\sigma}(x, T)|_{\mathcal{F},-\sigma} \leq C_{\Lambda,\sigma} C'_{\beta,\sigma}(\mathcal{F}) T^{-1/2} \beta(T).$$

The estimate on ergodic averages of general smooth functions follows from the orthogonal decomposition of Sobolev spaces into isotypical components, that is, subspaces which are (finite) direct sums of unitarily equivalent irreducible components. For any $\sigma \in \mathbb{R}$, let $\{W^\sigma_n(M) | n \in \mathbb{Z}\}$ denote the sequence of isotypical components of the Sobolev space $W^\sigma(M)$ which are orthogonal to the subspace of functions defined on the base (Abelianized) torus (or, equivalently, which have zero average along each fiber of the circle fibration $\pi : M \to \mathbb{T}^2$ of M over the base torus). By definition, for each $n \in \mathbb{Z}$ the space $W^\sigma_n(M)$ is unitarily equivalent to a finite direct sum of model spaces $W^\sigma(\pi^X_{\Lambda_n})$ with $\Lambda_n \in \mathfrak{h}^*$ such that $\Lambda_n(Z_0) \neq 0$, hence for any $\sigma > 2$ the estimates proved in Lemma 3.4.9 hold for functions in $W^\sigma_n(M)$. For any $\sigma \geq 0$, we have decomposition of the form

$$W^\sigma(M) = \pi^* W^\sigma(\mathbb{T}^2) \oplus \bigoplus_{n \in \mathbb{Z}} W^\sigma_n(M).$$

For any $f \in W^\sigma(M)$, let $\bar{f} \in W^\sigma(M)$ denote the projection onto the subspace $\pi^* W^\sigma(\mathbb{T}^2)$ of toral functions, and, for every $n \in \mathbb{N}$, let $f_n \in W^\sigma_n(M)$ denote the orthogonal projection on a non-toral isotypical component. By definition we have the decomposition

$$f = \bar{f} + \sum_{n\in\mathbb{Z}} f_n .$$

It can be proved that if $f \in W^\sigma(M)$ for any $\sigma > \tau$, the Sobolev norms of the functions $f_n \in W^\tau(M)$ decay polynomially. In fact, for all $a \in \mathbb{N}$ we have

$$\left(\sum_{n\in Z} n^{2a} \|f_n\|_\tau^2\right)^{1/2} = \|Z_0^a f\|_\tau \le \|f\|_{\tau+a} ,$$

which by interpolation implies that for all $\sigma > \tau$ we have

$$\|f_n\|_\tau \le (1 + n^{2(\sigma-\tau)})^{-1/2} \|f\|_\sigma .$$

Ergodic averages of the projection $\bar{f}$ on the subspace of toral functions can be written as ergodic averages with respect to the projected linear flows on the torus. Under our Diophantine condition all functions $\bar{f} \in W^\sigma(\mathbb{T}^2)$ for $\sigma > 5/2$ are continuous coboundaries, hence a stronger estimate holds, as in this case ergodic integrals are bounded. Finally, by Lemma 3.4.6 the constant $C_{\Lambda,\sigma} > 0$ in Lemma 3.4.9 can be estimated in terms of $\Lambda \in \mathfrak{h}^*$ as follows: for every $\tau > 2$ there exists $C_\tau > 0$ such that

$$C_{\Lambda_n,\tau} \le C_\tau (1 + n)^{\tau+2}.$$

The statement of Theorem 3.1.2 for $\sigma > 7$ then follows. In fact, since

$$\sum_{n\in\mathbb{Z}} C_{\Lambda_n,\tau} |f_n|_{\mathcal{F},\tau} \le C_\tau \left(\sum_{n\in\mathbb{Z}} (1+n)^{\tau+2} (1 + n^{2(\sigma-\tau)})^{-1/2}\right) \|f\|_\sigma ,$$

it follows from Lemma 3.4.9 that for all $\sigma > 2\tau+3 > 7$, there exists a constant $C_{\sigma,\tau} > 0$ such that

$$|\mathcal{A}^X_{\Lambda,\sigma}(x,T)(f - \bar{f})| \le \sum_{n\in\mathbb{Z}} |\mathcal{A}^X_{\Lambda,\sigma}(x,T)(f_n)| \le C_{\sigma,\tau} \|f\|_\sigma T^{-1/2} \beta(T).$$

By more careful estimates the result can be proved under the weaker regularity condition $\sigma > 5/2$ (see [FF2]). □

3.5 Higher-Step Filiform Nilflows

In this final section we describe a generalization of our approach to higher-step filiform nilflows. The main difficulty in the higher-step case comes from the fact that no renormalization dynamics seems to be available. However, our method can be generalized by introducing an appropriate *renormalization group* of scaling operators suggested by the harmonic analysis of the cohomological equation and invariant distributions. The renormalization group generalizes the renormalization introduced in the Heisenberg case, but (approximate) self-similarity, which is related to the recurrence of the renormalization dynamics, is lost. In fact, the dynamics of our renormalization group on the moduli space of nilflows is completely dissipative (hence it is actually pointless to introduce a moduli space). We discuss how to (partially) overcome these difficulties and outline an approach which we believe can be applied to other effective equidistribution problems in homogeneous unipotent dynamics.

Scaling of Invariant Distributions and of the Green Operator

In this section we introduce a scaling of (normalized) adapted bases of filiform Lie algebras which optimizes the scaling of the transverse Sobolev norms of invariant distributions.

Let $\mathcal{F} := (X, Y_1, \dots, Y_k)$ denote any *filiform basis* of a k-step filiform Lie algebra on 2 generators, that is, a basis satisfying the filiform commutation relations

$$\begin{aligned} &[X, Y_i] = Y_{i+1}\,, &&\text{for all } i < k \quad \text{and} \quad [X, Y_k] = 0\,; \\ &[Y_i, Y_j] = 0\,, &&\text{for all } i, j = 1, \dots, k. \end{aligned}$$

It can be proved that the *optimal* scaling for the renormalization group is given by the following formulas:

$$\rho_t \begin{pmatrix} X \\ Y_1 \\ \dots \\ Y_j \\ \dots \\ Y_k \end{pmatrix} = \begin{pmatrix} X(t) \\ Y_1(t) \\ \dots \\ Y_j(t) \\ \dots \\ Y_k(t) \end{pmatrix} := \begin{pmatrix} e^t X \\ e^{-\frac{2}{k}t} Y_1 \\ \dots \\ e^{-\frac{2(k-j)}{k(k-1)}t} Y_j \\ \dots \\ Y_k \end{pmatrix}. \tag{3.73}$$

The above scaling is optimal in the sense that any other (admissible) choice of the scaling gives a weaker result on the speed of equidistribution of ergodic averages (although it may give better or easier estimates on the deformation of the geometry of the nilmanifold).

We also remark that, in contrast with the Heisenberg case, in the higher-step case the scaling is not given by automorphisms of the Lie algebra. In fact, we have the following commutation relations:

$$[X(t), Y_j(t)] = e^{t-\frac{2}{k(k-1)}t} Y_{j+1}(t), \qquad \text{for all } j \in \{1, \dots, k-1\}.$$

From this formula for commutators it follows that the scaling group $\{\rho_t\}$ cannot be recurrent up to the action of diffeomorphisms (that is, on a moduli space) as diffeomorphisms preserve commutators. In particular, rescaled bases are not filiform bases.

Let $\mathcal{F} := (X, Y_1, \dots, Y_k)$ denote a filiform basis and, for all $t \geq 0$, let $\mathcal{F}(t)$ denote the rescaled basis $\mathcal{F}(t) = \rho_t(\mathcal{F})$ defined above.

In analogy with the Heisenberg case we introduce an intertwining (unitary) operator $U_h : L^2(\mathbb{R}) \to L^2(\mathbb{R})$ by the formula

$$U_h(f)(x) = e^{h/2} f(e^h x), \qquad \text{for all } f \in L^2(\mathbb{R}).$$

It follows by the definitions that, for all $V \in \mathfrak{f}_k$ we have

$$\pi_\Lambda^{X(t)}(V) \circ U_h = U_h \circ \pi_\Lambda^{X(t+h)}(V).$$

For all $\sigma \in \mathbb{R}$ and for all t and $h \in \mathbb{R}$, the operator U_h is unitary from Sobolev spaces $W^\sigma_{\mathcal{F}}(\pi_\Lambda^{X(t+h)})$ for the representation $\pi_\Lambda^{X(t+h)}$ onto Sobolev spaces $W^\sigma_{\mathcal{F}}(\pi_\Lambda^{X(t)})$ for the representation $\pi_\Lambda^{X(t)}$.

Lemma 3.5.1 *For all $t \in \mathbb{R}$, we have the following identity:*

$$|\mathcal{D}_\Lambda^{X(t+h)}|_{\mathcal{F}(t),-\sigma} = e^{-h/2} |\mathcal{D}_\Lambda^{X(t)}|_{\mathcal{F}(t),-\sigma} = e^{-h/2} I_\sigma(\Lambda, \mathcal{F}(t)).$$

Proof For all $f \in C^\infty(\pi_\Lambda^{X(t+h)}) \subset L^1(\mathbb{R})$, by intertwining and change of variables, we have

$$\mathcal{D}_\Lambda^{X(t+h)}(f) = \mathcal{D}_\Lambda^{X(t)}(U_h f) = e^{-h/2} \mathcal{D}_\Lambda^{X(t)}(f).$$

The statement then follows from Lemma 3.3.3 on the Sobolev norms of invariant distributions. □

Let $G_{X,\Lambda}^{X(t)}$ denote the Green operator for the cohomological equation $X(t)u = f$ in the representation π_Λ^X. We recall that, according to our definitions above, see formula (3.50), $G_\Lambda^{X(t)}$ denotes the Green operator for the same cohomological equation $X(t)u = f$ in the representation $\pi_\Lambda^{X(t)}$.

In the case of a filiform basis $\mathcal{F} := (X, Y_1, \dots, Y_k)$, the degree of the vectors $(Y_1, \dots, Y_k)$ are

$$(d_1, \dots, d_i, \dots, d_k) = (k-1, \dots, k-i, \dots, 0),$$

hence in particular the weight introduced in formula (3.56) takes the form

$$w_{\mathcal{F}}(\Lambda) := \min_{\{i\,:\,d_i\neq 0\}} \left|\frac{\Lambda_i^{(d_i)}(\mathcal{F})}{d_i!}\right|^{-\frac{1}{d_i}} = \min_{\{i=1,\dots,k\}} \left|\frac{\Lambda(Y_k)}{(k-i)!}\right|^{-\frac{1}{k-i}}$$

For convenience of notation we introduce the following modified norm: for all $\Lambda \in \mathfrak{a}^*_{k,\Lambda}$, let

$$\|\Lambda\|_{\mathcal{F}} := |\Lambda(\mathcal{F})|\,(1 + \frac{1}{|\Lambda(Y_k)|}). \tag{3.74}$$

The following Lemma generalizes Lemma 3.4.6 to the higher-step filiform case.

Lemma 3.5.2 *For all $k \geq 2$ and all $\sigma > 1/2$, there exists a constant $C_{k,\sigma} > 0$ such that, for all $\Lambda \in \mathfrak{a}^*_k$ such that $\Lambda(Y_k) \neq 0$, for all $t \in \mathbb{R}$, the following bounds hold:*

$$\frac{C_{k,\sigma}^{-1}}{(1+\|\Lambda(\mathcal{F})\|)^{\sigma}} \leq \frac{e^{(\frac{1}{2}-\frac{1}{k(k-1)})t}}{w_{\mathcal{F}}^{1/2}(\Lambda)} |\mathcal{D}_{\Lambda}^{X(t)}|_{\mathcal{F}(t),-\sigma} \leq C_{k,\sigma}(1+\|\Lambda(\mathcal{F})\|). \tag{3.75}$$

For $\sigma > \tau(k-1)+1$ there exists a constant $G_{k,\sigma,\tau} > 0$ such that, for all $t \in \mathbb{R}$ and for all $f \in \mathcal{K}^{\infty}(\pi_{\Lambda}^{X(t)})$, the following holds:

$$|G_{X,\Lambda}^{X(t)}(f)|_{\mathcal{F}(t),\tau} \leq G_{k,\sigma,\tau}(1+\|\Lambda\|_{\mathcal{F}})^{\tau k+2} e^{-(1-\frac{2}{k(k-1)})t} |f|_{\mathcal{F}(t),\sigma}. \tag{3.76}$$

Proof The argument proceeds as in the proof of Lemma 3.4.6. By Lemma 3.3.3 and Lemma 3.3.4, the bounds stated above are reduced to estimates given in Lemma 3.3.5 on the integrals $I_\sigma(\Lambda, \mathcal{F}(t))$ and $J_\sigma(\Lambda, \mathcal{F}(t))$ which were defined in formula (3.51). By Lemma 3.3.5 the above integrals can be estimated in terms of the weights $w_{\mathcal{F}(t)}(\Lambda)$ given in formula (3.56), and $\hat{\Lambda}(\mathcal{F}(t))$, given in formula (3.58). By definition, for all $t \in \mathbb{R}$ we have

$$w_{\mathcal{F}(t)}(\Lambda) = e^{-(1-\frac{2}{k(k-1)})t} w_{\mathcal{F}}(\Lambda)\,,$$

and, by an elementary estimate, for all $t \geq 0$ we have

$$|\hat{\Lambda}(\mathcal{F}(t))| \leq \|\Lambda\|_{\mathcal{F}}. \tag{3.77}$$

In fact, it follows from the definition in formula (3.57) that for any rescaled basis $\mathcal{F}(t)$, for all $i \in \{1,\dots,k\}$ and $j \in \{1,\dots,k-i\}$, we have

$$\hat{\Lambda}_i^{(j)}(\mathcal{F}(t)) = \Lambda_i^{(j)}(\mathcal{F}(t))w_{\mathcal{F}(t)}(\Lambda)^j = e^{\frac{2(j-k+i)}{k(k-1)}t}\hat{\Lambda}_i^{(j)}(\mathcal{F}) \leq \hat{\Lambda}_i^{(j)}(\mathcal{F}).$$

The bound in formula (3.77) follows from the definition of $\hat{\Lambda}(\mathcal{F})$ in formula (3.58).

The result then follows from Lemma 3.3.3, Lemma 3.3.4 and Lemma 3.3.5. □

Heuristically, following the method explained in the case of Heisenberg nilflow, we should be able to derive from Lemma 3.5.2 corresponding polynomial bounds for ergodic averages. However, for that we need to prove appropriate bounds on the scaled geometry of nilmanifolds. It does not seem to be possible to prove bounds on the injectivity radius of the rescaled metrics, as higher-step nilflows may have very close returns which correspond to very short loops in certain scaled metrics. However, such very short return should be quite rare. In the next section we introduce a notion of *average width* of an orbit segment and prove a corresponding generalization of the Sobolev trace theorem, which shows that the best Sobolev constant can be bounded in terms of the average width of orbits.

The Average Width and a Trace Theorem

The content of this section consists of the core technical novelties which make it possible to extend the renormalization approach from the Heisenberg case to the higher-step filiform case.

Let $\mathcal{F} := (X, Y_1, \dots, Y_k)$ be any normalized adapted basis of the filiform Lie algebra $\mathfrak{f}_k$. For any $x \in M$, let $\phi_x : \mathbb{R} \times \mathbb{R}^k \to M$ be the local embedding defined for all $(t, \mathbf{s}) \in \mathbb{R} \times \mathbb{R}^k$ by the formula

$$\phi_x(t, \mathbf{s}) = x \exp(tX) \exp(\mathbf{s} \cdot Y). \tag{3.78}$$

We have the following elementary results.

Lemma 3.5.3 *For any $x \in M$ and any $f \in C^\infty(M)$ we have*

$$\frac{\partial \phi_x^*(f)}{\partial t}(t, \mathbf{s}) = \phi_x^*(Xf)(t, \mathbf{s}) + \sum_j s_j \phi_x^*([X, Y_j]f)(t, \mathbf{s})\,;$$

$$\frac{\partial \phi_x^*(f)}{\partial s_j} = \phi_x^*(Y_j f), \quad \textit{for all } j = 1, \dots, k.$$

Lemma 3.5.4 *For any $x \in M$, we have*

$$\phi_x^*(\omega) = dt \wedge ds_1 \wedge \cdots \wedge ds_k.$$

Let Leb_k denote the k-dimensional Lebesgue measure on $\mathbb{R}^k$.

Definition 3.5.5 *For any open neighborhood of the origin $O \subset \mathbb{R}^k$, let $\mathcal{R}_O$ be the family of all k-dimensional symmetric (i.e. centered at the origin) rectangles $R \subset [-1/2, 1/2]^k$ such that $R \subset O$. The* inner width *of the open set $O \subset \mathbb{R}^k$ is the positive number*

$$w(O) := \sup\{Leb_k(R) \,|\, R \in \mathcal{R}_O\}.$$

The width function *of a set $\Omega \subset \mathbb{R} \times \mathbb{R}^k$ containing the line $\mathbb{R} \times \{0\}$ is the function $w_\Omega : \mathbb{R} \to [0, 1]$ defined as follows:*

$$w_\Omega(\tau) := w(\{\mathbf{s} \in \mathbb{R}^k \mid (\tau, \mathbf{s}) \in \Omega\}), \quad \textit{for all } \tau \in \mathbb{R}.$$

Definition 3.5.6 *Let $\mathcal{F} = (X, Y_1, \ldots, Y_k)$ be any normalized adapted basis. For any $x \in M$ and $T > 1$, we consider the family $\mathcal{O}_{x,T}$ of all open sets $\Omega \subset \mathbb{R} \times \mathbb{R}^k$ satisfying:*

- $[0, T] \times \{0\} \subset \Omega \subset \mathbb{R} \times [-1/2, 1/2]^k$;
- *the map*

$$\phi_x : \Omega \to M$$

defined by formula (3.78) *is injective.*

The average width *of the orbit segment*

$$\gamma^X(x, T) := \{x \exp(tX) \mid 0 \le t \le T\} = \{\phi_x(t, 0) \mid 0 \le t \le T\},$$

relative to the normalized adapted basis $\mathcal{F}$, is the positive real number

$$w_{\mathcal{F}}(x, T) := \sup_{\Omega \in \mathcal{O}_{x,T}} \left(\frac{1}{T} \int_0^T \frac{ds}{w_\Omega(s)} \right)^{-1}. \tag{3.79}$$

The average width *of the nilmanifold M, relative to the normalized adapted basis $\mathcal{F}$, at a point $y \in M$ is the positive real number*

$$w_{\mathcal{F}}(y) := \sup\{w_{\mathcal{F}}(x, 1) | y \in \gamma^X(x, 1)\}. \tag{3.80}$$

For any $\sigma \ge 0$, let $W^\sigma(M, \mathcal{F})$ denote the transverse Sobolev space (defined with respect to the transverse Laplacian) introduced in section 3.3.

The following generalization of the Sobolev embedding theorem and of the Sobolev trace theorem hold and can be proven by elementary methods (see Theorems 3.9 and 3.10 in [FF4]).

Theorem 3.5.7 *Let $\mathcal{F} = (X, Y_1, \ldots, Y_k)$ be any normalized adapted basis. For any $\sigma > k/2$, there exists a positive constant $C_{k,\sigma}$ such that, for all* *functions* $u \in W^{\sigma+1}(M, \mathcal{F})$ *such that* $Xu \in W^{\sigma}(M, \mathcal{F})$ *and for all* $y \in M$, *we have*

$$|u(y)| \le \frac{C_{k,\sigma}}{w_{\mathcal{F}}(y)^{1/2}} \Big\{ |u|_{\mathcal{F},\sigma} + |Xu|_{\mathcal{F},\sigma} + \sum_{j=1}^{k} \big|[X, Y_j]u\big|_{\mathcal{F},\sigma} \Big\}.$$

For a vector field X on M and $x \in M$ the ergodic average $\mathcal{A}^X(x, T)$ is defined as follows: for all $f \in L^2(M)$,

$$\mathcal{A}^X(x, T)(f) := \frac{1}{T} \int_0^T f \circ \phi_t^X(x) \, \mathrm{d}t , \qquad \text{for all } T \in \mathbb{R}_+,$$

where $\{\phi_t^X\}$ is the flow generated by the vector field X on M. The following generalized Sobolev trace theorem for the linear functional $\mathcal{A}^X(x, T)$ holds.

Theorem 3.5.8 *Let* $\mathcal{F} = (X, Y_1, \dots, Y_k)$ *be any normalized adapted basis. For any* $\sigma > k/2$, *there exists a positive constant* $C_{k,\sigma}$ *such that, for all functions* $f \in W^{\sigma}(M, \mathcal{F})$, *for all* $T \in [1, +\infty)$ *and all* $x \in M$ *we have*

$$|\mathcal{A}^X(x, T)(f)| \le \frac{C_{k,\sigma}}{T^{1/2} w_{\mathcal{F}}(x, T)^{1/2}} |f|_{\mathcal{F},\sigma}.$$

Spectral Estimates

Spectral estimates in the higher-step filiform case can be derived along the same lines as spectral estimates in the Heisenberg case.

For any adapted basis $\mathcal{F} := (X, Y_1, \dots, Y_k)$ and for any $(x, T) \in M \times \mathbb{R}^+$ we let

$$B_{\mathcal{F}}(x, T) := \frac{1}{w_{\mathcal{F}}^{1/2}(x)} + \frac{1}{w_{\mathcal{F}}^{1/2}(\phi_T^X(x))}.$$

For any $T \ge e$, let $[T]$ denote its integer part and let $h \in [1, 2]$ denote the ratio $\log T/[\log T]$. We define a sequence of rescaled basis $\mathcal{F}_j := (X^{(j)}, Y_1^{(j)}, \dots, Y_k^{(j)})$ given by the forward orbit of the renormalization group: for all $j \in \mathbb{N}$,

$$\begin{aligned}(X_j, Y_1^{(j)}, \dots, Y_i^{(j)}, \dots, Y_k^{(j)}) &:= \rho_{jh}(X, Y_1, \dots, Y_i, \dots, Y_k) \\ &= (e^{jh} X, e^{-\frac{2jh}{k}} Y_1, \dots, e^{-\frac{2(k-i)jh}{k(k-1)}} Y_i, \dots, Y_k).\end{aligned}$$

Let us define

$$\hat{B}_{\mathcal{F}}(x, T) := \sum_{j=1}^{[\log T]} e^{(1-\frac{1}{k(k-1)})jh} B_{\mathcal{F}_j}(x, T_j). \tag{3.81}$$

Lemma 3.5.9 *For all $\sigma > (k/2+1)(k-1)+1$ there exists $C_{\Lambda,\sigma} > 0$ such that, for any filiform basis $\mathcal{F} = (X, Y_1, \ldots, Y_k)$ and for all $(x, T) \in M \times \mathbb{R}^+$, we have*

$$|\mathcal{A}^X_{\Lambda,\sigma}(x,T)|_{\mathcal{F},-\sigma} \le C_{\Lambda,\sigma} \frac{\hat{B}_{\mathcal{F}}(x,T)}{T}.$$

Proof Let (T_j) be the finite sequence given by the formula

$$T_j := e^{-jh}T\,, \qquad \text{for all } \ j \in \mathbb{N}.$$

By change of variable we have the identities

$$\mathcal{A}^X_{\Lambda,\sigma}(x,T) = \mathcal{A}^{X_j}_{\Lambda,\sigma}(x,T_j)\,, \qquad \text{for all } \ j \in \mathbb{N}.$$

We then write the orthogonal decomposition of the distribution $\mathcal{A}^{X_j}_{\Lambda,\sigma}(x,T)$ in the Sobolev space $W^{\sigma}_{\mathcal{F}_j}(\pi^X_\Lambda)$: for all $j \in \mathbb{N}$ we have

$$\mathcal{A}^X_{\Lambda,\sigma}(x,T) = \mathcal{A}^{X_j}_{\Lambda,\sigma}(x,T_j) = c^{X_j}_{\Lambda,\sigma}(x,T_j)\mathcal{D}^{X_j}_\Lambda + R^{X_j}_{\Lambda,\sigma}(x,T_j)\,,$$

which in particular imply the following

$$c^{X_{j-1}}_{\Lambda,\sigma}(x,T_{j-1})\mathcal{D}^{X_{j-1}}_\Lambda + R^{X_{j-1}}_{\Lambda,\sigma}(x,T_{j-1}) = c^{X_j}_{\Lambda,\sigma}(x,T_j)\mathcal{D}^{X_j}_\Lambda + R^{X_j}_{\Lambda,\sigma}(x,T_j).$$

Let now π_j denote the orthogonal projection from $W^{-\sigma}_{\mathcal{F}_j}(\pi^X_\Lambda)$ onto the one-dimensional subspace generated by the invariant distribution $\mathcal{D}^{X_j}_\Lambda$. Since the normalized invariant distributions $\mathcal{D}^{X_j}_\Lambda$ and $\mathcal{D}^{X_{j-1}}_\Lambda$ are proportional, it follows that

$$c^{X_{j-1}}_{\Lambda,\sigma}(x,T_{j-1})\mathcal{D}^{X_{j-1}}_\Lambda = c^{X_j}_{\Lambda,\sigma}(x,T_j)\mathcal{D}^{X_j}_\Lambda + \pi_{j-1}R^{X_j}_{\Lambda,\sigma}(x,T_j)\,,$$

from which we derive the bound

$$\begin{aligned}|c^{X_{j-1}}_{\Lambda,\sigma}&(x,T_{j-1})||\mathcal{D}^{X_{j-1}}_\Lambda|_{\mathcal{F}_{j-1},-\sigma} \\ &\le |c^{X_j}_{\Lambda,\sigma}(x,T_j)||\mathcal{D}^{X_j}_\Lambda|_{\mathcal{F}_{j-1},-\sigma} + |R^{X_j}_{\Lambda,\sigma}(x,T_j)|_{\mathcal{F}_{j-1},-\sigma}.\end{aligned}$$

On the one hand, by the scaling of invariant distributions (see Lemma 3.5.1 and Lemma 3.5.2) there exists an explicit constant $K_{\Lambda,\sigma} > 0$ (given in Lemma 3.5.2) such that

$$|\mathcal{D}^{X_j}_\Lambda|_{\mathcal{F}_{j-1},-\sigma} = e^{-h/2}|\mathcal{D}^{X_{j-1}}_\Lambda|_{\mathcal{F}_{j-1},-\sigma} \ge K^{-1}_{\Lambda,\sigma}e^{-h/2}e^{-(\frac{1}{2}-\frac{1}{k(k-1)})(j-1)h}\,; \tag{3.82}$$

on the other hand, from the estimates on solutions of the cohomological equation (see Lemma 3.5.2) there exist constants $K_\sigma > 0$ and $K'_{\Lambda,\sigma} > 0$ such that

$$|R^{X_j}_{\Lambda,\sigma}(x, T_j)|_{\mathcal{F}_{j-1},-\sigma} \le K_\sigma |R^{X_j}_{\Lambda,\sigma}(x, T_j)|_{\mathcal{F}_j,-\sigma} \le K'_{\Lambda,\sigma} \frac{B_{\mathcal{F}_j}(x, T_j)}{T_j}. \tag{3.83}$$

In fact, by Lemma 3.5.2, for any $\sigma > \tau(k-1)+1$ and for any coboundary $f \in \mathcal{K}^\sigma(\pi^{X_j}_\Lambda)$ we have

$$|G^{X_j}_{X,\Lambda}(f)|_{\mathcal{F}_j,\tau} \le G_{k,\sigma,\tau}(1 + \|\Lambda\|_{\mathcal{F}})^{\tau k+2} e^{-(1-\frac{2}{k(k-1)})jh} \, |f|_{\mathcal{F}_j,\sigma} \,,$$

and, by Theorem 3.5.7, for $\tau > k/2$ there exist constants $C_{k,\tau}$, $C'_\tau > 0$ such that

$$|G^{X_j}_{X,\Lambda}(f)(y)| \le \frac{C_{k,\tau}}{w^{1/2}_{\mathcal{F}_j}(y)} \left(|f|_{\mathcal{F}_j,\tau} + C_\tau e^{(1-\frac{2}{k(k-1)})jh} |G^{X_j}_{X,\Lambda}(f)|_{\mathcal{F}_j,\tau+1} \right).$$

The bound in formula (3.83) then follows from the above estimates for the Green operator by an argument similar to the one given in the proof of Lemma 3.4.8.

From the bounds in formulas (3.82) and (3.83) we then derive that there exists a constant $K''_{\Lambda,\sigma} > 0$ such that we have the following estimate:

$$|c^{X_{j-1}}_{\Lambda,\sigma}(x, T_{j-1})| \le e^{-h/2} |c^{X_j}_{\Lambda,\sigma}(x, T_j)| + K''_{\Lambda,\sigma} \frac{B_{\mathcal{F}_j}(x, T_j)}{T_j} e^{(\frac{1}{2}-\frac{1}{k(k-1)})jh}.$$

By finite (backward) induction over the set $\{0, \ldots, [\log T]\}$ we find

$$|c^X_{\Lambda,\sigma}(x, T)| \le T^{-1/2} |c^{X_{[\log T]}}_{\Lambda,\sigma}(x, 1)| + \frac{K''_{\Lambda,\sigma}}{T} \sum_{j=1}^{[\log T]} B_{\mathcal{F}_j}(x, T_j) e^{(1-\frac{1}{k(k-1)})jh}.$$

By the a priori bound in Theorem 3.5.8 we derive that

$$|c^{X_{[\log T]}}_{\Lambda,\sigma}(x, 1)| |\mathcal{D}^{X_{[\log T]}}_\Lambda|_{\mathcal{F}_{[\log T]},\sigma} \le |\mathcal{A}^{X_{[\log T]}}_{\Lambda,\sigma}|_{\mathcal{F}_{[\log T]},\sigma} \le \frac{C_{k,\sigma}}{w^{1/2}_{\mathcal{F}_{[\log T]}}(x)}.$$

From the above bound and from Lemma 3.5.2 on the scaling of invariant distributions we derive that

$$|c^{X_{[\log T]}}_{\Lambda,\sigma}(x, 1)| \le K_{\Lambda,\sigma} B_{\mathcal{F}_{[\log T]}}(x, 1) T^{\frac{1}{2}-\frac{1}{k(k-1)}} \le K_{\Lambda,\sigma} \frac{\hat{B}_{\mathcal{F}}(x, T)}{T^{1/2}}.$$

Finally, the stated estimate is a consequence of the decomposition (3.69) of the distribution $\mathcal{A}^X_{\Lambda,\sigma}(x,T)$, of the bound on the remainder distribution $R^X_{\Lambda,\sigma}(x,T)$ proved in Lemma 3.4.8 and of the above bound. □

Bounds on the Average Width

From Lemma 3.5.9 it is possible to derive effective equidistribution bounds on higher-step filiform nilflows conditioned to bounds on the average width of orbits. In this section we formulate a conjecture on the average width of higher-step filiform nilflows, then describe results in support of the conjecture.

Conjecture 3.5.10 *For any filiform basis $\mathcal{F} = (X, Y_1, \ldots, Y_k)$ such that the frequency of the toral flow $\{\psi_t^{\bar{X}}\}$ generated by the projected vector field $\bar{X}$ on $\mathbb{T}^2$ is a Diophantine irrational of exponent $\nu \le k/2$ the following holds. For every $\epsilon > 0$ there exists a constant $C_\epsilon > 0$ such that*

$$w_{\mathcal{F}(t)}(x) \ge C_\epsilon^{-1} e^{-\epsilon t}, \qquad \textit{for all } (x,t) \in M \times \mathbb{R}^+.$$

From the above conjecture and from Lemma 3.5.9 we would then be able to derive the following result.

Theorem 3.5.11 (Conditional Theorem) *Let us assume that Conjecture 3.5.10 holds. Let $\{\phi_t^X\}$ be a nilflow on a k-step filiform nilmanifold M which projects to a toral linear flow on $\mathbb{T}^2$ with Diophantine frequency of exponent $\nu \le k/2$. For every $\sigma > (k/2+1)(k-1)+1$ and for every $\epsilon > 0$ there exists a constant $C_{\sigma,\epsilon}(X) > 0$ such that the following holds: for all function $f \in W^\sigma(M)$ and all $(x,T) \in M \times \mathbb{R}^+$, we have*

$$\left|\frac{1}{T}\int_0^T f(\phi_t^X(x))dt - \int_M f dvol\right| \le C_{\sigma,\epsilon}(X)\|f\|_\sigma T^{-\frac{1}{k(k-1)}+\epsilon}.$$

By Lemma 3.2.5 the above conditional theorem implies (conditional) estimates on Weyl sums for polynomials of degree $k \ge 3$ under a Diophantine condition on the leading coefficient.

We introduce a definition of *good points*, that is, points on the nilmanifold for which we can prove bounds on the width of sufficiently many orbit segments to derive by our method bounds on ergodic averages. The set of good points is of full measure, as stated below.

Definition 3.5.12 *Let $\mathcal{F} = (X, Y_1, \ldots, Y_k)$ be a filiform basis. For any increasing sequence (T_i) of positive real numbers, let $N_i := [\log T_i/\log 2]$*

and $T_{j,i} := T_i^{j/N_i}$*, for all* $j = 0, \ldots, N_i$*. Let* $\epsilon > 0$ *and* $w > 0$*. We say that a point* $x \in M$ *is a* $(w, (T_i), \epsilon)$-good point for the basis $\mathcal{F}$ *if having set* $y_i = \phi^X_{T_i}(x)$*, for all* $i \in \mathbb{N}$ *and for all* $0 \leq j \leq N_i$*, we have*

$$w_{\mathcal{F}(\log T_{j,i})}(x, 1) \geq w/T_i^{\epsilon}, \qquad w_{\mathcal{F}(\log T_{j,i})}(y_i, 1) \geq w/T_i^{\epsilon}.$$

By a rather technical Borel–Cantelli argument based on the maximal ergodic theorem, we have proved that the set of good points has full measure (see [FF4], Lemma 5.18).

Lemma 3.5.13 *Let* $\epsilon > 0$ *be fixed and let* (T_i) *be an increasing sequence of positive real numbers satisfying the condition*

$$\Sigma\big((T_i), \epsilon\big) := \sum_{i \in \mathbb{N}} (\log T_i)^2 T_i^{-\epsilon} < +\infty. \tag{3.84}$$

For any filiform basis $\mathcal{F} = (X, Y_1, \ldots, Y_k)$ *such that the frequency of the toral flow* $\{\psi_t^{\bar{X}}\}$ *generated by the projected vector field* $\bar{X}$ *on* $\mathbb{T}^2$ *is a Diophantine irrational of exponent* $\nu \leq k/2$ *the following holds. The Lebesgue measure of the complement of the set* $\mathcal{G}\big(w, (T_i), \epsilon\big)$ *of* $\big(w, (T_i), \epsilon\big)$*-good points is bounded above as follows: there exists a constant* $C_{\mathcal{F}} := C_{\mathcal{F}}(k, \nu) > 0$ *such that*

$$\text{meas}\big(M \setminus \mathcal{G}(w, (T_i), \epsilon)\big) \leq C_{\mathcal{F}}\, \Sigma\big((T_i), \epsilon\big)\, w.$$

Our main effective equidistribution result for higher-step nilflows (see Theorem 3.1.3 in the Introduction) can be derived from the above almost everywhere bounds on average width functions along the lines of the proof of Lemma 3.5.9. Bounds on Weyl sums then follow immediately by Lemma 3.2.5.

We conclude these lectures with the proof of a uniform bound for the average width in the case of 3-step filiform nilflows. The lemma establishes a weak form of our Conjecture 3.5.10 in the case of 3-step filiform nilflows and by Lemma 3.5.9 it implies Theorem 3.1.5 stated in the Introduction.

The argument is essentially based only on the *linear divergence* of nearby orbits that takes place in this case. It can be generalized to other similar cases (see for instance [FFT]).

Lemma 3.5.14 *Let* M *be a* 3*-step filiform nilmanifold. For every nilflow* $\{\phi_s^X\}$ *such that the frequency of the projected linear flow* $\{\psi_s^{\bar{X}}\}$ *on* $\mathbb{T}^2$ *is Diophantine of Roth type and for any filiform basis* $\mathcal{F} = (X, Y_1, Y_2, Y_3)$ *the following holds. For every* $\epsilon > 0$ *there exists a constant* $C_\epsilon(X) > 0$ *such that, for all* $t \geq 0$ *and for all* $(x, T) \in M \times \mathbb{R}^+$ *we have*

$$w_{\mathcal{F}(t)}(x, T) \geq C_\epsilon(X)^{-1} e^{-\epsilon t}.$$

Proof Let $\mathcal{F} = (X, Y_1, Y_2, Y_3)$ be a filiform basis. It is not restrictive to assume (up to a change of basis) that the system (Y_1, Y_2, Y_3) is rational with respect to the lattice Γ which defines the nilmanifold M. For all $t \geq 0$, let

$$\mathcal{F}(t) = (X(t), Y_1(t), Y_2(t), Y_3) := (e^t X, e^{-2t/3} Y_1, e^{-t/3} Y_2, Y_3)$$

denote the optimally rescaled basis introduced above, written in the special case of a 3-step filiform nilmanifold. We prove below a uniform bound on the average width for all such rescaled bases.

By definition of the average width (see Definition 3.5.6), for any $t \geq 0$ and for any $(x, T) \in M \times [1, +\infty)$ we have to construct an open set $\Omega_t(x, T) \subset \mathbb{R}^4$ which contains the segment $\{(s, 0, 0, 0) | 0 \leq s \leq T\}$, such that the map

$$\phi_x(s, y_1, y_2, y_3) = \Gamma x \exp(s e^t X) \exp(e^{-2t/3} y_1 Y_1 + e^{-t/3} y_2 Y_2 + y_3 Y_3)$$

is injective on $\Omega_t(x, T)$. Injectivity fails if and only if there exist vectors

$$(s, y_1, \ldots, y_3) \neq (s', y_1', \ldots, y_3') \in \Omega_t(x, T)$$

such that

$$\begin{aligned} &\Gamma x \exp(s' e^t X) \exp(e^{-\frac{2t}{3}} y_1' Y_1 + e^{-\frac{t}{3}} y_2' Y_2 + y_3' Y_3) \\ &\qquad = \Gamma x \exp(s e^t X) \exp(e^{-\frac{2t}{3}} y_1 Y_1 + e^{-\frac{t}{3}} y_2 Y_2 + y_3 Y_3). \end{aligned} \tag{3.85}$$

Let us denote for convenience $r = s' - s$ and $z_i := y_i' - y_i$ for all $i \in \{1, 2, 3\}$. Since $s, s' \in [0, T]$ it follows that $r \in [-T, T]$. Let $c_\Gamma > 0$ denote the distance from the identity of the smallest non-zero element of the lattice Γ.

Let us assume that

$$(z_1, z_2, z_3) \in [-c_\Gamma/4, c_\Gamma/4]^3. \tag{3.86}$$

For all $t \geq 0$ and all $s \in [0, T]$, let us adopt the notation

$$z_2(t, s) := z_2 + z_1 s e^{2t/3} \quad \text{and} \quad z_3(t, s) := z_3 + \frac{1}{2} z_1 s^2 e^{4t/3} + z_2 s e^{2t/3}.$$

From the identity in formula (3.85) we derive the identity

$$\exp(r e^t X) \exp\left(e^{-2t/3} z_1 Y_1 + e^{-t/3} z_2(t, s) Y_2 + z_3(t, s) Y_3\right) \in x^{-1} \Gamma x. \tag{3.87}$$

By projecting the above identity on the base torus we have

$$\exp(r e^t \bar{X}) \exp(e^{-2t/3} z_1 \bar{Y}_1) \in \bar{\Gamma}, \tag{3.88}$$

which implies that re^t is a return time for the projected toral linear flow $\{\psi_s^{\bar{X}}\}$ at distance at most $e^{-2t/3} c_\Gamma / 2$. By the Diophantine condition (of exponent

$\nu \geq 1$) on the projected toral flow $\{\psi_s^{\bar{X}}\}$ it follows that there exists a constant $C(\bar{X}) > 0$ such that all solutions of formula (3.88) satisfy the following lower bound:

$$|z_1| \geq \frac{C(\bar{X})}{r^\nu} e^{(\frac{2}{3}-\nu)t}.$$

Let $\mathcal{R}_t(x, T)$ denote the set of $r \in [-T, T]$ such that the equation in formula (3.88) has a solution $z_1 \in [-c_\Gamma/2, c_\Gamma/2]$. By construction for every $r \in \mathcal{R}_t(x, T)$, the solution $z_1 := z_1(r)$ of the identities in formulas (3.87) and (3.88) is unique. Let then $\mathcal{S}(r)$ denote the set of $s \in [0, T]$ such that there exists a solution of identity (3.87) satisfying the condition in formula (3.86). By its definition the set $\mathcal{S}(r)$ is a union of intervals I^* of length at most $c_\Gamma |z_1(r)|^{-1} e^{-2t/3}/2$. As long as $|z_1(r)| \geq e^{-2t/3}$, for each connected component I^* of $\mathcal{S}(r)$ there exists $s^* \in I^*$ solution of the equation $z_2(t, s) - z_1(r) s e^{2t/3} = 0$. Let $\mathcal{S}^*(r)$ denote the set of all such solutions. Its cardinality can be estimated from above by counting lattice points on central fibers after projection on the quotient Heisenberg manifold:

$$\#\mathcal{S}^*(r) \leq c_\Gamma^{-1} e^{t/3} |z_1(r)| T.$$

For every $r \in \mathcal{R}_t(x, T)$ and every $s \in [0, T]$, we now define the function

$$\delta_r(t, s) = \begin{cases} \frac{1}{10} \max\{|z_1(r)|, |z_1(r)(s - s^*) e^{2t/3}|\}, & \text{for all } s \in I^*; \\ \frac{c_\Gamma}{10}, \quad \text{for all } s \in [0, T] \setminus \mathcal{S}(r). \end{cases}$$

We then define the set $\Omega_t(r) \subset [0, T] \times \mathbb{R}^3$ as follows

$$\Omega_t(r) := \{(s, y_1, y_2, y_3) | \max\{|y_1|, |y_2|\} < \delta_r(t, s), |y_3| < c_\Gamma/10\},$$

and finally we define the set $\Omega_t(x, T)$ as the intersection

$$\Omega_t(x, T) := \cap_{r \in \mathcal{R}_t(x,T)} \Omega_t(r).$$

It can be verified that by the above construction the above map $\phi_x : \mathbb{R}^4 \to M$ is injective on $\Omega_t(x, T)$. In fact, the open sets $\Omega_t(r) \cap \Omega_t(-r)$ are narrowed near both endpoints of the return orbits of return time r so that their images in M have no self-intersections given by return times r and $-r$.

Let us compute the average width function associated with the sets $\Omega_t(r)$ for each $r \in \mathcal{R}_t(x, T)$. By definition we have

$$\delta_r(t, s) = \begin{cases} \frac{1}{10}|z_1(r)|, \quad \text{for } s \in I^* \text{ with } |s - s^*| \leq e^{-2t/3}; \\ \frac{1}{10}|z_1(r)(s - s^*) e^{2t/3}|, \quad \text{for } s \in I^* \text{ with } |s - s^*| \geq e^{-2t/3}; \\ \frac{c_\Gamma}{10}, \quad \text{for all } s \in [0, T] \setminus \mathcal{S}(r). \end{cases}$$

By the definition of inner width and by construction of the set $\Omega_t(r)$ we have that

$$w_{\Omega_t(r)}(s) = \frac{4}{5} c_\Gamma \, \delta_r(t,s)^2 \,, \qquad \text{for all } s \in [0,T] \,,$$

from which it follows that for every subinterval $I^* \subset \mathcal{S}(r)$ we have

$$\int_{I^*} \frac{ds}{w_{\Omega_t(r)}(s)} \le \frac{500 c_\Gamma^{-1}}{e^{2t/3}|z_1(r)|^2} .$$

By the upper bound on the length of the intervals I^* and on the cardinality of the set $\mathcal{S}^*(r)$ we finally derive

$$\frac{1}{T}\int_0^T \frac{ds}{w_{\Omega_t(r)}(s)} \le \frac{10^3}{c_\Gamma^3} \frac{1}{e^{t/3}|z_1(r)|} .$$

Finally it can be verified that we have the straightforward estimate

$$\frac{1}{T}\int_0^T \frac{ds}{w_{\Omega_t(x,T)}(s)} \le \frac{1}{T} \sum_{r \in \mathcal{R}_t(x,T)} \int_0^T \frac{ds}{w_{\Omega_t(r)}(s)} .$$

For every $n \in \mathbb{N}$, let $\mathcal{R}_t^{(n)}(x,T)$ denote the subset of $\mathcal{R}_t(x,T)$ characterized by the condition that

$$|z_1(r)| \in (\frac{c_\Gamma}{2^{n+1}}, \frac{c_\Gamma}{2^n}] .$$

By a Diophantine condition of exponent $\nu \ge 1$ on the frequency of the projection of the nilflow to a linear flow on the base torus we have the following estimates which can be proved by methods from the theory of continued fractions (see [FF4], Lemma 5.13) :

$$\#\mathcal{R}_t^{(n)}(x,T) \le C_\nu(X) \max\{(Te^t)^{1-\frac{1}{\nu}}, Te^t \frac{c_\Gamma}{2^n} e^{-2t/3}\} .$$

It follows from the above estimates that, if the frequency of the projected linear flow is Diophantine of Roth type (that is, it is Diophantine of exponent $\nu = 1+\epsilon$ for all $\epsilon > 0$), then for all $\epsilon > 0$ there exists $C_\epsilon(\bar{X}) > 0$ such that

$$\#\mathcal{R}_t^{(n)}(x,T) \le C_\epsilon(\bar{X}) T \frac{c_\Gamma}{2^n} e^{t/3+\epsilon t/2} ,$$

hence we conclude that there exists a constant $C_\epsilon(\Gamma, X) > 0$ such that

$$\frac{1}{T}\int_0^T \frac{ds}{w_{\Omega_t(x,T)}(s)} \le C_\epsilon(\Gamma, X) T e^{\epsilon t} .$$

□

Acknowledgements. The author would like to thank D. Badziahin, A. Gorodnik, N. Peyerimhoff and T. Ward, organizers of the School on *Dynamics and Analytic Number Theory* held at Durham University, UK, in April 2014, for the opportunity to present his work there and for the pleasant and stimulating environment provided by the school.

Giovanni Forni
Department of Mathematics, University of Maryland
4176 Campus Drive – Mathematics Building
College Park, MD 20742-4015, USA
email: gforni@math.umd.edu

References

[Ad] R. Adams. *Sobolev Spaces*, Boston, MA: Academic Press, 1975.

[AGH] L. Auslander, L. Green and F. Hahn. *Flows on Homogeneous Spaces*, Princeton, NJ: Princeton University Press, 1963.

[Bo] E. Bombieri. On Vinogradov's mean value theorem and Weyl sums, in: *Automorphic Forms and Analytic Number Theory*, Univ. de Montréal, Montréal, 1990, 7–24.

[BF] A. Bufetov and G. Forni. Limit theorems for Horocycle flows, *Annales Scientifiques de l'ENS*, **47** (2014), 851–903.

[CM] F. Cellarosi and J. Marklof. Quadratic Weyl sums, automorphic functions, and invariance principles, Preprint (arXiv:1501.07661v2), to appear in Proceedings of the London Mathematical Society.

[CF] S. Cosentino and L. Flaminio. Equidistribution for higher-rank Abelian actions on Heisenberg nilmanifolds, *J. Mod. Dyn.* **9** (2015), 305–353.

[Da] S. G. Dani. Bounded orbits of flows on homogeneous spaces, *Comm. Math. Helv.* **61** (1986), 636–660.

[DT] M. Drmota and R. F. Tichy. *Sequences, Discrepancies and Applications*, Lecture Notes in Mathematics, vol. **1651**, Springer-Verlag, Berlin, 1997.

[FK] A. Fedotov and F. Klopp. An exact renormalization formula for Gaussian exponential sums and applications, *American Journal of Mathematics* **134** (2012), 711–748.

[FJK] H. Fiedler, W. Jurkat and O. Körner. Asymptotic expansions of finite theta series, *Acta Arith.* **32** (2) (1977), 129–146.

[FF1] L. Flaminio and G. Forni. Invariant distributions and time averages for horocycle flows, *Duke Math. J.* **119** (2003), 465–526.

[FF2] L. Flaminio and G. Forni. Equidistribution of nilflows and applications to theta sums, *Erg. Th. Dyn. Sys.* **26** (2006), 409–433.

[FF3] L. Flaminio and G. Forni. On the cohomological equation for nilflows, *J. Mod. Dyn.* **1** (2007), 37–60.

[FF4] L. Flaminio and G. Forni. On the effective equidistribution of a class of nilflows. Preprint (arXiv:1407.3640v1).

[FFT] L. Flaminio, G. Forni and J. Tanis. Effective equidistribution of twisted horocycle flows and horocycle maps. Preprint (arXiv:1507.05147v1), to appear in GAFA.

[Fu] H. Furstenberg. *Recurrence in Ergodic Theory and Combinatorial Number Theory*, Princeton University Press, 1981.

[Gr] L. W. Green. Spectra of nilflows, *Bull. Amer. Math. Soc.* **67** (1961), 414–415.

[GM] J. Griffin and J. Marklof. Limit theorems for skew translations, *J. Mod. Dyn.* **8** (2014), 177–189.

[GT1] B. Green and T. Tao. The quantitative beahaviour of polynomial orbits on nilmanifolds, *Ann. of Math.* **175** (2) (2012), 465–540.

[GT2] B. Green and T. Tao. The Möbius function is strongly orthogonal to nilsequences, *Ann. of Math.* **175** (2) (2012), 541–566.

[HL] G. H. Hardy and J. E. Littlewood. The trigonometrical series associated with elliptic theta θ-functions, *Acta Math.* **37** (1914), 193–238.

[He] E. Hebey. *Nonlinear Analysis on Manifolds: Sobolev Spaces and Inequalities*, CIMS Lecture Notes, Courant Institute of Mathematical Sciences, New York University, Volume 5, 1999. Second edition in AMS/CIMS Lecture Notes, Volume 5, 2000.

[Hu] L.-K. Hua. An improvement of Vinogradov's mean value theorem and several applications, *Quart. J. Math. Oxford* **20** (1949), 48–61.

[Ma] J. Marklof. Limit theorems for theta sums, *Duke Mathematical Journal* **97** (1999), 127–153.

[Su] D. Sullivan. Disjoint spheres, approximation by imaginary quadratic numbers, and the logarithm law for geodesics, *Acta Math.* **149** (1982), 215–237.

[Va] R. C. Vaughan. *The Hardy-Littlewood Method*, 2nd edition, Cambridge University Press, 1997.

[Vi] I. M. Vinogradov. New estimates for Weyl sums, *Dokl. Akad. Nauk SSSR* **8** (1935), 195–198.

[We] H. Weyl. Über die Gleichverteilung von Zahlen mod Eins, *Math. Ann.* **77** (1916), 313–352.

[Wo1] T. D. Wooley. New estimates for Weyl sums, *Quart. J. Math.* Oxford **46** (1995), 119–127.

[Wo2] T. D. Wooley. Vinogradov's mean value theorem via efficient congruencing, *Ann. of Math.* **175** (2012), 1575–1627.

[Wo3] T. D. Wooley. Vinogradov's mean value theorem via efficient congruencing, *II, Duke Math. J.* **162** (2013), 673–730.

[Wo4] T. D. Wooley. Multigrade efficient congruencing and Vinogradov's mean value theorem, *Proc. London Math. Soc.* (3) **111** (2015), no. 3, 519–560.

[Wo5] T. D. Wooley. Translation invariance, exponential sums and Waring's problem, *Proceedings of the International Congress of Mathematicians*, Seoul, 2014, 505–529.

[Wo6] T. D. Wooley. Perturbations of Weyl sums, *Internat. Math. Res. Notices* **2016** (9), (2016), 2632–2646. doi: 10.1093/imrn/rnv225, 15 pp.

4

Multiple Recurrence and Finding Patterns in Dense Sets

Tim Austin

Abstract

Szemerédi's theorem asserts that any positive-density subset of the integers must contain arbitrarily long arithmetic progressions. It is one of the central results of additive combinatorics. After Szemeredi's original combinatorial proof, Furstenberg noticed the equivalence of this result to a new phenomenon in ergodic theory that he called 'multiple recurrence'. Furstenberg then developed some quite general structural results about probability-preserving systems to prove the multiple recurrence theorem directly. Furstenberg's ideas have since given rise to a large body of work around multiple recurrence and the associated 'non-conventional' ergodic averages, and to further connections with additive combinatorics.

This course is an introduction to multiple recurrence and some of the ergodic theoretic structure that lies behind it. We begin by explaining the correspondence observed by Furstenberg, and then give an introduction to the necessary background from ergodic theory. We emphasize the formulation of multiple recurrence in terms of joinings of probability-preserving systems. The next step is a proof of Roth's theorem (the first nontrivial case of Szemeredi's theorem), which illustrates the general approach. We finish with a proof of a more recent convergence theorem for some non-conventional ergodic averages, showing some of the newer ideas in use in this area.

The classic introduction to this area of combinatorics and ergodic theory is Furstenberg's book [Fur81], but the treatment below has a more modern point of view.

4.1 Szemerédi's Theorem and Its Relatives

In 1927, van der Waerden gave a clever combinatorial proof of the following surprising fact:

Theorem 4.1 (Van der Waerden's Theorem [vdW27]) *For any fixed integers $c, k \geq 1$, if the elements of $\mathbb{Z}$ are coloured using c colours, then there is a non-trivial k-term arithmetic progression which is monochromatic: that is, there are some $a \in \mathbb{Z}$ and $n \geq 1$ such that*

$$a, a+n, \ldots, a+(k-1)n.$$

all have the same colour.

This result now fits into a whole area of combinatorics called Ramsey theory. The classic account of this Theory is the book by [GRS90].

It is crucial to allow both the start point a and the common difference $n \geq 1$ to be chosen freely. This theorem has some more difficult relatives which allow certain restrictions on the choice of n, but if one tries to fix a single value of n a priori then the conclusion is certainly false.

In 1936, Erdős and Turán realized that a deeper phenomenon might lie beneath van der Waerden's theorem. Observe that for any c-colouring of $\mathbb{Z}$ and for any finite subinterval of $\mathbb{Z}$, at least one of the colour-classes must occupy at least a fraction $1/c$ of the points in that subinterval. In [ET36] they asked whether any subset of $\mathbb{Z}$ which has 'positive density' in arbitrarily long subintervals must contain arithmetic progressions of any finite length.

This turns out to be true. The formal statement requires the following definition. We give it for subsets of $\mathbb{Z}^d$, $d \geq 1$, for the sake of a coming generalization. Let $[N] := \{1, 2, \ldots, N\}$.

Definition 4.2 (Upper Banach Density) For $E \subseteq \mathbb{Z}^d$, its **upper Banach density** is the quantity

$$\bar{\mathrm{d}}(E) := \limsup_{N \longrightarrow \infty} \sup_{\mathbf{v} \in \mathbb{Z}^d} \frac{|E \cap (\mathbf{v} + [N]^d)|}{N^d}.$$

That is, $\bar{\mathrm{d}}(E)$ is the supremum of those $\delta > 0$ such that one can find cubical boxes in $\mathbb{Z}^d$ with arbitrarily long sides such that E contains at least a proportion δ of the lattice points in those boxes.

Exercise Prove that Definition 4.2 is equivalent to

$$\bar{\mathrm{d}}(E) = \limsup_{L \longrightarrow \infty} \sup \left\{ \frac{\left|E \cap \prod_{i=1}^{d} [M_i, N_i]\right|}{\prod_{i=1}^{d} (N_i - M_i)} : N_i \geq M_i + L \ \forall i = 1, 2, \ldots, d \right\}.$$

◁

Theorem 4.3 (Szemerédi's Theorem) *If $E \subseteq \mathbb{Z}$ has $\bar{\mathrm{d}}(E) > 0$, then for any $k \geq 1$ there are $a \in \mathbb{Z}$ and $n \geq 1$ such that*

$$\{a, a+n, \ldots, a+(k-1)n\} \subseteq E.$$

The special case $k = 3$ of this theorem was proved by Roth in [Rot53], so it is called Roth's theorem. The full theorem was finally proved by Szemerédi in [Sze75].

As already remarked, Szemerédi's theorem implies van der Waerden's theorem, because if $\mathbb{Z}$ is coloured using c colours then at least one of the colour-classes must have upper Banach density at least $1/c$.

Szemerédi's proof of Theorem 4.3 is one of the virtuoso feats of modern combinatorics. It is also the earliest major application of several tools that have since become workhorses of that area, particularly the Szemerédi regularity lemma in graph theory. However, shortly after Szemerédi's proof appeared, Furstenberg gave a new and very different proof using ergodic theory. In [Fur77], he showed the equivalence of Szemerédi's theorem to an ergodic-theoretic phenomenon called 'multiple recurrence', and proved some new structural results in ergodic theory which imply that multiple recurrence always occurs.

Multiple recurrence is introduced in the next subsection. First we bring the combinatorial side of the story closer to the present. Furstenberg and Katznelson quickly realized that Furstenberg's ergodic-theoretic proof could be considerably generalized, and in [FK78] they obtained a multidimensional version of Szemerédi's theorem as a consequence:

Theorem 4.4 (Furstenberg–Katznelson Theorem) *If $E \subseteq \mathbb{Z}^d$ has $\bar{\mathrm{d}}(E) > 0$, and if $\mathbf{e}_1, \ldots, \mathbf{e}_d$, is the standard basis in $\mathbb{Z}^d$, then there are some $\mathbf{a} \in \mathbb{Z}^d$ and $n \geq 1$ such that*

$$\{\mathbf{a} + n\mathbf{e}_1, \ldots, \mathbf{a} + n\mathbf{e}_d\} \subseteq E$$

(so 'dense subsets contain the set of outer vertices of an upright right-angled isosceles simplex').

This easily implies Szemerédi's theorem, because if $k \geq 1$, $E \subseteq \mathbb{Z}$ has $\bar{\mathrm{d}}(E) > 0$, and we define

$$\Pi : \mathbb{Z}^{k-1} \longrightarrow \mathbb{Z} : (a_1, a_2, \ldots, a_{k-1}) \mapsto a_1 + 2a_2 + \cdots + (k-1)a_{k-1},$$

then the pre-image $\Pi^{-1}(E)$ has $\bar{\mathrm{d}}(\Pi^{-1}(E)) > 0$, and an upright isosceles simplex found in $\Pi^{-1}(E)$ projects under Π to a k-term progression in E. Similarly, by projecting from higher dimensions to lower, one can prove that Theorem 4.4 actually implies the following:

Corollary 4.5 *If $F \subset \mathbb{Z}^d$ is finite and $E \subseteq \mathbb{Z}^d$ has $\bar{\mathrm{d}}(E) > 0$, then there are some $\mathbf{a} \in \mathbb{Z}^d$ and $n \geq 1$ such that $\{\mathbf{a} + n\mathbf{b} : \mathbf{b} \in F\} \subseteq E$.* □

For about twenty years, the ergodic-theoretic proof of Furstenberg and Katznelson was the only known proof of Theorem 4.4. That changed when a new approach using hypergraph theory was developed roughly in parallel by Gowers [Gow06], Nagle, Rödl, and Schacht [NRS06] and Tao [Tao06b]. These works gave the first 'finitary' proofs of this theorem, implying somewhat effective bounds: unlike the ergodic-theoretic approach, the hypergraph approach gives an explicit value $N = N(\delta)$ such that any subset of $[N]^d$ containing at least δN^d points must contain a whole simplex. (In principle, one could extract such a bound from the Furstenberg–Katznelson proof, but it would be extremely poor: see Tao [Tao06a] for the one-dimensional case.)

The success of Furstenberg and Katznelson's approach gave rise to a new sub-field of ergodic theory sometimes called 'ergodic Ramsey theory'. It now contains several other results asserting that positive-density subsets of some kind of combinatorial structure must contain a copy of some special pattern. Some of these have been re-proven by purely combinatorial means only very recently. We will not state these in detail here, but only mention by name the IP Szemerédi theorem of [FK85], the density Hales–Jewett theorem of [FK91] (finally given a purely combinatorial proof by the members of the 'Polymath 1' project in [Pol09]), the polynomial Szemerédi theorem Bergelson and Leibman [BL96] and the nilpotent Szemerédi theorem of Leibman [Lei98].

4.2 Multiple Recurrence

4.2.1 The Setting of Ergodic Theory

Ergodic theory studies the 'statistical' properties of dynamical systems. The following treatment is fairly self-contained, but does assume some standard facts from functional analysis and probability, at the level of advanced textbooks such as Folland [Fol99] or Royden [Roy88] and Billingsley [Bil95].

Let G be a countable group; later we will focus on $\mathbb{Z}$ or $\mathbb{Z}^d$. A **G-space** is a pair (X, T) in which X is a compact metrizable topological space, and $T = (T^g)_{g\in G}$ is an action of G on X by Borel measurable transformations: thus,

$$T^e = \mathrm{id}_X \quad \text{and} \quad T^g \circ T^h = T^{gh} \quad \forall g, h \in G,$$

where e is the identity of G. A $\mathbb{Z}$-action T is specified by the single transformation T^1 which generates it. Similarly, a $\mathbb{Z}^d$-action T may be identified with the commuting d-tuple of transformations $T^{\mathbf{e}_i}$, where $\mathbf{e}_1, \ldots, \mathbf{e}_d$ are the standard basis vectors of $\mathbb{Z}^d$.

The set of Borel probability measures on a compact metrizable space X is denoted by $\Pr(X)$. This set is compact in the weak* topology: for instance, this can be seen by the Riesz Representation Theorem ([Fol99, Theorem 7.17]), which identifies $\Pr(X)$ with a closed, bounded, convex subset of the Banach-space dual $C(X)^\star$, followed by an application of Alaoglu's Theorem ([Fol99, Theorem 5.18]).

Next, let $(X, \mathscr{S})$ and $(Y, \mathscr{T})$ be any two measurable spaces, let μ be a probability measure on $\mathscr{S}$, and let $\pi : X \longrightarrow Y$ be measurable. Then the **image measure of μ under π** is the measure $\pi_*\mu$ on $\mathscr{T}$ defined by setting

$$(\pi_*\mu)(B) := \mu(\pi^{-1}(B)) \quad \forall B \in \mathscr{T}.$$

If X and Y are compact metrizable spaces, then π_* defines a map $\Pr(X) \longrightarrow \Pr(Y)$. If, in addition, π is continuous, then π_* is continuous for the weak* topologies.

Finally, a **G-system** is a triple (X, μ, T) in which (X, T) is a G-space and μ is a T-invariant member of $\Pr(X)$, meaning that $T^g_*\mu = \mu$ for every $g \in G$. When needed, the Borel σ-algebra of X will be denoted by $\mathscr{B}_X$. We often denote a G-system (X, μ, T) by a single boldface letter such as $\mathbf{X}$. A $\mathbb{Z}$-system will sometimes be called just a **system**.

The definitions above ignore a host of other possibilities, such as dynamics with an infinite invariant measure, or with a non-invertible transformation. Ergodic theory has branches for these too, but they do not appear in this course.

Examples

1. Let $X = \mathbb{T} = \mathbb{R}/\mathbb{Z}$ with its usual topology, let μ be Lebesgue measure, and let T be the rotation by a fixed element $\alpha \in X$:

$$Tx := x + \alpha.$$

This is called a **circle rotation**.

2. Let $\mathbf{p} = (p_1, \ldots, p_m)$ be a stochastic vector: that is, a probability distribution on the set $\{1, 2, \ldots, m\}$. Let $X := \{1, 2, \ldots, m\}^{\mathbb{Z}}$ with the product topology, let $\mu := \mathbf{p}^{\otimes \mathbb{Z}}$ (the law of an i.i.d. random sequence of numbers each chosen according to $\mathbf{p}$), and let T be the leftward coordinate-shift:

$$T((x_n)_n) := (x_{n+1})_n.$$

This is called the **Bernoulli shift over p**. ◁

An important subtlety concerns the topology on X. In most of ergodic theory, no particular compact topology on X is very important, except through the resulting Borel σ-algebra: it is this measurable structure that underlies the theory. This is why we allow arbitrary Borel measurable transformations T^g, rather than just homeomorphisms. However, general measurable spaces can exhibit certain pathologies which Borel σ-algebras of compact metrizable spaces cannot. The real assumption we need here is that our measurable spaces be 'standard Borel', but the assumption of a compact metric is a convenient way to guarantee this.

Having explained this, beware that many authors restrict the convenient term 'G-space' to actions by homeomorphisms.

4.2.2 The Phenomenon of Multiple Recurrence

In order to introduce multiple recurrence, it is helpful to start with the probability-preserving version of Poincaré's classical Recurrence Theorem.

Theorem 4.6 (Poincaré Recurrence) *If (X, μ, T) is a system and $A \in \mathscr{B}_X$ has $\mu(A) > 0$, then there is some $n \neq 0$ such that $\mu(A \cap T^{-n}A) > 0$.*

Proof The pre-images $T^{-n}A$ are all subsets of the probability space X of equal positive measure, so some two of them must overlap in positive measure. Once we have $\mu(T^{-n}A \cap T^{-m}A) > 0$ for some $n \neq m$, the invariance of μ under T^n implies that also $\mu(A \cap T^{n-m}A) > 0$. □

Furstenberg's main result from [Fur77] strengthens this conclusion. He shows that in fact one may find several of the sets $T^{-n}A$, $n \in \mathbb{Z}$, that simultaneously overlap in a positive-measure set, where the relevant times n form an arithmetic progression.

Theorem 4.7 (Multiple Recurrence Theorem) *If (X, μ, T) is a system and $A \in \mathscr{B}_X$ has $\mu(A) > 0$, then for any $k \geq 1$ there is some $n \geq 1$ such that*

$$\mu(T^{-n}A \cap \cdots \cap T^{-kn}A) > 0.$$

The Multidimensional Multiple Recurrence Theorem from [FK78] provides an analog of this for several commuting transformations.

Theorem 4.8 (Multidimensional Multiple Recurrence Theorem) *If (X, μ, T) is a $\mathbb{Z}^d$-system and $A \in \mathscr{B}_X$ has $\mu(A) > 0$, then there is some $n \geq 1$ such that*

$$\mu(T^{-n\mathbf{e}_1}A \cap \cdots \cap T^{-n\mathbf{e}_d}A) > 0.$$

Note that for $d = 2$, simply applying the Poincaré recurrence theorem for the transformation $T^{\mathbf{e}_1-\mathbf{e}_2}$ gives the conclusion of Theorem 4.8.

This course will include a proof of Theorem 4.7 in the first case beyond Poincaré Recurrence: $k = 3$. Two different ergodic-theoretic proofs of the full Theorem 4.8 can be found in [Fur81] and [Aus10c]. These are too long to be included in this course, but we will formulate and prove a related convergence result which gives an introduction to some of the ideas.

First, let us prove the equivalence of Theorem 4.8 and Theorem 4.4. This equivalence is often called the 'Furstenberg correspondence principle'. Although easy to prove, it has turned out to be a hugely fruitful insight into the relation between ergodic theory and combinatorics. The version we give here essentially follows [Ber87].

Proposition 4.9 (Furstenberg Correspondence Principle) *If $E \subseteq \mathbb{Z}^d$, then there are a $\mathbb{Z}^d$-system (X, μ, T) and a set $A \in \mathscr{B}_X$ such that $\mu(A) = \bar{\mathrm{d}}(E)$, and such that for any $\mathbf{v}_1, \mathbf{v}_2, \ldots, \mathbf{v}_k \in \mathbb{Z}^d$ one has*

$$\bar{\mathrm{d}}((E - \mathbf{v}_1) \cap (E - \mathbf{v}_2) \cap \cdots \cap (E - \mathbf{v}_k)) \geq \mu(T^{-\mathbf{v}_1}A \cap \cdots \cap T^{-\mathbf{v}_k}A).$$

In order to visualize this, observe that

$$(E - \mathbf{v}_1) \cap (E - \mathbf{v}_2) \cap \cdots \cap (E - \mathbf{v}_k)$$

is the set of those $\mathbf{a} \in \mathbb{Z}^d$ such that $\mathbf{a} + \mathbf{v}_i \in E$ for each $i \leq k$. Its density may be seen as the 'density of the set of translates of the pattern $\{\mathbf{v}_1, \mathbf{v}_2, \ldots, \mathbf{v}_k\}$ that lie entirely inside E'. In these terms, the above propositions asserts that one can synthesize a $\mathbb{Z}^d$-system which provides a lower bound on this density for any given pattern in terms of the intersection of the corresponding shifts of the subset A.

Proof Choose a sequence of boxes $R_j := \prod_{i=1}^d [M_{j,i}, N_{j,i}]$ such that

$$\min_{i\in\{1,2,\ldots,d\}} (N_{j,i} - M_{j,i}) \longrightarrow \infty \quad \text{as} \quad j \longrightarrow \infty$$

and

$$\frac{|E \cap R_j|}{|R_j|} \longrightarrow \bar{\mathrm{d}}(E) \quad \text{as } j \longrightarrow \infty.$$

We can regard the set E as a point in the space $X := \mathcal{P}(\mathbb{Z}^d)$ of subsets of $\mathbb{Z}^d$, on which $\mathbb{Z}^d$ naturally acts by translation: $T^{\mathbf{n}}B := B - \mathbf{n}$. This X can be identified with the Cartesian product $\{0, 1\}^{\mathbb{Z}^d}$ by associating to each subset its indicator function. It therefore carries a compact metrizable product topology which makes (X, T) a $\mathbb{Z}^d$-space.

Let

$$\nu_j := \frac{1}{|R_j|}\sum_{\mathbf{n}\in R_j} \delta_{T^{\mathbf{n}}(E)} \quad \text{for each } j,$$

the uniform measure on the piece of the T-orbit of E indexed by the large box R_j. Because the side-lengths of these boxes all tend to ∞, these measures are approximately invariant: that is, $\|T_*^{-\mathbf{m}}\nu_j - \nu_j\|_{\mathrm{TV}} \longrightarrow 0$ as $j \longrightarrow \infty$ for any fixed $\mathbf{m} \in \mathbb{Z}^d$, where $\|\cdot\|_{\mathrm{TV}}$ is the total variation norm (see [Fol99, Section 7.3]).

Since $\Pr(X)$ is weak* compact, we may let $\mu \in \Pr(X)$ be a subsequential weak* limit of the measures ν_j. By passing to a subsequence, we may assume that in fact $\nu_j \longrightarrow \mu$ (weak*). Since the measures ν_j are approximately T-invariant, and each $T_*^{-\mathbf{m}}$ acts continuously for the weak* topology on $\Pr(X)$, μ itself is strictly T-invariant.

Finally, let $A := \{H \in X : H \ni \mathbf{0}\}$. This corresponds to the cylinder set $\{(\omega_{\mathbf{n}})_{\mathbf{n}} : \omega_{\mathbf{0}} = 1\} \subseteq \{0,1\}^{\mathbb{Z}^d}$. We will show that (X, μ, T) and A have the desired properties. By our initial choice of the sequence of boxes R_j, we have

$$\mu(A) = \lim_{j\longrightarrow\infty} \nu_j(A) = \lim_{j\longrightarrow\infty} \frac{1}{|R_j|}\sum_{\mathbf{n}\in R_j} 1_{T^{\mathbf{n}}(E)}(\mathbf{0})$$

$$= \lim_{j\longrightarrow\infty} \frac{|E\cap R_j|}{|R_j|} = \bar{\mathrm{d}}(E).$$

The first convergence here holds because 1_A is a continuous function for the product topology on X, and so weak* convergence applies to it.

On the other hand, for any $\mathbf{v}_1, \mathbf{v}_2, \ldots, \mathbf{v}_k \in \mathbb{Z}^d$, the indicator function $1_{T^{-\mathbf{v}_1}A\cap\cdots\cap T^{-\mathbf{v}_k}A}$ is also continuous on X, and so

$$\begin{aligned}\mu(T^{-\mathbf{v}_1}A\cap\cdots\cap T^{-\mathbf{v}_k}A) &= \lim_{j\longrightarrow\infty} \nu_j(T^{-\mathbf{v}_1}A\cap\cdots\cap T^{-\mathbf{v}_k}A)\\ &= \lim_{j\longrightarrow\infty} \frac{1}{|R_j|}\sum_{\mathbf{n}\in R_j} 1_{T^{\mathbf{n}+\mathbf{v}_1}E\cap\cdots\cap T^{\mathbf{n}+\mathbf{v}_k}E}(\mathbf{0})\\ &= \lim_{j\longrightarrow\infty} \frac{1}{|R_j|}\sum_{\mathbf{n}\in R_j} 1_{T^{\mathbf{v}_1}E\cap\cdots\cap T^{\mathbf{v}_k}E}(\mathbf{n})\\ &\leq \bar{\mathrm{d}}(T^{\mathbf{v}_1}E\cap\cdots\cap T^{\mathbf{v}_k}E),\end{aligned}$$

since the upper Banach density is defined by a lim sup over *all* box-sequences with increasing side-lengths. □

Corollary 4.10 *Theorems 4.4 and 4.8 are equivalent.*

Proof ($\Longrightarrow$) Let (X,μ,T) be a $\mathbb{Z}^d$-system and $A \in \mathscr{B}_X$ with $\mu(A) > 0$. For each $x \in X$ let $E_x := \{\mathbf{n} \in \mathbb{Z}^d : T^{\mathbf{n}}x \in A\}$, and for each $N \in \mathbb{N}$ let

$$Y_N := \{x \in X : |E_x \cap [N]^d| \geq \mu(A)N^d/2\}.$$

A simple calculation gives

$$\mu(Y_N)N^d + \mu(X \setminus Y_N)\mu(A)N^d/2 \geq \int_X |E_x \cap [N]^d|\,\mu(\mathrm{d}x)$$
$$= \sum_{\mathbf{n}\in[N]^d} \int_X |E_x \cap \{\mathbf{n}\}|\,\mu(\mathrm{d}x) = \sum_{\mathbf{n}\in[N]^d} \int_X 1_{T^{-\mathbf{n}}(A)}\,\mathrm{d}\mu = \mu(A)N^d,$$

and therefore $\mu(Y_N) \geq \mu(A)/2$ for every N. Therefore, by the Borel–Cantelli Lemma, the set

$$Y := \big\{x \in X : \limsup_{N\longrightarrow\infty} |E_x \cap [N]^d|/N^d \geq \mu(A)/2\big\}$$

has positive measure.

By Theorem 4.4, if $x \in Y$ then E_x contains some pattern of the form $\{\mathbf{a} + n\mathbf{e}_1, \ldots, \mathbf{a}+n\mathbf{e}_d\}$. Since there are only countably many such patterns, it follows that for some choice of $\mathbf{a}$ and n, one has

$$\mu\big\{x \in X : \{\mathbf{a}+n\mathbf{e}_1, \mathbf{a}+n\mathbf{e}_2, \ldots, \mathbf{a}+n\mathbf{e}_d\} \subseteq E_x\big\} > 0.$$

By the definition of E_x, this measure is equal to

$$\mu\big(T^{-\mathbf{a}}(T^{-n\mathbf{e}_1}A \cap \cdots \cap T^{-n\mathbf{e}_d}A)\big) = \mu(T^{-n\mathbf{e}_1}A \cap \cdots \cap T^{-n\mathbf{e}_d}A),$$

so this completes the proof.

($\Longleftarrow$) Given E with $\bar{\mathrm{d}}(E) > 0$, Proposition 4.9 produces a $\mathbb{Z}^d$-system (X,μ,T) and positive-measure set A such that the positivity of $\mu(T^{-n\mathbf{e}_1}A \cap \cdots \cap T^{-n\mathbf{e}_d}A)$ for some $n \geq 1$ implies that

$$(E - n\mathbf{e}_1) \cap (E - n\mathbf{e}_2) \cap \cdots \cap (E - n\mathbf{e}_d)$$

has positive upper density, and so is certainly nonempty. □

4.3 Background from Ergodic Theory

This section covers most of the general theory that will be used later. Much of it overlaps with [Fur81, Chapter 5], which is also very accessible.

4.3.1 Factors

'Factors' are the 'morphisms' of ergodic theory: maps from one system to another which preserve the basic structure. They can be introduced in two distinct ways.

First, a **factor** of the G-system (X, μ, T) is a σ-subalgebra $\mathscr{S} \leq \mathscr{B}_X$ which is **globally T-invariant**, meaning that, for each $g \in G$,

$$A \in \mathscr{S} \quad \Longleftrightarrow \quad T^g(A) \in \mathscr{S}.$$

Secondly, given two G-systems $\mathbf{X} = (X, \mu, T)$ and $\mathbf{Y} = (Y, \nu, S)$, a **factor map** from $\mathbf{X}$ to $\mathbf{Y}$ is a Borel map $\pi : X \longrightarrow Y$ such that:

- (map is measure-respecting) $\pi_*\mu = \nu$;
- (map is equivariant a.e.) $\pi \circ T^g(x) = S^g \circ \pi(x)$ for μ-a.e. $x \in X$, for all $g \in G$.

This will be the meaning of the notation $\pi : \mathbf{X} \longrightarrow \mathbf{Y}$.

A factor map π is an **isomorphism** if there is another factor map $\phi : \mathbf{Y} \longrightarrow \mathbf{X}$ such that

$$\phi \circ \pi = \mathrm{id}_X \text{ a.s.} \quad \text{and} \quad \pi \circ \phi = \mathrm{id}_Y \text{ a.s.}$$

As for real-valued functions, we generally identify two factor maps that agree a.e., and may be sloppy about distinguishing individual maps and a.e.-equivalence classes of maps. Given two factors $\mathscr{S}$ and $\mathscr{T}$, we write that $\mathscr{S} \subseteq \mathscr{T}$ **modulo** μ if for every $A \in \mathscr{S}$ there is some $B \in \mathscr{T}$ such that $\mu(A \triangle B) = 0$. Equality modulo μ is defined from this in the obvious way.

In the definition above, it is important that the equivariance of π be allowed to fail on a μ-negligible set. Otherwise the theory is too rigid for many applications. It can be helpful to throw away some negligible part of one system in order to pass to another, and our intuition is that this does not alter the 'statistical' properties of the domain system. For instance, let (Y, ν, S) be any system with no fixed points (such as a circle rotation), and define (X, μ, T) so that $X = Y \sqcup \{x_0\}$ for some T-fixed point x_0 which carries zero measure, and with $T|_Y = S$. 'Statistically' we should like to consider these two systems isomorphic, but no map $X \longrightarrow Y$ can be equivariant at x_0.

If a factor map exists as above, one sometimes also refers to $\mathbf{Y}$ as a **factor** of $\mathbf{X}$, or to $\mathbf{X}$ as an **extension** of $\mathbf{Y}$.

Beware that factor maps are not assumed to be continuous. This is in keeping with our remark that it is $\mathscr{B}_X$, rather than the topology on X, that really matters. Insisting on homeomorphism-actions and continuous factor maps would lead to the area of topological dynamics, a rich but very different theory.

Given a factor map as above, the σ-subalgebra $\pi^{-1}(\mathscr{B}_Y)$ is a factor: this is the relation between these notions. The map π is said to **generate** this factor. Since we work among compact metric spaces, some routine measure theory shows that for any G-system (X, μ, T) and factor $\mathscr{S} \leq \mathscr{B}_X$, there are another G-system (Y, ν, S) and a factor map $\pi : (X, \mu, T) \longrightarrow (Y, \nu, S)$ such that

$$\mathscr{S} = \pi^{-1}(\mathscr{B}_Y) \quad \text{modulo } \mu.$$

Example If (X, μ, T) is a G-system and $H \trianglelefteq G$, then the σ-algebra of H-invariant sets,

$$\mathscr{B}_X^{T^H} := \{A \in \mathscr{B}_X : T^h A = A \; \forall h \in H\},$$

is a factor. It is called the **H-partially invariant factor** of (X, μ, T). An easy exercise characterizes it as follows: $\mathscr{B}_X^{T^H}$ is the largest factor of (X, μ, T), modulo μ, which can be generated modulo μ by a factor map $\pi : (X, \mu, T) \longrightarrow (Y, \nu, S)$ to a system in which $S^h = \mathrm{id}_Y$ for all $h \in H$. We usually abbreviate $\mathscr{B}_X^{T^G}$ to $\mathscr{B}_X^T$. $\triangleleft$

4.3.2 Ergodicity and Disintegration

Suppose that $\mathbf{X} = (X, \mu, T)$ is a G-system, and that $A \in \mathscr{B}_X$ is T-invariant and has $0 < \mu(A) < 1$. Then its complement $X \setminus A$ has the same properties, and now the partition

$$X = A \sqcup (X \setminus A)$$

gives a decomposition of (X, μ, T) into two disjoint subsystems, each with its own dynamics under T, re-weighted so that their measures have total mass $\mu(A)$ and $1 - \mu(A)$.

Definition 4.11 (Ergodic System) A system $\mathbf{X}$ is **ergodic** if it is not decomposable in this way: that is, if any T-invariant $A \in \mathscr{B}_X$ has $\mu(A) \in \{0, 1\}$.

Equivalently, $\mathbf{X}$ is ergodic if and only if

$$\mathscr{B}_X^T = \{\emptyset, X\} \quad \text{modulo } \mu.$$

In case $\mathbf{X}$ is not ergodic, $\mathscr{B}_X^T$ may be generated modulo μ by a nontrivial factor map

$$\pi : (X, \mu, T) \longrightarrow (Y, \nu, \mathrm{id}_Y). \tag{4.1}$$

For many purposes, this enables a reduction from arbitrary to ergodic systems, by virtue of the following classical result from measure theory.

Theorem 4.12 (Measure Disintegration) *Let $\pi : X \longrightarrow Y$ be a Borel map between compact metric spaces, let $\mu \in \Pr(X)$, and let $\nu := \pi_*\mu \in \Pr(Y)$. Then there is a map*

$$Y \longrightarrow \Pr(X) : y \mapsto \mu_y$$

with the following properties:

- *for every $A \in \mathscr{B}_X$, the real-valued map $y \mapsto \mu_y(A)$ is Borel measurable;*
- *$\mu_y(\pi^{-1}\{y\}) = 1$ for ν-a.e. y;*
- *$\mu = \int_Y \mu_y\,\nu(\mathrm{d}y)$, in the sense that $\mu(A) = \int_Y \mu_y(A)\,\nu(\mathrm{d}y)$ for every $A \in \mathscr{B}_X$.*

Moreover, this map $y \mapsto \mu_y$ is essentially unique: if $y \mapsto \mu'_y$ is another such, then $\mu_y = \mu'_y$ for ν-a.e. y. □

A map $y \mapsto \mu_y$ as above is referred to as a **disintegration** of μ over π.

Many textbooks stop short of measure disintegration in this generality, but it can be found, for instance, in [Bre68, Section 4.3] or [Fur81, Theorem 5.8].

If (X, μ) is as above and $\mathscr{S} \leq \mathscr{B}_X$ is a σ-subalgebra, then for any $f \in L^2(\mu)$ there is an $\mathscr{S}$-measurable function $f' \in L^2(\mu)$ with the property that

$$\int_X fh\,\mathrm{d}\mu = \int_X f'h\,\mathrm{d}\mu \quad \forall\, \mathscr{S}\text{-measurable } h \in L^2(\mu).$$

This f' may be obtained as the orthogonal projection of f onto the subspace of $\mathscr{S}$-measurable functions in $L^2(\mu)$. It follows that it is unique up to μ-a.e. equality. It is called the **conditional expectation** of f on $\mathscr{S}$, and denoted by $\mathsf{E}_\mu(f \mid \mathscr{S})$. See, for instance, [Bil95, Section 34] for more on this important construction in probability.

In the setting above, if $\mathscr{S} := \pi^{-1}(\mathscr{B}_Y)$, then a disintegration gives a 'formula' for the conditional expectation:

$$\mathsf{E}_\mu(f \mid \mathscr{S})(x) = \int_X f\,\mathrm{d}\mu_{\pi(x)} \quad \forall f \in L^2(\mu). \tag{4.2}$$

Exercise Prove (4.2) from the properties in Theorem 4.12. ◁

Now consider a G-system $\mathbf{X}$, and let us apply this machinery to the factor $\mathscr{B}_X^T$. Let $\pi : \mathbf{X} \longrightarrow \mathbf{Y}$ generate $\mathscr{B}_X^T$ modulo μ, where $\mathbf{Y} = (Y, \nu, \mathrm{id}_Y)$. Theorem 4.12 gives a disintegration

$$\mu = \int_Y \mu_y\,\nu(\mathrm{d}y).$$

For any $g \in G$, applying the transformation T^g and recalling that μ is T^g-invariant gives

$$\mu = T^g_* \mu = \int_Y T^g_* \mu_y \, \nu(dy).$$

Using this, one can now check that the map $y \mapsto T^g_* \mu_y$ is also a disintegration of μ over π. By the essential uniqueness of disintegration, this implies that $\mu_y = T^g_* \mu_y$ for ν-a.e. y. Since G is countable, this holds simultaneously for all g for ν-a.e. y, and so μ has been represented as an integral of T-invariant measures.

The crucial property that one gains from this construction is the following.

Proposition 4.13 *In the disintegration above, the system (X, μ_y, T) is ergodic for ν-a.e. y.*

There are many approaches to this proposition. We base ours on the following lemma. It gives a quantitative relation between the failure of some $f \in L^2(\mu)$ to be G-invariant and the distance from f to $\mathsf{E}(f \mid \mathscr{B}_X^T)$.

Lemma 4.14 *If (X, μ, T) is any G-system and $f \in L^2(\mu)$, then*

$$\sup_{g \in G} \|f - f \circ T^g\|_2 \geq \|f - \mathsf{E}(f \mid \mathscr{B}_X^T)\|_2.$$

Proof On $L^2(\mu)$, the function

$$\psi(h) := \sup_{g \in G} \|h - f \circ T^g\|_2^2$$

is continuous, nonnegative, and T-invariant. It is also strictly convex, since this is true of the squared norm $\|\cdot\|_2^2$. Therefore ψ has a unique minimizer h, which is also T-invariant owing to its uniqueness. On the other hand, for any T-invariant function h' one has

$$\psi(h') = \sup_{g \in G} \|h' \circ T^{g^{-1}} - f\|_2^2 = \|h' - f\|_2^2 \quad \forall g \in G,$$

and on the subspace of T-invariant functions this is minimized by the orthogonal projection $h' = \mathsf{E}(f \mid \mathscr{B}_X^T)$. Therefore h must equal $\mathsf{E}(f \mid \mathscr{B}_X^T)$, and so

$$\psi(f) \geq \psi(\mathsf{E}(f \mid \mathscr{B}_X^T)) = \|f - \mathsf{E}(f \mid \mathscr{B}_X^T)\|_2^2.$$

□

Proof of Proposition 4.13 We will show that if some positive-ν-measure subset of the measures μ_y failed to be ergodic, then we could synthesize a single function which contradicts Lemma 4.14 for the measure μ. The key facts to use are that $C(X)$ is both (i) separable for the uniform topology and (ii) dense in $L^2(\theta)$ for any $\theta \in \Pr(X)$.

Let $h_1, h_2, \ldots$ be a uniformly dense sequence in $C(X)$. Now consider some $\theta \in \Pr(X)$ which is T-invariant but not ergodic. Then $L^2(\theta)$ contains an invariant function which is not a.s. constant. Approximating this function sufficiently well in $\|\cdot\|_{L^2(\theta)}$ by some h_i, it follows that for any such θ there is some $i \geq 1$ for which

$$\sup_{g\in G} \|h_i - h_i \circ T^g\|_{L^2(\theta)} < \left\| h_i - \int_X h_i \,\mathrm{d}\theta \right\|_{L^2(\theta)}.$$

In the setting of our disintegration, it therefore suffices to prove that each of the sets

$$A_i := \left\{ y \in Y : \sup_{g\in G} \|h_i - h_i \circ T^g\|_{L^2(\mu_y)} < \left\| h_i - \int_X h_i \,\mathrm{d}\mu_y \right\|_{L^2(\mu_y)} \right\}$$

has $\nu(A_i) = 0$, since it follows that μ_y is ergodic for every y outside the negligible set $\bigcup_{i\geq 1} A_i$.

The proof is completed by contradiction: suppose that $\nu(A_i) > 0$ for some i. On X, consider the function $h(x) := 1_{A_i}(\pi(x))h_i(x)$. Using the properties of disintegration and (4.2), this function satisfies

$$\begin{aligned}
\sup_{g\in G} \|h - h \circ T^g\|_{L^2(\mu)} &= \sup_{g\in G} \left(\int_Y \|h - h \circ T^g\|^2_{L^2(\mu_y)} \,\nu(\mathrm{d}y) \right)^{1/2} \\
&\leq \left(\int_{A_i} \sup_{g\in G} \|h_i - h_i \circ T^g\|^2_{L^2(\mu_y)} \,\nu(\mathrm{d}y) \right)^{1/2} \\
&< \left(\int_{A_i} \left\| h_i - \int h_i \,\mathrm{d}\mu_y \right\|^2_{L^2(\mu_y)} \nu(\mathrm{d}y) \right)^{1/2} \\
&= \left(\int_Y \left\| h - \int h \,\mathrm{d}\mu_y \right\|^2_{L^2(\mu_y)} \nu(\mathrm{d}y) \right)^{1/2} \\
&= \|h - \mathsf{E}(h \mid \mathscr{B}_X^T)\|_{L^2(\mu)}.
\end{aligned}$$

This contradicts Lemma 4.14. □

Definition 4.15 (Ergodic Decomposition) A disintegration of μ into ergodic T-invariant measures as above is called an **ergodic decomposition** of μ.

For many purposes in ergodic theory, an ergodic decomposition quickly permits one to restrict attention to ergodic systems, by arguing about the disintegrands μ_y individually.

4.3.3 Joinings

Factors give us a notion of the 'parts' of a probability-preserving system. It can also be important to study the ways in which some 'ingredient' systems can be combined into a new, 'larger' system.

First, let us recall some useful nomenclature from probability. Suppose that $(X_i, \mathscr{S}_i)$, $i \in I$, is a countable collection of measurable spaces and that μ is a probability measure on the product space $\prod_{i \in I} X_i$ with the product σ-algebra $\bigotimes_{i \in I} \mathscr{S}_i$ (see, for instance, [Fol99, Section 1.2]). If $J \subseteq I$ and $\pi : \prod_{i \in I} X_i \longrightarrow \prod_{j \in J} X_j$ is the corresponding coordinate projection, then the image measure $\pi_* \mu$ is the **marginal** of μ on the coordinates indexed by J. In particular, if π is the projection onto the coordinate-copy of X_i, then $\pi_* \mu$ is the **marginal of μ on the ith coordinate**.

Definition 4.16 Suppose that $(X_i, \mathscr{S}_i, \mu_i)$, $i \in I$, is a countable collection of probability spaces. A **coupling** of them is a probability measure λ on $\prod_{i \in I} X_i$ with the product σ-algebra such that the marginal of λ on the ith coordinate is μ_i for each $i \in I$: that is,

$$\lambda\{(x_i)_{i \in I} : x_i \in A\} = \mu_j(A) \quad \forall j \in I, \ A \in \mathscr{S}_j.$$

The set of such couplings is denoted $\mathrm{Cpl}((\mu_i)_{i \in I})$.

Now suppose that $\mathbf{X}_i = (X_i, \mu_i, T_i)$, $i \in I$, is a countable collection of G-systems. A **joining** of them is a coupling $\lambda \in \mathrm{Cpl}((\mu_i)_{i \in I})$ which is invariant under the diagonal G-action defined by

$$\prod_{i \in I} T_i^g \quad \text{for } g \in G.$$

The set of such joinings is denoted $\mathrm{J}((\mathbf{X}_i)_{i \in I})$.

A joining of d copies of the same system $\mathbf{X}$ is called a **d-fold self-joining**, and the set of these is denoted $\mathrm{J}^{(d)}(\mathbf{X})$.

Joinings were introduced into ergodic theory in Furstenberg's classic paper [Fur67], which still makes delightful reading.

A joining of some systems (X_i, μ_i, T_i), $i = 1, 2, \ldots$, arises when those systems appear together as factors of some other system $\widetilde{\mathbf{X}} = (\widetilde{X}, \widetilde{\mu}, \widetilde{T})$: that is, when one has a diagram

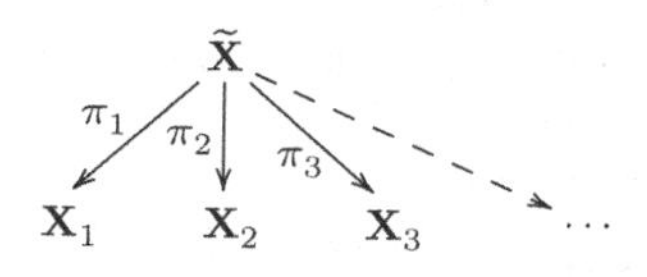

In this case, the corresponding joining is the pushforward measure

$$(\pi_1, \pi_2, \ldots)_* \widetilde{\mu}.$$

This is called the **joint distribution of $\pi_1, \pi_2, \ldots$ in $\widetilde{\mathbf{X}}$.**

Examples

1. Any collection of systems $(X_i, \mu_i, T_i)_{i \in I}$ always has at least one joining: the product measure $\bigotimes_{i \in I} \mu_i$.
2. For any system (X, μ, T) and $d \in \mathbb{N}$, another d-fold self-joining of (X, μ, T) is given by the **diagonal measure**

$$\mu^{\Delta(d)} := \int_X \delta_{(x,x,\ldots,x)}\, \mu(\mathrm{d}x),$$

which may be viewed as a copy of μ living on the diagonal in X^d. ◁

The following is obvious given example (1) above.

Lemma 4.17 *For any G and countable family $(\mathbf{X}_i)_{i \in I}$ of G-systems, the space $\mathrm{J}((\mathbf{X}_i)_{i \in I})$ is nonempty and convex.* □

We will henceforth focus on couplings and joinings of finitely many spaces or systems, for simplicity.

Given any spaces X_i for $i \leq m$ and functions $f_i : X_i \longrightarrow \mathbb{R}$, we will often write $f_1 \otimes \cdots \otimes f_m$ for the function

$$X_1 \times \cdots \times X_m \longrightarrow \mathbb{R} : (x_1, \ldots, x_m) \mapsto f_1(x_1) \cdots f_m(x_m).$$

Using such functions, coupling-spaces and joining-spaces can be endowed with a natural topology. This extra structure will play a crucial rôle later.

Definition 4.18 Given a tuple $(X_i, \mathcal{S}_i, \mu_i)$, $i \leq m$, of probability spaces, the **coupling topology** on $\mathrm{Cpl}(\mu_1, \ldots, \mu_m)$ is the weakest topology for which the evaluation functionals

$$\mathrm{ev}_{f_1,\ldots,f_m} : \mathrm{Cpl}(\mu_1, \ldots, \mu_m) \longrightarrow \mathbb{R} : \lambda \mapsto \int f_1 \otimes \cdots \otimes f_m \, \mathrm{d}\lambda$$

are continuous for all tuples $f_i \in L^\infty(\mu_i)$.

Given G-systems $\mathbf{X}_i = (X_i, \mu_i, T_i)$ for $i \leq m$, the restriction to $\mathrm{J}(\mathbf{X}_1, \ldots, \mathbf{X}_m)$ of the coupling topology on $\mathrm{Cpl}(\mu_1, \ldots, \mu_m)$ is called the **joining topology**.

It is worth recording the following easy consequence immediately.

Lemma 4.19 *Let $(X_i, \mathscr{S}_i, \mu_i)$ and $(Y_i, \mathscr{T}_i, \nu_i)$ for $i \leq m$ be two tuples of probability spaces, and let $\pi_i : X_i \longrightarrow Y_i$ be measurable maps satisfying $\pi_{i*}\mu_i = \nu_i$ for each i. Then the map*

$$\mathrm{Cpl}(\mu_1, \ldots, \mu_m) \longrightarrow \mathrm{Cpl}(\nu_1, \ldots, \nu_m) : \lambda \mapsto (\pi_1 \times \cdots \times \pi_m)_*\lambda$$

is continuous for the coupling topologies on domain and target. □

It is obvious from Definition 4.18 that the joining topology depends only on the Borel σ-algebras of the spaces X_i, not on their specific topologies. Nevertheless, it turns out that the joining topology can also be characterized using only continuous functions for those topologies.

Lemma 4.20 *If each (X_i, μ_i) for $i \leq m$ is a compact metric space with a Borel probability measure, then the coupling topology agrees with the restriction to $\mathrm{Cpl}(\mu_1, \ldots, \mu_m)$ of the weak* topology on $\mathrm{Pr}(X_1 \times \cdots \times X_m)$.*

Proof *Step 1.* The weak* topology is the weakest for which the functionals

$$\mathrm{ev}_F : \theta \mapsto \int F \, \mathrm{d}\theta$$

are continuous for all $F \in C(X_1 \times \cdots \times X_m)$. If $F = f_1 \otimes \cdots \otimes f_m$ for some tuple $f_i \in C(X_i)$, then ev_F is clearly also continuous for the coupling topology, since $C(X_i) \subseteq L^\infty(\mu_i)$. It follows that ev_F is continuous in case F is a linear combination of such product-functions.

By the Stone–Weierstrass theorem, the algebra of these linear combinations is dense in $C(X_1 \times \cdots \times X_m)$. Therefore, for any continuous F and $\varepsilon > 0$, there is some such linear combination G for which $\|F - G\|_\infty < \varepsilon$, and this implies

$$|\mathrm{ev}_F(\theta) - \mathrm{ev}_G(\theta)| = \left|\int (F - G) \, \mathrm{d}\theta\right| < \varepsilon \quad \forall \theta \in \mathrm{Cpl}(\mu_1, \ldots, \mu_m).$$

Therefore ev_F is a uniform limit of functionals that are continuous for the coupling topology, so is itself continuous for the coupling topology.

Step 2. To prove the reverse, fix $f_i \in L^\infty(\mu_i)$ for $i \leq m$. We will prove that $\mathrm{ev}_{f_1,\ldots,f_m}$ is continuous on $\mathrm{Cpl}(\mu_1, \ldots, \mu_m)$ for the restriction of the weak* topology.

By rescaling, we may clearly assume $\|f_i\|_\infty \leq 1$ for each i. Let $\varepsilon > 0$. For each $i \leq m$, choose $g_i \in C(X_i)$ such that $\|g_i\|_\infty \leq 1$ and $\|f_i - g_i\|_{L^1(\mu_i)} < \varepsilon/m$. Then for any $\theta \in \mathrm{Cpl}(\mu_1, \ldots, \mu_m)$, one has

$$\begin{aligned}
&|\mathrm{ev}_{f_1,\dots,f_m}(\theta) - \mathrm{ev}_{g_1,\dots,g_m}(\theta)| \\
&= \Big| \int (f_1 \otimes \cdots \otimes f_m - g_1 \otimes \cdots \otimes g_m)\, \mathrm{d}\theta \Big| \\
&= \Big| \sum_{i=1}^m \int g_1 \otimes \cdots \otimes g_{i-1} \otimes (f_i - g_i) \otimes f_{i+1} \otimes \cdots \otimes f_m \, \mathrm{d}\theta \Big| \\
&\leq \sum_{i=1}^m \|g_1\|_\infty \cdots \|g_{i-1}\|_\infty \cdot \|f_i - g_i\|_{L^1(\mu_i)} \cdot \|f_{i+1}\|_\infty \cdots \|f_m\|_\infty \\
&< m(\varepsilon/m) = \varepsilon,
\end{aligned}$$

because the marginal of θ on X_i is assumed to equal μ_i.

Thus, the functionals $\mathrm{ev}_{f_1,\dots,f_m}$ may be uniformly approximated by functionals $\mathrm{ev}_{g_1,\dots,g_m}$ with $g_i \in C(X_i)$ for each i. Since the latter are all continuous for the weak* topology by definition, and a uniform limit of continuous functions is continuous, this completes the proof. □

Corollary 4.21 *Given G-systems* $\mathbf{X}_i$ *for* $i \leq m$, *the joining topology on* $\mathrm{J}(\mathbf{X}_1, \dots, \mathbf{X}_m)$ *is compact.*

Proof By the preceding lemma, the coupling topology on $\mathrm{Cpl}(\mu_1, \dots, \mu_m)$ is the restriction of the weak* topology to the further subset

$$\bigcap_{i \leq m} \bigcap_{f_i \in C(X_i)} \Big\{\theta \in \Pr(X_1 \times \cdots \times X_m) : \int f_i(x_i)\, \theta(\mathrm{d}x_1, \dots, \mathrm{d}x_m) = \int f_i \, \mathrm{d}\mu_i \Big\},$$

because the ith marginal of any $\theta \in \Pr(X_1 \times \cdots \times X_m)$ is uniquely determined by the integrals of all continuous functions. This is therefore a weak*-closed subset of the weak*-compact space $\Pr(X_1 \times \cdots \times X_m)$, hence also weak*-compact.

Finally, the joining topology on $\mathrm{J}(\mathbf{X}_1, \dots, \mathbf{X}_m)$ is the further restriction of the weak* topology to the subset

$$\bigcap_{g \in G} \bigcap_{f_1 \in L^\infty(\mu_1), \dots, f_m \in L^\infty(\mu_m)} \Big\{\theta \in \mathrm{Cpl}(\mu_1, \dots, \mu_m) : \mathrm{ev}_{f_1,\dots,f_m}(\theta) = \mathrm{ev}_{f_1 \circ T_1^g, \dots, f_m \circ T_m^g}(\theta)\Big\}.$$

Since each of the functionals $\mathrm{ev}_{f_1,\dots,f_m}$ and $\mathrm{ev}_{f_1 \circ T_1^g, \dots, f_m \circ T_m^g}$ has been shown to be continuous for the restriction of the weak* topology to $\mathrm{J}(\mathbf{X}_1, \dots, \mathbf{X}_m)$, this is a further weak*-closed subset, hence weak*-compact. □

Exercise Generalize Definition 4.18, Lemma 4.20 and Corollary 4.21 to couplings or joinings of countable collections of spaces or systems. [Hint: Work with functions of only finitely many coordinates.] $\lhd$

4.3.4 Relative Products

With the use of factor maps and disintegrations, we can introduce a considerable generalization of product joinings. Suppose that $\mathbf{X}_i$ are systems for $i \leq m$, that $\pi_i : \mathbf{X}_i \longrightarrow \mathbf{Y}_i$ are factor maps, that $\mathscr{S}_i := \pi_i^{-1}(\mathscr{B}_{Y_i})$, and that $\theta \in \mathrm{J}(\mathbf{Y}_1, \dots, \mathbf{Y}_m)$. For each $i \leq m$, let $y \mapsto \mu_{i,y}$ be a disintegration of μ_i over π_i. Then the **relative product** of the systems $\mathbf{X}_i$ over the maps π_i and joining θ is the measure

$$\lambda := \int_{\prod_i Y_i} \mu_{1,y_1} \otimes \cdots \otimes \mu_{m,y_m} \, \theta(\mathrm{d}y_1, \dots, \mathrm{d}y_m).$$

One checks easily that $\lambda \in \mathrm{J}(\mathbf{X}_1, \dots, \mathbf{X}_m)$.

If $\lambda \in \mathrm{J}(\mathbf{X}_1, \dots, \mathbf{X}_m)$ is given, then it is **relatively independent** over $(\pi_1, \dots, \pi_m)$ or over $(\mathscr{S}_1, \dots, \mathscr{S}_m)$ if it is of the above form for some θ. In that case, of course, one must have

$$\theta = (\pi_1 \times \cdots \times \pi_m)_* \lambda.$$

Using (4.2), this conclusion is equivalent to

$$\int \bigotimes_{i \leq m} f_i \, \mathrm{d}\lambda = \int \bigotimes_{i \leq m} \mathsf{E}_{\mu_i}(f_i \mid \mathscr{S}_i) \, \mathrm{d}\lambda \tag{4.3}$$

for all 'test functions' $f_1 \in L^\infty(\mu_1), \dots, f_m \in L^\infty(\mu_m)$.

As an important special case of the above, consider a single factor map $\pi : \mathbf{X} \longrightarrow \mathbf{Y}$. Let θ be the diagonal self-joining of two copies of $\mathbf{Y}$, and let $\mu \otimes_\pi \mu$ denote the relative product of two copies of $\mathbf{X}$ over π and θ. For this joining one obtains

$$(\mu \otimes_\pi \mu)(A) = \int_Y (\mu_y \otimes \mu_y)(A) \, \nu(\mathrm{d}y) \quad \text{for } A \in \mathscr{B}_{X^2}.$$

This is supported on the Borel subset $\{(x, x') \in X^2 : \pi(x) = \pi(x')\}$. As a result, the two coordinate projections $\pi_1, \pi_2 : X^2 \longrightarrow X$ satisfy

$$\pi \circ \pi_1 = \pi \circ \pi_2 \quad (\mu \otimes_\pi \mu)\text{-a.s.},$$

and so there is a commutative diagram of factor maps

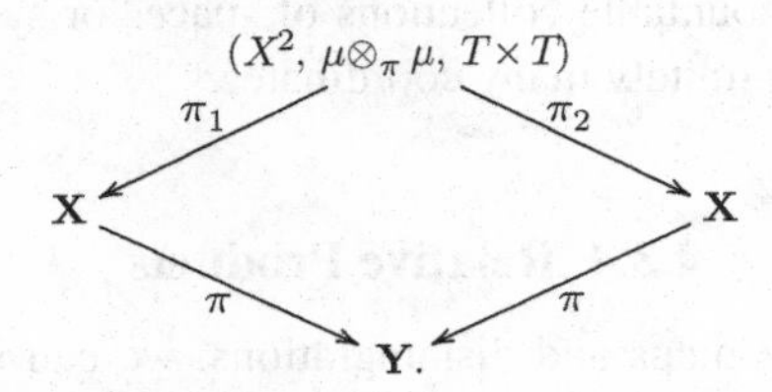

An equivalent description of $\mu \otimes_\pi \mu$ is that for any bounded measurable functions $f, g : X \longrightarrow \mathbb{R}$, one has

$$\int_{X^2} f \otimes g \, \mathrm{d}(\mu \otimes_\pi \mu) = \int_X \mathsf{E}(f \mid \pi^{-1}(\mathscr{B}_Y))\mathsf{E}(g \mid \pi^{-1}(\mathscr{B}_Y)) \, \mathrm{d}\mu.$$

From this, it follows at once that $\mu \otimes_\pi \mu = \mu \otimes_{\pi'} \mu$ whenever the factor maps π and π' generate the same factor of $\mathbf{X}$ modulo μ.

4.3.5 Inverse Limits

The last general construction we need can be viewed as giving 'limits' of 'increasing' sequences of systems.

To be precise, suppose that we have a sequence of G-systems $\mathbf{X}_n := (X_n, \mu_n, T_n)$ for $n \geq 0$, together with **connecting factor maps** π_n:

$$\cdots \xrightarrow{\pi_3} \mathbf{X}_3 \xrightarrow{\pi_2} \mathbf{X}_2 \xrightarrow{\pi_1} \mathbf{X}_1 \xrightarrow{\pi_0} \mathbf{X}_0.$$

Proposition 4.22 *In this situation, there is a G-system $\mathbf{X} = (X, \mu, T)$ together with a sequence of factor maps $\phi_n : \mathbf{X} \longrightarrow \mathbf{X}_n$ such that*

- *(the factor maps are consistent) $\pi_n \circ \phi_{n+1} = \phi_n$ μ-a.e. for all $n \geq 0$;*
- *(the factor maps generate everything) $\mathscr{B}_X$ is generated modulo μ by the union of the factors $\phi_n^{-1}(\mathscr{B}_{X_n})$ over $n \geq 0$; and*
- *(universality) given any other system $\mathbf{Y} = (Y, \nu, S)$ and factor maps $\psi_n : \mathbf{Y} \longrightarrow \mathbf{X}_n$ such that $\pi_n \circ \psi_{n+1} = \psi_n$ for all n, there is an essentially unique factor map $\alpha : \mathbf{Y} \longrightarrow \mathbf{X}$ with $\psi_n = \phi_n \circ \alpha$ for all n.*

Proof Let $X := \prod_{n\geq 0} X_n$ with its product topology, and let $T^g := \prod_{n\geq 0} T_n^g$ for each $g \in G$.

For $m > n \geq 0$, let

$$\pi_n^m := \pi_n \circ \pi_{n+1} \circ \cdots \circ \pi_{m-1} : \mathbf{X}_m \longrightarrow \mathbf{X}_n,$$

so $\pi_n^{n+1} = \pi_n$ for each n. These are all still factor maps.

Now, for each $m \geq 1$, consider the measure μ'_m on $\prod_{n=0}^m X_n$ given by

$$\mu'_m = \int_{X_m} \delta_{(\pi_0^m(x),\ \pi_1^m(x),\ \dots,\ \pi_{m-1}^m(x),\ x)}\ \mu_m(\mathrm{d}x).$$

These measures are consistent, in the sense that μ'_{m_1} is the image of μ'_{m_2} under coordinate projection whenever $m_2 \geq m_1$. Therefore, by the Kolmogorov extension theorem, they are the finite-dimensional marginals of a unique measure $\mu \in \Pr(X)$ (see [Fol99, Theorem 10.18] or [Bil95, Section 36] – both of those sources treat the case of measures on $\mathbb{R} \times \mathbb{R} \times \cdots$, but the same proofs work in general). This μ is supported on the Borel subset

$$\widetilde{X} := \{(x_n)_{n\geq 0} \in X : \pi_n(x_{n+1}) = x_n\ \forall n \geq 0\}.$$

Since each μ'_m is invariant under the diagonal G-action, and they specify μ uniquely, μ is also invariant under the diagonal G-action.

This defines the system $\mathbf{X} = (X, \mu, T)$. For each n, let $\phi_n : X \longrightarrow X_n$ be the projection onto the nth coordinate. An easy check gives that ϕ_n defines a factor map $\mathbf{X} \longrightarrow \mathbf{X}_n$, and the μ-a.s. relation $\phi_n = \pi_n \circ \phi_{n+1}$ follows from the fact that μ is supported on $\widetilde{X}$. Clearly the union of the σ-algebras $\phi_n^{-1}(\mathscr{B}_{X_n})$ contains all finite-dimensional Borel subsets of X, and so generates the whole of $\mathscr{B}_X$.

Finally, suppose that $\mathbf{Y}$ and $\psi_n : \mathbf{Y} \longrightarrow \mathbf{X}_n$ are as posited, and define $\alpha : \mathbf{Y} \longrightarrow \mathbf{X}$ by

$$\alpha(y) := (\psi_0(y), \psi_1(y), \dots).$$

The assumption that $\pi_n \circ \psi_{n+1} = \psi_n$ implies that $\alpha(y) \in \widetilde{X}$ for almost all y, and moreover it is clear by definition that α intertwines S with T (since this holds coordinate-wise in X). Finally, $\alpha_*\nu$ has the same marginal as μ on any finite-dimensional projection of X, so in fact $\alpha_*\nu = \mu$. □

Definition 4.23 (Inverse Limit) A choice of system $\mathbf{X}$ together with the factor maps $(\phi_n)_{n\geq 0}$ as constructed above is an **inverse limit** of the tower of systems $(\mathbf{X}_n)_{n\geq 0}$, $(\pi_n)_{n\geq 0}$.

4.3.6 Idempotent Classes

This subsection represents a further level of abstraction, since it concerns whole classes of systems. However, the following definition will be of great value in organizing several later results.

Definition 4.24 (Idempotent Class) Let G be a countable group, and let $\mathfrak{C}$ be a class of G-systems. Then $\mathfrak{C}$ is **idempotent** if it has the following closure properties:

1. (Isomorphism) If $\mathbf{X} \in \mathfrak{C}$ and $\mathbf{Y}$ is isomorphic to $\mathbf{X}$, then $\mathbf{Y} \in \mathfrak{C}$.
2. (Finite joinings) If $\mathbf{X} = (X, \mu, T)$ and $\mathbf{Y} = (Y, \nu, S)$ are both in $\mathfrak{C}$, and $\lambda \in \mathrm{J}(\mathbf{X}, \mathbf{Y})$, then
$$(X \times Y, \lambda, T \times S) \in \mathfrak{C}.$$
3. (Inverse limits) If
$$\cdots \longrightarrow \mathbf{X}_3 \longrightarrow \mathbf{X}_2 \longrightarrow \mathbf{X}_1$$
is an inverse sequence consisting of members of $\mathfrak{C}$, then some choice of inverse limit is a member of $\mathfrak{C}$ (and hence so are all such choices).

Example For a subset $S \subseteq G$, let $\mathfrak{C}^S$ be the class of those G-systems (X, μ, T) in which $T^s = \mathrm{id}_X$ μ-a.e. for all $s \in S$. This class is easily checked to be idempotent. It is called the ***S*-partially invariant class**. A simple argument shows that $\mathfrak{C}^S = \mathfrak{C}^H$, where H is the normal subgroup of G generated by S. ◁

Lemma 4.25 *Let $\mathfrak{C}$ and $\mathfrak{D}$ be idempotent classes, and let $\mathfrak{C} \vee \mathfrak{D}$ denote the class of G-systems that are joinings of a member of $\mathfrak{C}$ and a member of $\mathfrak{D}$, up to isomorphism. Then $\mathfrak{C} \vee \mathfrak{D}$ is idempotent.* □

Exercise Prove this lemma. [Hint: Prove that an inverse limit of joinings is a joining of inverse limits, etc.] ◁

The class $\mathfrak{C} \vee \mathfrak{D}$ constructed above is called the **join** of $\mathfrak{C}$ and $\mathfrak{D}$.

Example If $G = \mathbb{Z}^d$, then the idempotent class
$$\mathfrak{C}^{\mathbf{e}_1 - \mathbf{e}_2} \vee \cdots \vee \mathfrak{C}^{\mathbf{e}_1 - \mathbf{e}_d}$$
consists of all systems of the form
$$(X_2 \times \ldots \times X_d, \lambda, T_2 \times \cdots \times T_d),$$
where

- for each $i \in \{2, 3, \ldots, d\}$, $\mathbf{X}_i = (X_i, \mu_i, T_i)$ is a $\mathbb{Z}^d$-system with the property that $T_i^{\mathbf{e}_1} = T_i^{\mathbf{e}_i}$ a.e..
- $\lambda \in \mathrm{J}(\mathbf{X}_2, \ldots, \mathbf{X}_d)$.

This example will play a key rôle later. ◁

Definition 4.26 For any G-system (X, μ, T), a factor $\mathscr{S} \leq \mathscr{B}_X$ is a **$\mathfrak{C}$-factor** if it is generated modulo μ by a factor map whose target system is a member of $\mathfrak{C}$.

Lemma 4.27 *Any G-system* $\mathbf{X}$ *has an essentially unique $\mathfrak{C}$-factor $\mathscr{S}$ which contains every other $\mathfrak{C}$-factor modulo μ.*

Proof Let C be the subset of $L^1(\nu)$ containing all functions that are measurable with respect to some $\mathfrak{C}$-factor of $\mathbf{X}$. Since $L^1(\nu)$ is separable, we may choose a dense sequence $(f_n)_{n\geq 1}$ in C. For each f_n, there are a factor map

$$\pi_n : (X, \mu, T) \longrightarrow (Y_n, \nu_n, S_n) = \mathbf{Y}_n \in \mathfrak{C}$$

and a function $g_n \in L^1(\nu_n)$ such that $f_n = g_n \circ \pi_n$ a.s.

From these, assemble the maps

$$\psi_n : X \longrightarrow \prod_{m=1}^{n} Y_m : x \mapsto (\pi_1(x), \ldots, \pi_n(x)).$$

Each ψ_n is a factor map from $\mathbf{X}$ to the system

$$\mathbf{W}_n := (Y_1 \times \cdots \times Y_n, \psi_{n*}\mu, S_1 \times \cdots \times S_n),$$

which is a joining of $\mathbf{Y}_1, \ldots, \mathbf{Y}_n$ and hence still a member of $\mathfrak{C}$.

Next, projecting out the last coordinate defines factor maps

$$\alpha_n : \mathbf{W}_{n+1} \longrightarrow \mathbf{W}_n$$

for each $n \geq 1$, so these systems $\mathbf{W}_n$ form an infinite tower of factors of $\mathbf{X}$. Letting $\mathbf{W} = (W, \theta, R)$ be an inverse limit, one still has a factor map $\psi : \mathbf{X} \longrightarrow \mathbf{W}$, owing to the universality of inverse limits. Also, $\mathbf{W} \in \mathfrak{C}$, since $\mathfrak{C}$ is closed under inverse limits. Finally, each f_n is measurable with respect to this factor map ψ. By the density of $\{f_n : n \geq 1\}$ among all $\mathfrak{C}$-factor-measurable members of $L^1(\nu)$, it follows that any function which is measurable with respect to a $\mathfrak{C}$-factor can be lifted through ψ, up to a negligible set. Now let $\mathscr{S} := \psi^{-1}(\mathscr{B}_W)$.

Essential uniqueness follows at once from maximality. □

Definition 4.28 A choice modulo μ of the factor constructed in the previous lemma is called a **maximal $\mathfrak{C}$-factor** of $\mathbf{X}$, and denoted $\mathfrak{C}_{\mathbf{X}}$. In view of its uniqueness modulo μ, we sometimes abusively refer to *the* maximal $\mathfrak{C}$-factor.

4.4 Multiple Recurrence in Terms of Self-Joinings

4.4.1 Reformulation

A key aspect of Furstenberg and Katznelson's approach to Theorems 4.7 and 4.8 is that it gives more than just the existence of one suitable time $n \geq 1$. In the multidimensional case, it actually proves the following.

Theorem 4.29 (Multidimensional Multiple Recurrence Theorem) *If (X, μ, T) is a $\mathbb{Z}^d$-system, and $A \in \mathscr{B}_X$ has $\mu(A) > 0$, then*

$$\liminf_{N \longrightarrow \infty} \frac{1}{N} \sum_{n=1}^{N} \mu(T^{-n\mathbf{e}_1} A \cap \cdots \cap T^{-n\mathbf{e}_d} A) > 0.$$

An analogous assertion for a single transformation lies behind Theorem 4.7. Similarly to many other applications of ergodic theory, one finds that these averages behave more regularly than the summand that appears for any particular value of n, and so are more amenable to analysis.

In these notes, we will prove only one simple special case of Theorem 4.29. However, we will answer a related and slightly easier question in general: whether the averages appearing in Theorem 4.29 actually converge, so that 'lim inf' may be replaced with 'lim'.

Furstenberg and Katznelson's original proof in [FK78] shows that these non-negative sequences stay bounded away from zero, but does not show their convergence. Naturally, ergodic theorists quickly went in pursuit of this question in its own right. It was fully resolved only recently, by Host and Kra [HK05b] for the averages associated to a single transformation and then by Tao [Tao08] for those appearing in Theorem 4.29. Both are challenging proofs, and they are quite different from one another.

Host and Kra's argument builds on a long sequence of earlier works, including [CL84, CL88a, CL88b, Rud95, Zha96, FW96, HK01] and several others referenced there. An alternative approach to the main results of Host and Kra has been given by Ziegler in [Zie07], also resting on much of that earlier work. In addition to proving convergence, these efforts have led to a very detailed description of the limiting behaviour of these averages, in terms of certain very special factors of an arbitrary $\mathbb{Z}$-system.

Tao's proof of convergence for the multidimensional averages in Theorem 4.29 departs significantly from those earlier works. He proceeds by first formulating a finitary, quantitative analog of the desired convergence, and then making contact with the hypergraph-regularity theory developed for the new combinatorial proofs of Szemerédi's theorem in [NRS06, Gow06, Tao06b] (as well as using several new insights in that finitary world).

In these notes, we will recount a more recent ergodic-theoretic proof of convergence in the general case. It avoids the transfer to a finitary setting that underlies Tao's proof, but also avoids the need for a very detailed description of the averages that is part of the earlier approaches. Although such a description would be of great interest in its own right, it is still mostly incomplete for the multidimensional averages.

Our next step will be to reformulate these questions in terms of a certain sequence of d-fold self-joinings of (X, μ, T). First, recall the diagonal self-joining:

$$\mu^{\Delta(d)} := \int_X \delta_{(x,x,\dots,x)}\,\mu(\mathrm{d}x).$$

Let $\vec{T} := T^{\mathbf{e}_1} \times \cdots \times T^{\mathbf{e}_d}$ be the **off-diagonal** transformation associated to the $\mathbb{Z}^d$-action T. Since different group elements act on different coordinates under $\vec{T}$, a self-joining of (X, μ, T) need not be $\vec{T}$-invariant.

Now, for each $N \geq 1$, let

$$\lambda_N := \frac{1}{N}\sum_{n=1}^{N} \vec{T}_*^n \mu^{\Delta(d)}, \tag{4.4}$$

the averages of $\mu^{\Delta(d)}$ under the off-diagonal transformation. A simple rearrangement gives

$$\frac{1}{N}\sum_{n=1}^{N}\int_X (f_1 \circ T^{n\mathbf{e}_1}) \cdot \dots \cdot (f_d \circ T^{n\mathbf{e}_d})\,\mathrm{d}\mu = \int_{X^d} f_1 \otimes \cdots \otimes f_d\,\mathrm{d}\lambda_N$$

for any functions $f_i \in L^\infty(\mu)$. In particular,

$$\frac{1}{N}\sum_{n=1}^{N}\mu(T^{-n\mathbf{e}_1}A \cap \cdots \cap T^{-n\mathbf{e}_d}A) = \lambda_N(A^d).$$

Thus, our questions of interest may be re-formulated as follows:

- (Convergence) Do the self-joinings λ_N converge to some limit λ_∞ in the joining topology?
- (Multiple Recurrence) If such a limit λ_∞ exists, does it have the property that

$$\mu(A) > 0 \quad \Longrightarrow \quad \lambda_\infty(A^d) > 0 \quad ?$$

The main results are that both answers are Yes. These notes will prove the first of these, and a special case of the second.

Theorem 4.30 *In the setting above, the joinings λ_N converge to some limit in the joining topology.*

A complete description of the limit joining is still lacking in general, although we will mention a few special cases where it is known precisely.

Before beginning the rest of our work on Theorems 4.30 and 4.29, it is worth mentioning another convergence theorem. Theorem 4.30 is central to the approach taken in these notes, but it is weaker than the usual formulation of 'non-conventional average convergence', which concerns the functional averages

$$A_N(f_1, f_2, \ldots, f_d) := \frac{1}{N}\sum_{n=1}^{N}(f_1 \circ T^{n\mathbf{e}_1}) \cdot (f_2 \circ T^{n\mathbf{e}_2}) \cdot \cdots \cdot (f_d \circ T^{n\mathbf{e}_d})$$

for $f_1, f_2, \ldots, f_d \in L^\infty(\mu)$.

Theorem 4.31 (Convergence of Non-Conventional Ergodic Averages) *If (X, μ, T) is a $\mathbb{Z}^d$-system and $f_1, f_2, \ldots, f_d \in L^\infty(\mu)$, then the averages $A_N(f_1, f_2, \ldots, f_d)$ converge in $\|\cdot\|_2$ as $N \longrightarrow \infty$.*

This is the result that was actually proved in [Tao08], rather than its joining analog. The earlier works [CL84, CL88a, CL88b, FW96, HK01, HK05b, Zie07] on special cases also proved convergence for functional averages. This functional-average convergence is stronger than Theorem 4.30, because

$$\int_{X^d} f_1 \otimes \cdots \otimes f_d \, \mathrm{d}\lambda_N = \int_X A_N(f_1, f_2, \ldots, f_2) \, \mathrm{d}\mu.$$

On the other hand, the approach to Theorem 4.30 given in this course can be turned into a proof of Theorem 4.31 with just a little extra work: the exercise at the end of Section 4.8 sketches something close to the alternative proof of Theorem 4.31 from [Aus09].

It is also interesting to ask whether the above functional averages converge pointwise a.e. as $N \longrightarrow \infty$. This question is still open in almost all cases. Aside from when $d = 1$, which reduces to the classical pointwise ergodic theorem, pointwise convergence is known only for the averages

$$\frac{1}{N}\sum_{n=1}^{N}(f_1 \circ T^n)(f_2 \circ T^{2n}).$$

This is a very tricky result of Bourgain [Bou90].

4.4.2 Brief Sketch of the Remaining Sections

The above questions about the joinings λ_N concern how the orbit $\vec{T}_*^n \mu^{\Delta(d)}$, $n = 1, 2, \ldots$, of the diagonal measure 'moves around' in X^d.

One basic intuition about this question is that if the system (X, μ, T) behaves very 'randomly', then these off-diagonal image measures will appear to 'spread out' over the whole of X^d, and in the limit will simply converge to the product measure $\mu^{\otimes d}$. The prototypical example of this situation is the following.

Example One-dimensional Bernoulli shifts were among the examples in Subsection 4.2.1. To obtain a d-dimensional version, let $\mathbf{p} = (p_1, \ldots, p_m)$ be a stochastic vector, let $X = \{1, 2, \ldots, m\}^{\mathbb{Z}^d}$, let $\mu := \mathbf{p}^{\otimes \mathbb{Z}^d}$, and let T be the action of $\mathbb{Z}^d$ by coordinate-translation: that is,

$$T^{\mathbf{n}}\big((x_{\mathbf{k}})_{\mathbf{k}}\big) = (x_{\mathbf{k}+\mathbf{n}})_{\mathbf{k}}.$$

The product X^d may be identified with the set $(\{1, 2, \ldots, m\} \times \cdots \times \{1, 2, \ldots, m\})^{\mathbb{Z}^d}$. To obtain a random element of X^d with law $\vec{T}_*^n \mu^{\Delta(d)}$, let $(x_{\mathbf{k}})_{\mathbf{k}}$ be a random element of X with law μ, and form the tuple

$$(x_{\mathbf{k}+n\mathbf{e}_1}, \ldots, x_{\mathbf{k}+n\mathbf{e}_\mathbf{d}})_{\mathbf{k}\in\mathbb{Z}^d}. \tag{4.5}$$

Now consider a large finite 'window' $W \subseteq \mathbb{Z}^d$, such as a large box around the origin. If n is sufficiently large compared with W, then one has $(W + n\mathbf{e}_i) \cap (W + n\mathbf{e}_j) = \emptyset$ whenever $i \neq j$. It follows that, for sufficiently large n, if one views the tuple (4.5) only in the finite window W, then its components

$$(x_{\mathbf{k}+n\mathbf{e}_1})_{\mathbf{k}\in W},\ (x_{\mathbf{k}+n\mathbf{e}_2})_{\mathbf{k}\in W},\ \ldots,\ (x_{\mathbf{k}+n\mathbf{e}_d})_{\mathbf{k}\in W}$$

have the same joint distribution as d independent draws from the probability measure $\mathbf{p}^{\otimes W}$. Thus, when n is large, the measure $\vec{T}_*^n \mu^{\Delta(d)}$ 'resembles' the product measure $\mu^{\otimes d}$: more formally, one can prove that $\vec{T}_*^n \mu^{\Delta(d)} \longrightarrow \mu^{\otimes d}$ weakly* as $n \longrightarrow \infty$. This, of course, is even stronger than asserting that $\lambda_N \longrightarrow \mu^{\otimes d}$ weakly*. $\lhd$

Exercise Let $\mathbf{p} = (1 - p, p)$ be a stochastic vector on $\{0, 1\}$. Consider again the proof of Proposition 4.9 (the Furstenberg correspondence principle). Fix a sequence of boxes R_j with all side-lengths tending to ∞. Let $(x_{\mathbf{k}})_{\mathbf{k}\in\mathbb{Z}^d} \in \{0, 1\}^{\mathbb{Z}^d}$ be drawn at random from the measure $\mathbf{p}^{\otimes\mathbb{Z}^d}$, and now let $E = \{\mathbf{k} \in \mathbb{Z}^d : x_{\mathbf{k}} = 1\}$. Show that with probability one in the choice of $(x_{\mathbf{k}})_{\mathbf{k}\in\mathbb{Z}^d} \in \{0, 1\}^{\mathbb{Z}^d}$, if one implements the proof of Proposition 4.9 with this choice of E then the resulting measure μ is equal to $\mathbf{p}^{\otimes\mathbb{Z}^d}$. (This requires either the law of large numbers or the pointwise ergodic theorem, both of which lie outside these notes.) $\lhd$

Of course, we cannot expect the simple behaviour of the Bernoulli example to hold for all systems. In the first place, if one happens to have a $\mathbb{Z}^d$-system (X, μ, T) in which $T^{\mathbf{e}_i} = T^{\mathbf{e}_j}$ for some $i \neq j$, then all the image measures $\vec{T}^n_* \mu^{\Delta(d)}$ will be supported on the subset

$$\{(x_1, \ldots, x_d) \in X^d : x_i = x_j\},$$

and hence so will any limit of the joinings λ_N. However, it turns out that the ways in which the 'random' intuition can fail are somewhat limited. More specifically, if $(\lambda_N)_N$ fails to converge to $\mu^{\otimes d}$ in the joining topology, then the system (X, μ, T) must exhibit some special extra 'structure', which can itself be exploited to show that $(\lambda_N)_N$ converges to some other limit λ_∞ instead, and to deduce something about the structure of that λ_∞[1].

One of the main discoveries in this area is the right way to describe that 'structure'. If $\lambda_N \not\longrightarrow \mu^{\otimes d}$ in the joining topology, then one finds that there must be a particular 'part' of (X, μ, T) which is responsible for this non-convergence. Specifically, the template for the results we will prove is the following: if $\lambda_N \not\longrightarrow \mu^{\otimes d}$, then there are d factor maps $\pi_i : \mathbf{X} \longrightarrow \mathbf{Y}_i$, $i = 1, 2, \ldots, d$, such that

- each $\mathbf{Y}_i$ is a system of a special kind, indicated by its membership of some relevant idempotent class $\mathfrak{C}_i$, and
- the 'random' intuition does hold *relative to these factors*, in the sense that λ_N converges to some limit joining λ_∞ which is relatively independent over the maps π_i (see Subsection 4.3.4).

These factor maps π_i are called **characteristic factors** for the limit joining λ_∞. This terminology originates with the 'partially characteristic' factors introduced in [FW96], and used in various ways since. The correct choice of the idempotent classes $\mathfrak{C}_i$ depends on the case of convergence that one is studying: that is, on the value of d, and whether one is interested in powers of a single transformation or in arbitrary $\mathbb{Z}^d$-systems. The goal is to obtain classes $\mathfrak{C}_i$ whose members are so highly structured that the limiting behaviour of the image joinings $(\pi_1 \times \cdots \times \pi_d)_* \lambda_N$ can be analysed fairly explicitly.

In the case of powers of a single transformation, the actual results can be made to match the above template exactly, with the added simplification that

[1] In fact, the non-averaged sequence $\vec{T}^n_* \mu^{\Delta(d)}$ can fail to converge much more easily, and there seems to be no good structure theory for this failure.

$\mathbf{Y}_i$ and $\mathfrak{C}_i$ are the same for every i. This will be the subject of Section 4.6, after some more necessary machinery has been developed in Section 4.5.

For general $\mathbb{Z}^d$-systems, one can prove results which fit into the template above, but it is easier to modify the template slightly. It turns out that a certain subclass of $\mathbb{Z}^d$-systems admits a considerably simpler analysis of the possible limit joinings λ_∞, for which the relevant idempotent classes $\mathfrak{C}_i$ in our template are easily described. A general $\mathbb{Z}^d$-system $\mathbf{X}$ may not have characteristic factors lying in those classes $\mathfrak{C}_i$, but one finds that $\mathbf{X}$ always admits an *extension* for which that simpler analysis can be carried out. Thus, in the general case we first extend $\mathbf{X}$ to a system whose analysis is easier, and then look for suitable characteristic factors of that extended system. This will be done in Sections 4.7 and 4.8, again in tandem with developing some necessary general tools.

Once this structural information about the limit joining λ_∞ has been obtained, it can also be used to prove Theorem 4.8. That deduction requires several extra steps, so it will not be completed in this course: see [Aus10a] or [Aus10c]. The original proofs of multiple recurrence in [Fur77, FK78] were different, but the proof in [Fur77] also involved some structural analysis of these limit joinings.

A key feature of the proofs below is that we need to work with 'candidate' limit joinings λ_∞ before showing that the sequence $(\lambda_N)_N$ actually converges. This is where we make crucial use of the compactness of the joining topology (Corollary 4.21). Given that compactness, we may start our proofs by taking a subsequential limit λ_∞ of the sequence $(\lambda_N)_N$. This subsequential limit must have the following extra property which distinguishes it among arbitrary joinings.

Lemma 4.32 (Off-Diagonal Invariance) *Any subsequential limit of the sequence $(\lambda_N)_N$ in the joining topology is invariant under $\vec{T}$, as well as under $T^{\times d}$.*

Proof Substituting from Definition (4.4) and observing that most terms cancel, one obtains

$$\vec{T}_*\lambda_N - \lambda_N = \frac{1}{N}(\vec{T}_*^{N+1}\mu^{\Delta(d)} - \vec{T}_*\mu^{\Delta(d)}).$$

These measures have total variation tending to 0 as $N \longrightarrow \infty$, so if $\lambda_{N_i} \longrightarrow \lambda$ is a subsequential limit in the joining topology, then also $\vec{T}_*\lambda_{N_i} \longrightarrow \lambda$. Since $\vec{T}_*$ acts continuously on the joining topology (Lemma 4.19), this implies $\lambda = \vec{T}_*\lambda$. □

It turns out that this extra invariance is already enough to force λ_∞ to be a relative product over some special factors of $\mathbf{X}$: that is, we can prove a result that matches the above template for any joining having this extra invariance. Having done so, one can deduce that the sequence $(\lambda_N)_N$ has only one possible subsequential limit, and hence that the sequence actually converges. We refer to this as the 'joining rigidity' approach to Theorem 4.30. It is much like the strategy for studying equidistribution under unipotent flows enabled by Ratner's theorems: see, for instance, [Sta00], or Manfred Einsiedler's and Tom Ward's chapter (Chapter 5) in the present volume.

In the remainder of this section we take some first steps towards the above analysis: reducing our work to the case of ergodic systems, and handling the 'trivial' case $d = 2$.

4.4.3 Reduction to Ergodic Systems

Next we show that Theorems 4.29 and 4.30 both follow for general systems if they are known for ergodic systems.

Indeed, if $\mathbf{X} = (X, \mu, T)$ is an arbitrary $\mathbb{Z}^d$-system with ergodic decomposition

$$\mu = \int_Y \mu_y \, \nu(\mathrm{d}y),$$

then for each d and N one has

$$\mu^{\Delta(d)} = \int_Y \mu_y^{\Delta(d)} \, \nu(\mathrm{d}y) \quad \text{and} \quad \lambda_N = \int_Y \lambda_{N,y} \, \nu(\mathrm{d}y),$$

where λ_N is as in (4.4) and $\lambda_{N,y}$ is its analog for the system (X, μ_y, T).

If convergence as in Theorem 4.30 is known for any ergodic system, this fact may be applied to the systems (X, μ_y, T) to give joining-topology limits $\lambda_{N,y} \longrightarrow \lambda_{\infty,y}$ for each y. Now the dominated convergence theorem implies that

$$\begin{aligned} \int_{X^d} f_1 \otimes \cdots \otimes f_d \, \mathrm{d}\lambda_N &= \int_Y \int_{X^d} f_1 \otimes \cdots \otimes f_d \, \mathrm{d}\lambda_{N,y} \, \nu(\mathrm{d}y) \\ &\longrightarrow \int_Y \int_{X^d} f_1 \otimes \cdots \otimes f_d \, \mathrm{d}\lambda_{\infty,y} \, \nu(\mathrm{d}y) \end{aligned}$$

for any functions $f_i \in L^\infty(\mu_i)$, and hence

$$\lambda_N \xrightarrow{\text{join}} \lambda_\infty := \int_Y \lambda_{\infty,y} \, \nu(\mathrm{d}y).$$

Having shown this, Theorem 4.29 also follows easily from the ergodic case. If $A \in \mathscr{B}_X$ has $\mu(A) > 0$, then the set

$$B := \{y \in Y : \mu_y(A) > 0\}$$

must have $\nu(B) > 0$. For each $y \in B$, the ergodic case of Theorem 4.29 gives $\lambda_{\infty,y}(A^d) > 0$, so integrating in y gives

$$\lambda_\infty(A^d) \geq \int_B \lambda_{\infty,y}(A^d)\,\nu(\mathrm{d}y) > 0.$$

We will henceforth freely assume that $\mathbf{X}$ is ergodic.

4.4.4 The 'Trivial' Case

In our theorems of interest, the 'trivial' case means $k = 2$ or $d = 2$. In that case 'multiple' recurrence is just Poincaré recurrence, as in Theorem 4.6. Nevertheless, it is worth describing the joining rigidity approach in this simple case, both to introduce some of the ideas and because some of the auxiliary results will be re-used later.

Proposition 4.33 *Let G be a countable group, let (X, μ, T) and (Y, ν, S) be G-systems, and suppose that $\lambda \in \Pr(X \times Y)$ is invariant under both $T \times S$ and $\mathrm{id} \times S$. Then λ is relatively independent over $(\mathscr{B}_X^T, \mathscr{B}_Y^S)$.*

Proof The assumptions imply that μ is also invariant under

$$(T \times S)^g \circ (\mathrm{id} \times S)^{g^{-1}} = T^g \times \mathrm{id}$$

for any g, so the assumptions are actually symmetric in X and Y.

We will prove the 'test-function' formula

$$\int f \otimes f'\,\mathrm{d}\lambda = \int \mathsf{E}_\mu(f \mid \mathscr{B}_X^T) \otimes \mathsf{E}_\nu(f' \mid \mathscr{B}_Y^S)\,\mathrm{d}\lambda \quad \forall f \in L^\infty(\mu), f' \in L^\infty(\nu):$$

see (4.3). By symmetry, it actually suffices to prove that

$$\int f \otimes f'\,\mathrm{d}\lambda = \int \mathsf{E}_\mu(f \mid \mathscr{B}_X^T) \otimes f'\,\mathrm{d}\lambda \quad \forall f \in L^\infty(\mu),\; f' \in L^\infty(\nu)$$

and then repeat the argument with the coordinates swapped.

Let $\pi_X : X \times Y \longrightarrow X$ denote the first coordinate projection, and π_Y the second. Fix $f' \in L^\infty(\nu)$, and consider the conditional expectation $\mathsf{E}_\lambda(1_X \otimes f' \mid \pi_X^{-1}(\mathscr{B}_X))$: that is, we lift f' to $X \times Y$ through the second coordinate, and then condition onto the first coordinate. Since it is $\pi_X^{-1}(\mathscr{B}_X)$-measurable, this conditional expectation is a.e. equal to $h \otimes 1_Y$ for some essentially unique $h \in L^\infty(\mu)$.

By the definition of conditional expectation, for any $f \in L^\infty(\mu)$ one has

$$\int f \otimes f' \,\mathrm{d}\lambda = \int (f \otimes 1_Y) \cdot \mathsf{E}_\lambda(1_X \otimes f' \mid \pi_X^{-1}(\mathscr{B}_X)) \,\mathrm{d}\lambda$$
$$= \int fh \otimes 1_Y \,\mathrm{d}\lambda = \int fh \,\mathrm{d}\mu.$$

Now let $g \in G$, and observe from the above that

$$\int f \cdot (h \circ T^g) \,\mathrm{d}\mu = \int (f \circ T^{g^{-1}}) \cdot h \,\mathrm{d}\mu = \int (f \otimes f') \circ (T^{g^{-1}} \times \mathrm{id}) \,\mathrm{d}\lambda$$
$$= \int f \otimes f' \,\mathrm{d}\lambda = \int fh \,\mathrm{d}\mu,$$

by the invariance of λ. Since this holds for every $f \in L^\infty(\mu)$, h must be essentially T^g-invariant for every g, and so h agrees a.e. with a $\mathscr{B}_X^T$-measurable function. Now the definition of conditional expectation gives

$$\int f \otimes f' \,\mathrm{d}\lambda = \int fh \,\mathrm{d}\mu = \int \mathsf{E}_\mu(f \mid \mathscr{B}_X^T) h \,\mathrm{d}\mu = \int \mathsf{E}_\mu(f \mid \mathscr{B}_X^T) \otimes f' \,\mathrm{d}\lambda,$$

as required. □

Let us see how this quickly handles the trivial case of Theorem 4.8.

Corollary 4.34 *Let (X, μ, T) be a $\mathbb{Z}^d$-system, let $\mu^\Delta \in \mathrm{J}(\mathbf{X}, \mathbf{X})$ be the 2-fold diagonal joining, and let $\pi : \mathbf{X} \longrightarrow \mathbf{Y}$ be a factor map to some other system which generates the invariant factor $\mathscr{B}_X^T$ modulo μ. Let R_i be a sequence of boxes in $\mathbb{Z}^d$ with all side-lengths tending to ∞. Then*

$$\lambda_i := \frac{1}{|R_i|} \sum_{\mathbf{n} \in R_i} (\mathrm{id} \times T^{\mathbf{n}})_* \mu^\Delta \xrightarrow{\text{join}} \mu \otimes_\pi \mu$$

and also

$$\frac{1}{|R_i|^2} \sum_{\mathbf{n}, \mathbf{m} \in R_i} (T^{\mathbf{n}} \times T^{\mathbf{m}})_* \mu^\Delta \xrightarrow{\text{join}} \mu \otimes_\pi \mu.$$

Proof We prove the first result and leave the second as an exercise.

By compactness of $\mathrm{J}(\mathbf{X}, \mathbf{X})$, we may pass to a subsequence and assume that $\lambda = \lim_{i \longrightarrow \infty} \lambda_i$ exists. It remains to show that $\lambda = \mu \otimes_\pi \mu$ for any such subsequence.

For any $\mathbf{m} \in \mathbb{Z}^d$, we obtain

$$\lambda - (\mathrm{id} \times T^{\mathbf{m}})_*\lambda = \lim_{i \longrightarrow \infty} \frac{1}{|R_i|} \sum_{\mathbf{n} \in R_i} \left((\mathrm{id} \times T^{\mathbf{n}})_*\mu^{\Delta} - (\mathrm{id} \times T^{\mathbf{m}+\mathbf{n}})_*\mu^{\Delta} \right)$$

in the joining topology. Some terms on the right-hand side here cancel, leaving

$$\left\| \frac{1}{|R_i|} \sum_{\mathbf{n} \in R_i} \left((\mathrm{id} \times T^{\mathbf{n}})_*\mu^{\Delta} - (\mathrm{id} \times T^{\mathbf{m}+\mathbf{n}})_*\mu^{\Delta} \right) \right\|_{\mathrm{TV}} \leq \frac{|R_i \triangle (R_i + \mathbf{m})|}{|R_i|}.$$

Since all side lengths of R_i are tending to ∞, this last estimate tends to 0 for any fixed $\mathbf{m}$. Hence the above joining limit must also be 0, and so, since $\mathbf{m}$ was arbitrary, λ is $(\mathrm{id} \times T)$-invariant. Therefore, by Proposition 4.33, λ is relatively independent over $(\mathscr{B}_X^T, \mathscr{B}_X^T)$.

Finally, suppose that $f, f' \in L^2(\mu)$, and let $h := \mathsf{E}(f \mid \mathscr{B}_X^T)$ and $h' := \mathsf{E}(f' \mid \mathscr{B}_X^T)$. Using the relative independence proved above and the T-invariance of h', we obtain

$$\int f \otimes f' \,\mathrm{d}\lambda = \int h \otimes h' \,\mathrm{d}\lambda \stackrel{\mathrm{def}}{=} \lim_{i \longrightarrow \infty} \frac{1}{|R_i|} \sum_{\mathbf{n} \in R_i} \int h(x) h'(T^{\mathbf{n}}x) \,\mu(\mathrm{d}x)$$

$$= \int hh' \,\mathrm{d}\mu = \int f \otimes f' \,\mathrm{d}(\mu \otimes_\pi \mu).$$

Hence $\lambda = \mu \otimes_\pi \mu$. □

As a digression, we can now derive from the preceding result a slightly unconventional proof of the norm ergodic theorem, which has not been used in these notes so far.

Theorem 4.35 (Norm Ergodic Theorem) *If (X, μ, T) is a $\mathbb{Z}^d$-system, the boxes R_i are as above, and $f \in L^2(\mu)$, then*

$$\left\| \frac{1}{|R_i|} \sum_{\mathbf{n} \in R_i} f \circ T^{\mathbf{n}} - \mathsf{E}(f \mid \mathscr{B}_X^T) \right\|_2 \longrightarrow 0 \quad \textit{as } N \longrightarrow \infty.$$

Proof Replacing f by $f - \mathsf{E}(f \mid \mathscr{B}_X^T)$, we may assume that $\mathsf{E}(f \mid \mathscr{B}_X^T) = 0$ and show that

$$A_i := \frac{1}{|R_i|} \sum_{\mathbf{n} \in R_i} f \circ T^{\mathbf{n}} \longrightarrow 0$$

in $\| \cdot \|_2$.

Squaring and expanding this norm gives

$$\|A_i\|_2^2 = \frac{1}{|R_i|^2} \sum_{\mathbf{n}, \mathbf{m} \in R_i} \langle f \circ T^{\mathbf{n}}, f \circ T^{\mathbf{m}} \rangle = \int f \otimes f \,\mathrm{d}\lambda_i$$

with

$$\lambda_i := \frac{1}{|R_i|^2} \sum_{\mathbf{n},\mathbf{m}\in R_i} (T^{\mathbf{n}} \times T^{\mathbf{m}})_* \mu^{\Delta}.$$

By the second part of Corollary 4.34, these joinings converge to $\mu \otimes_\pi \mu$ for any factor map π generating $\mathscr{B}_X^T$, and hence

$$\|A_i\|_2^2 \longrightarrow \int f \otimes f \, \mathrm{d}(\mu \otimes_\pi \mu) = \int \mathsf{E}(f \mid \mathscr{B}_X^T) \otimes \mathsf{E}(f \mid \mathscr{B}_X^T) \, \mathrm{d}(\mu \otimes_\pi \mu) = 0.$$

□

Remark The above connection notwithstanding, we do not need the ergodic theorem at any point in this course. ◁

4.5 Weak Mixing

Our next milestone will be Roth's theorem, but before reaching it we must introduce some more general machinery.

Definition 4.36 (Weak Mixing) For a countable group G, a G-system (X, μ, T) is **weakly mixing** if the Cartesian product system $(X \times Y, \mu \otimes \nu, T \times S)$ is ergodic for any ergodic G-system (Y, ν, S).

This property was studied long before Furstenberg introduced multiple recurrence. Clearly it requires that (X, μ, T) itself be ergodic, but not all ergodic systems are weakly mixing.

Example Let $(\mathbb{T}, m, R_\alpha)$ be an ergodic circle rotation (using Corollary 4.47, this ergodicity holds if and only if α is irrational). Its Cartesian square is $(\mathbb{T}^2, m^{\otimes 2}, R_{(\alpha,\alpha)})$, in which any set of the form

$$\{(u, v) : u - v \in A\}, \quad A \in \mathscr{B}_{\mathbb{T}}$$

is invariant. ◁

However, it turns out that examples similar to the above are the *only* way in which a product of ergodic systems can fail to be ergodic. To explain this, we must first formalize that class of examples.

4.5.1 Isometric Systems and the Kronecker Factor

Definition 4.37 For any countable group G, a G-system (X, μ, T) is **concretely isometric** if the topology of X can be generated by a T-invariant compact metric d. A G-system is simply **isometric** if it is isomorphic to a concretely isometric G-system.

Given a general G-system (X, μ, T), a factor $\mathscr{S} \leq \mathscr{B}_X$ is **isometric** if it is generated modulo μ by a factor map from $\mathbf{X}$ to an isometric G-system.

Lemma 4.38 *The class of all isometric G-systems is idempotent.*

Proof Closure under isomorphisms is written into the definition. Given this, it suffices to check that the concretely isometric G-systems are closed under joinings and inverse limits, up to isomorphism.

Suppose that $\mathbf{X} = (X, \mu, T)$ and $\mathbf{Y} = (Y, \nu, S)$ are concretely isometric G-systems with invariant compact metrics d_X and d_Y. Then any $\lambda \in \mathrm{J}(\mathbf{X}, \mathbf{Y})$ is a $(T \times S)$-invariant measure on the space $X \times Y$. On this space, any of the usual compact product-space metrics is $(T \times S)$-invariant, such as

$$d((x, y), (x', y')) := d_X(x, x') + d_Y(y, y').$$

Similarly, given an inverse sequence

$$\cdots \longrightarrow \mathbf{X}_3 \longrightarrow \mathbf{X}_2 \longrightarrow \mathbf{X}_1$$

of concretely isometric systems, it is easily checked that the inverse limit constructed in the proof of Proposition 4.22 is concretely isometric, where again $\prod_{n \geq 1} X_n$ is given any standard compact product-space metric. □

Definition 4.39 The idempotent class defined above is called the **Kronecker class**[2] of G-systems, and denoted $\mathfrak{K}^G$, or just $\mathfrak{K}$ if G is understood.

Having defined the Kronecker class, Lemma 4.27 immediately gives the following.

Proposition 4.40 *Every G-system (X, μ, T) has an essentially unique isometric factor $\mathscr{S}$ with the property that any other isometric factor is contained in $\mathscr{S}$ modulo μ.* □

Definition 4.41 The factor given by Proposition 4.40 is called the **Kronecker factor** of (X, μ, T).

[2] This is not standard terminology.

4.5.2 Describing the Failure of Weak Mixing

Theorem 4.42 *Let (X, μ, T) be an ergodic G-system with Kronecker factor $\mathscr{S}$, and let (Y, ν, S) be another G-system. Suppose that $F : X \times Y \longrightarrow \mathbb{R}$ is measurable and $(T \times S)$-invariant. Then F is $(\mu \otimes \nu)$-essentially measurable with respect to $\mathscr{S} \otimes \mathscr{B}_Y$.*

In particular, $\mathbf{X}$ *is weakly mixing if and only if its Kronecker factor is trivial.*

The proof of this will require the following lemma.

Lemma 4.43 *Let (X, μ, T) be an ergodic G-system, let (Y, d) be a Polish (that is, complete and separable) metric space, and let $S = (S^g)_{g \in G}$ be an action of G on Y by isometries for the metric d. Suppose that $F : X \longrightarrow Y$ is an equivariant Borel map: thus,*

$$F \circ T^g = S^g \circ F \quad \mu\text{-a.s. } \forall g \in G.$$

Then there is an S-invariant compact subset $K \subseteq Y$ such that $\mu(F^{-1}(K)) = 1$.

Proof Let $\nu := F_*\mu$, an S-invariant Borel measure on Y. For each $r > 0$, consider the non-negative function

$$f_r(x) := \nu\big(B_{r/2}(F(x))\big).$$

One checks easily that f_r is a measurable function on X (exercise!). It is also T-invariant, because F intertwines T with S, and S preserves both the metric d and the measure ν. Therefore, by ergodicity, $f_r(x)$ is a.s. equal to some constant, say c_r.

We next show that $c_r > 0$. Consider the set

$$U_r := \{y \in Y : \ \nu(B_{r/2}(y)) = 0\}.$$

Since Y is separable, so is U_r. Letting $y_1, y_2, \dots$ be a dense sequence in U_r, it follows that

$$\nu(U_r) \leq \sum_{n \geq 1} \nu(B_{r/2}(y_n)) = 0.$$

Hence

$$\mu\{f_r = 0\} = \nu(U_r) = 0,$$

and so $c_r > 0$ for every $r > 0$.

It follows that F a.s. takes values in the set

$$K := \bigcap_{m \geq 1} \{y \in Y : \nu(B_{1/2m}(y)) \geq c_m\}.$$

This is an intersection of closed sets (exercise!), and each of them is S-invariant, since S preserves both d and ν. The proof is completed by showing that K is totally bounded: more specifically, that each of the 'packing numbers'

$$\max\{|F| : F \subseteq K, \ F \text{ is } r\text{-separated}\}$$

is finite for any $r > 0$, where we recall that F is 'r-separated' if

$$d(x, y) \geq r \text{ whenever } x, y \in F \text{ are distinct.}$$

To see this, suppose $F \subseteq K$ is r-separated, and let $m \geq 1/r$. Then the balls $B_{1/2m}(y)$ for $y \in F$ are pairwise disjoint and all have ν-measure at least c_m, so $|F|$ must be at most $1/c_m$. □

Proof of Theorem 4.42 By considering a sequence of truncations, it suffices to prove this for $F \in L^\infty(\mu \otimes \nu)$. For such F, its slices $F(x, \cdot)$ are also bounded, hence all lie in $L^1(\nu)$. Now some routine measure theory shows that the resulting map

$$\pi : X \longrightarrow L^1(\nu) : x \mapsto F(x, \cdot)$$

is measurable. Let $\widehat{S}$ be the isometric action of G on $L^1(\nu)$ given by

$$\widehat{S}^g f := f \circ S^{g^{-1}}.$$

Then the $(T \times S)$-invariance of F translates into

$$\pi(T^g x)(y) = F(T^g x, y) = F(x, S^{g^{-1}} y) = \widehat{S}^g(\pi(x))(y).$$

We may therefore apply Lemma 4.43 to find a compact, $\widehat{S}$-invariant subset $K \subseteq L^1(\nu)$ such that $\pi(x) \in K$ for a.e. x. Now π defines a factor map

$$(X, \mu, T) \longrightarrow (K, \pi_*\mu, \widehat{S})$$

with target a concretely isometric system.

Let $\mathscr{S}_1 := \pi^{-1}(\mathscr{B}_K)$, so by definition we have $\mathscr{S}_1 \leq \mathscr{S}$. The map π is μ-essentially $\mathscr{S}_1$-measurable and takes values in the compact metric space K, so we may find a sequence of $\mathscr{S}_1$-measurable maps $\pi_n : X \longrightarrow K$ which take only finitely many values and such that $\|\pi_n(x) - \pi(x)\|_1 \longrightarrow 0$ for μ-a.e. x. Since the level-sets of π_n form a finite, $\mathscr{S}_1$-measurable partition of X, the functions

$$F_n(x, y) := \pi_n(x)(y)$$

are all $(\mathscr{S}_1 \otimes \mathscr{B}_Y)$-measurable. On the other hand, Fubini's theorem and the dominated convergence theorem give

$$\|F - F_n\|_{L^1(\mu\otimes\nu)} = \int_X \|\pi(x) - \pi_n(x)\|_{L^1(\nu)}\,\mu(\mathrm{d}x) \longrightarrow 0.$$

Therefore F is essentially $(\mathscr{S}_1\otimes\mathscr{B}_Y)$-measurable, hence also essentially $(\mathscr{S}\otimes\mathscr{B}_Y)$-measurable. □

Exercise After seeing the structural results of the next subsection, generalize the example at the beginning of this section to prove the 'only if' part of Theorem 4.42. ◁

Later we will need the following slight generalization of Theorem 4.42 in the case $G = \mathbb{Z}$.

Corollary 4.44 *Let (X, μ, T) be an ergodic system with Kronecker factor $\mathscr{S}$, and let (Y, ν, S) be another system. Suppose that $F : X \times Y \longrightarrow \mathbb{R}$ is measurable and $(T^2 \times S)$-invariant. Then F is $(\mu\otimes\nu)$-essentially measurable with respect to $\mathscr{S} \otimes \mathscr{B}_Y$.*

Proof Define a new system $(\widetilde{Y}, \widetilde{\nu}, \widetilde{S})$ by setting

$$\widetilde{Y} := Y \times \{0, 1\} \quad \text{and} \quad \widetilde{\nu} := \nu \otimes \frac{\delta_0 + \delta_1}{2}.$$

Consider the transformation $\widetilde{S}$ on $(\widetilde{Y}, \widetilde{\nu})$ defined by

$$\widetilde{S}(y, 0) := (y, 1) \quad \text{and} \quad \widetilde{S}(y, 1) := (Sy, 0) \quad \forall y \in Y.$$

Observe that $\widetilde{S}^2 = S \times \mathrm{id}$. The intuition here is there we have enlarged Y in order to create a 'square root' for the transformation S.

Now consider the product system $(X \times \widetilde{Y}, \mu \otimes \widetilde{\nu}, T \times \widetilde{S})$, and on it the function defined by

$$\widetilde{F}(x, y, 0) := F(x, y) \quad \text{and} \quad \widetilde{F}(x, y, 1) := F(T^{-1}x, y).$$

On the one hand,

$$(\widetilde{F} \circ (T \times \widetilde{S}))(x, y, 0) = \widetilde{F}(Tx, y, 1) = F(T^{-1}Tx, y) = \widetilde{F}(x, y, 0),$$

and on the other, the invariance of F under $(T^2 \times S)$ gives

$$\begin{aligned}(\widetilde{F} \circ (T \times \widetilde{S}))(x, y, 1) = \widetilde{F}(Tx, Sy, 0) &= F(Tx, Sy)\\ &= F(T^{-1}x, y) = \widetilde{F}(x, y, 1).\end{aligned}$$

So $\widetilde{F}$ is $(T \times \widetilde{S})$-invariant, and thus by Theorem 4.42 it is measurable with respect to $\mathscr{S} \otimes \mathscr{B}_{\widetilde{Y}}$. Since F is just the restriction of $\widetilde{F}$ to $X \times (Y \times \{0\})$, it is measurable with respect to $\mathscr{S} \otimes \mathscr{B}_Y$. □

Exercise Generalize Corollary 4.44 by showing that any $(T^n \times S)$-invariant function for any $n \geq 1$ is essentially $(\mathscr{S} \otimes \mathscr{B}_Y)$-measurable. ◁

4.5.3 The Structure of Isometric $\mathbb{Z}^d$-Systems

Theorem 4.42 explains the importance of isometric systems. We next give a much more precise description of their structure.

Recall that a **compact metric group** is a group G endowed with a compact metric D such that

- multiplication and inversion are both continuous for the topology defined by D, and
- D is invariant under left- and right-translation and under inversion:

$$D(g,h) = D(kg,kh) = D(gk,hk) = D(g^{-1},h^{-1}) \quad \forall g,h,k \in G.$$

Examples The following are all easily equipped with compact group metrics:

1. finite groups;
2. tori;
3. other compact Lie groups, such as closed subgroups of $\mathrm{O}(d)$;
4. countable Cartesian products of other compact metric groups. ◁

We will soon need the following important fact: If G is a compact metric group, then there is a unique Borel probability measure on G (as a compact metric space) which is invariant under both left- and right-translation. It is called the **Haar probability measure**, and is denoted by m_G. This classical fact can be found, for instance, in [Fol99, Section 11.1]. From translation-invariance and the compactness of G, it follows that $m_G(U) > 0$ for any nonempty open $U \subseteq G$ (since finitely many translates of U cover G, and $m_G(G) = 1$).

Lemma 4.45 *Let (X,d) be a compact metric space, let* $\mathrm{Isom}(X,d)$ *be the group of its isometries, and endow this group with the metric*

$$D(S,T) := \sup_{x \in X} d(Sx,Tx).$$

Then this is a compact metric group.

Proof Exercise. [Hint: On route to compactness, let $\{x_k : k \geq 1\}$ be a countable dense subset of X, and prove that if $(S_n)_{n\geq 1}$ is a sequence in $\mathrm{Isom}(X, d)$ such that $(S_n x_k)_{n\geq 1}$ tends to some limit $y_k \in X$ for each k, then the map $x_k \mapsto y_k$ is the restriction to $\{x_k : k \geq 1\}$ of some limiting isometry $T : X \longrightarrow X$.] □

Proposition 4.46 *Let (X, μ, T) be an ergodic and concretely isometric $\mathbb{Z}^d$-system with invariant metric d.*

1. *Let*
$$Z := \overline{\{T^{\mathbf{n}} : \mathbf{n} \in \mathbb{Z}^d\}} \leq \mathrm{Isom}(X, d),$$
where the closure is in the metric D. Then (Z, D) is a compact metric Abelian group, and there is some $x \in X$ such that
$$\mu = \int_Z \delta_{R(x)}\, m_Z(\mathrm{d}R), \tag{4.6}$$
where m_Z is the Haar probability measure on Z.
2. *There are a compact metric Abelian group Z and an action S of $\mathbb{Z}^d$ by rotations on Z such that (X, μ, T) is isomorphic to (Z, m_Z, S).*

Proof *Part 1.* The subgroup Z is closed and therefore compact since $\mathrm{Isom}(X, d)$ is compact. It is Abelian because it contains the dense Abelian subgroup $\{T^{\mathbf{n}} : \mathbf{n} \in \mathbb{Z}^d\}$.

For any $R \in Z$, the definition of Z gives a sequence $(\mathbf{n}_i)_{i\geq 1}$ in $\mathbb{Z}^d$ such that $T^{\mathbf{n}_i} \longrightarrow R$ in the metric D. This implies that
$$R_*\mu = \lim_{i\longrightarrow\infty} T_*^{\mathbf{n}_i}\mu = \mu$$
in the weak* topology (exercise!), and hence that μ is Z-invariant.

For any $f \in C(X)$, we may average over the action of Z to define the new function
$$\Phi(f)(x) := \int_Z f(R(x))\, m_Z(\mathrm{d}R).$$
One checks easily that $\Phi(f)$ is still a continuous functions on Z. It is Z-invariant as an average over the action of Z, hence also invariant under $T^{\mathbf{n}}$ for every $\mathbf{n} \in \mathbb{Z}^d$. Therefore, by ergodicity, for every $f \in C(X)$ there is a Borel subset $X_f \subseteq X$ with $\mu(X_f) = 1$ on which $\Phi(f)$ is equal to the constant $\int \Phi(f)\, \mathrm{d}\mu$. Since μ is Z-invariant, this constant may be evaluated as follows:
$$\int_X \Big(\int_Z f(R(x'))\, m_Z(\mathrm{d}R) \Big)\, \mu(\mathrm{d}x')$$
$$\overset{\text{Fubini}}{=} \int_Z \Big(\int_X f(x')\, (R_*\mu)(\mathrm{d}x') \Big)\, m_Z(\mathrm{d}R) = \int f\, \mathrm{d}\mu.$$

Now let $f_1, f_2, \dots$ be a countable and uniformly dense sequence in $C(X)$. The resulting intersection $\bigcap_n X_{f_n}$ still has μ-measure 1, hence is nonempty; let x be an element of this intersection. Having done so, the definition of the sets X_{f_n} gives

$$\int f_n \, \mathrm{d}\mu = \Phi(f_n)(x) = \int_X f_n(R(x)) \, m_Z(\mathrm{d}R) \quad \forall n \geq 1.$$

Since $\{f_n : n \geq 1\}$ is dense in $C(X)$, this must actually hold for every $f \in C(X)$, and this now implies that

$$\mu = \int_Z \delta_{R(x)} \, m_Z(\mathrm{d}R).$$

Part 2. Now reconsider the situation of Part 1, but replace X with the compact orbit Zx for some fixed x satisfying (4.6). This is T-invariant and has full μ-measure, so it gives the same isometric $\mathbb{Z}^d$-system up to isomorphism. We may therefore simply assume $X = Zx$. We now treat Z as a subgroup of $\mathrm{Isom}(X, d)$ with this extra assumption.

However, if X is itself equal to the single orbit Zx, then the action of Z on X must be free. Indeed, if $R(x') = x'$ for some $R \in Z$ and $x' \in X$, then also $R(R'(x')) = R'(R(x')) = R'(x')$ for any $R' \in Z$, because Z is Abelian; since the Z-orbit of any point $x' \in X$ equals the whole of X, this implies that $R = \mathrm{id}_X$. Therefore the orbit map

$$\pi : Z \longrightarrow X : R \mapsto R(x)$$

is continuous and injective, and hence is a homeorphism. Equation (4.6) asserts precisely that μ is equal to $\pi_* m_Z$. Finally, letting $S^{\mathbf{n}}$ be the rotation of the group Z by its element $T^{\mathbf{n}}$ for each $\mathbf{n} \in \mathbb{Z}^d$, it follows that π defines an isomorphism

$$(Z, m_Z, S) \longrightarrow (X, \mu, T).$$

□

Corollary 4.47 (Abstract Weyl Equidistribution) *If (X, d) is a compact metric space, $T : \mathbb{Z}^d \longrightarrow \mathrm{Isom}(X, d)$ is an action by isometries, and $x \in X$, then*

$$\frac{1}{N^d} \sum_{\mathbf{n} \in [N]^d} \delta_{T^{\mathbf{n}} x} \overset{\text{weak}^*}{\longrightarrow} \int_Z \delta_{R(x)} \, m_Z(\mathrm{d}R),$$

where $Z = \overline{\{T^{\mathbf{n}} : \mathbf{n} \in \mathbb{Z}^d\}} \leq \mathbb{Z}^d$.

Exercise Prove Corollary 4.47 from Proposition 4.46. ◁

4.6 Roth's Theorem

This section proves the special case of Theorem 4.7 with $k = 3$. This is rather easier to handle than the general case, but it already exhibits some nontrivial structure. By the Furstenberg correspondence principle, it implies Roth's theorem (the case $k = 3$ of Theorem 4.3).

Example Suppose that $(\mathbb{T}, m_{\mathbb{T}}, R_\alpha)$ is an ergodic circle rotation. Let $f_1 : t \mapsto e^{4\pi i t} \in \mathbb{C}$ and $f_2 : t \mapsto e^{-2\pi i t}$. Then for any n we have

$$f_1(t + n\alpha) f_2(t + 2n\alpha) = e^{4\pi i t + 4n\pi i\alpha} e^{-2\pi i t - 4n\pi i\alpha} = e^{2\pi i t},$$

and so, even though T is ergodic, the averages

$$\frac{1}{N}\sum_{n=1}^{N}(f_1 \circ T^n)(f_2 \circ T^{2n})$$

are all equal to the non-constant function $e^{2\pi i t}$. By letting A be a set such that 1_A has nonzero Fourier coefficients corresponding to both of the characters f_1 and f_2, and using a little care, one can produce a set A such that

$$\frac{1}{N}\sum_{n=1}^{N} m_{\mathbb{T}}(A \cap T^{-n} A \cap T^{-2n} A)$$

tends to some limit other than $m_{\mathbb{T}}(A)^3$. $\lhd$

By analogy with Theorem 4.42, we might hope that the above examples represent essentially the only way in which the limit joining can fail to be the product measure. This turns out to be true for the case $k = 3$ of Theorem 4.7. The story is more complicated for $k \geq 4$ or in higher dimensions.

Proposition 4.48 *Let (X, μ, T) be an ergodic $\mathbb{Z}$-system and let $\mathscr{S} \leq \mathscr{B}_X$ be its Kronecker factor. Let λ be any 3-fold self-joining of $\mathbf{X}$ which is also invariant under $\vec{T} = T \times T^2 \times T^3$. Then λ is relatively independent over $(\mathscr{S}, \mathscr{S}, \mathscr{S})$.*

Proof Suppose that $f_i \in L^\infty(\mu)$ for $i = 1, 2, 3$.

Consider the projection $\pi_{23} : X^3 \longrightarrow X^2$ onto the second and third coordinates. The measure $\pi_{23*}\lambda$ has both marginals equal to μ and is invariant under both $T \times T$ and $T^2 \times T^3$. It is therefore also invariant under $\mathrm{id} \times T$. Since $\mathbf{X}$ is ergodic, this implies that $\pi_{23*}\lambda = \mu^{\otimes 2}$, by the ergodic case of Proposition 4.33.

By writing $X^3 = X \times X^2$, we may now regard λ as a joining of $\mathbf{X}$ and $\mathbf{X}^2$ which is invariant under both $T \times (T \times T)$ and under $T \times (T^2 \times T^3)$, hence also under $\mathrm{id} \times (T \times T^2)$. By another appeal to Proposition 4.33, with this interpretation λ is relatively independent over $(\mathscr{B}_X, \mathscr{B}_{X^2}^{T\times T^2})$.

However, now Theorem 4.42 (for T) and Corollary 4.44 (for T^2) give that $\mathscr{B}_{X^2}^{T\times T^2} \subseteq \mathscr{S} \otimes \mathscr{S}$. Suppose that $f_1, f_2, f_3 \in L^\infty(\mu)$. A simple exercise (for instance, based on the formula (4.2)) shows that

$$\mathsf{E}_{\mu\otimes\mu}(f_2 \otimes f_3 \,|\, \mathscr{S} \otimes \mathscr{S}) = \mathsf{E}_\mu(f_2 \,|\, \mathscr{S}) \otimes \mathsf{E}_\mu(f_3 \,|\, \mathscr{S}).$$

Using this, the law of iterated conditional expectation ([Bil95, Theorem 34.4]) gives

$$\begin{aligned}\mathsf{E}_{\mu\otimes\mu}(f_2 \otimes f_3 \,|\, \mathscr{B}_{X^2}^{T\times T^2}) &= \mathsf{E}_{\mu\otimes\mu}\big(\mathsf{E}_{\mu\otimes\mu}(f_2 \otimes f_3 \,|\, \mathscr{S} \otimes \mathscr{S}) \,\big|\, \mathscr{B}_{X^2}^{T\times T^2}\big)\\ &= \mathsf{E}_{\mu\otimes\mu}\big(\mathsf{E}_\mu(f_2 \,|\, \mathscr{S}) \otimes \mathsf{E}_\mu(f_3 \,|\, \mathscr{S}) \,\big|\, \mathscr{B}_{X^2}^{T\times T^2}\big),\end{aligned}$$

and so we obtain

$$\begin{aligned}\int f_1 \otimes f_2 \otimes f_3 \,\mathrm{d}\lambda &= \int f_1 \otimes \mathsf{E}_{\mu\otimes\mu}(f_2 \otimes f_3 \,|\, \mathscr{B}_{X^2}^{T\times T^2}) \,\mathrm{d}\lambda\\ &= \int f_1 \otimes \mathsf{E}_{\mu\otimes\mu}\big(\mathsf{E}_\mu(f_2 \,|\, \mathscr{S}) \otimes \mathsf{E}_\mu(f_3 \,|\, \mathscr{S}) \,\big|\, \mathscr{B}_{X^2}^{T\times T^2}\big) \,\mathrm{d}\lambda\\ &= \int f_1 \otimes \mathsf{E}_\mu(f_2 \,|\, \mathscr{S}) \otimes \mathsf{E}_\mu(f_3 \,|\, \mathscr{S}) \,\mathrm{d}\lambda.\end{aligned}$$

Alternatively, we may regard λ as a joining of $\mathbf{X}^2$ and $\mathbf{X}$ via the first two coordinates and the last coordinate, giving the analog of the above with $\mathscr{S}$ in the first and second positions. Combining these equalities, we obtain

$$\int f_1 \otimes f_2 \otimes f_3 \,\mathrm{d}\lambda = \int \mathsf{E}_\mu(f_1 \,|\, \mathscr{S}) \otimes \mathsf{E}_\mu(f_2 \,|\, \mathscr{S}) \otimes \mathsf{E}_\mu(f_3 \,|\, \mathscr{S}) \,\mathrm{d}\lambda.$$

Since the f_is were arbitrary, this is the desired relative independence. □

Given $A \in \mathscr{B}_X$, the above proposition and Lemma 4.32 imply that

$$\lambda(A^3) = \int \mathsf{E}_\mu(1_A \,|\, \mathscr{S})^{\otimes 3} \,\mathrm{d}\lambda$$

for any limit λ of the joinings λ_N. If $\mu(A) > 0$ and we let $f := \mathsf{E}_\mu(1_A \,|\, \mathscr{S})$, then this is a $[0,1]$-valued function with $\int f \,\mathrm{d}\mu > 0$. Therefore the set $B := \{f > 0\}$ also has positive measure, and the case $k = 3$ of Theorem 4.7 will be proved if we show that $\lambda(B^3) > 0$. However, B is $\mathscr{S}$-measurable, so is lifted through a factor map from some isometric system. Therefore the desired conclusion will follow from a final description of λ in the case of an isometric system.

Lemma 4.49 *Let Z be a compact Abelian group, m its Haar probability measure, and R_α an ergodic rotation. Consider the off-diagonal-averaged joinings λ_N in the case $k = 3$. These converge to the Haar probability measure of the subgroup*

$$W := \{(z, z + w, z + 2w) : z, w \in Z\} \leq Z^3.$$

Proof For any $f \in C(Z^3)$ and fixed $z \in Z$, Corollary 4.47 gives

$$\frac{1}{N}\sum_{n=1}^{N} f(z, z + n\alpha, z + 2n\alpha) \longrightarrow \int_Z f(z, z + w, z + 2w)\, m(\mathrm{d}w).$$

Integrating with respect to $m(\mathrm{d}z)$ gives

$$\frac{1}{N}\sum_{n=1}^{N}\int_Z \delta_{(z,z+n\alpha,n+2\alpha)}\, m(\mathrm{d}z) \overset{\text{weak}^*}{\longrightarrow} m_W.$$

This is the desired joining convergence, by Lemma 4.20. □

Lemma 4.50 *If (Z, m) are as above, then the rotation-action R of Z on $L^1(m)$ is continuous.*

Proof Exercise. □

Proof of Theorem 4.7 for $k = 3$ As argued above, Proposition 4.48 reduces our task to the proof for a concretely isometric system, whose structure may be described using the second part of Proposition 4.46. So let (Z, m) be a compact Abelian group with its Haar probability measure. Let R be the rotation action of Z on itself, and let $T = R_\alpha$ be some particular ergodic rotation. Let $A \in \mathscr{B}_X$ with $\delta := m(A) > 0$, and let $f := 1_A$. Let λ be the limit joining described in Lemma 4.49.

Applying Lemma 4.50 to f, let U be a neighbourhood of 0 in Z such that

$$\|f - f \circ R_{-u}\|_1,\ \|f - f \circ R_{-2u}\|_1 < \delta/3 \quad \forall u \in U.$$

Now we have

$$\lambda(A^3) = \int f \otimes f \otimes f\, \mathrm{d}\lambda = \int_Z\int_Z f(z) f(z + u) f(z + 2u)\, m(\mathrm{d}z)\, m(\mathrm{d}u)$$

$$\geq \int_U \Big(\int_Z f \cdot (f \circ R_{-u}) \cdot (f \circ R_{-2u})\, \mathrm{d}m\Big)\, m(\mathrm{d}u)$$

$$\geq \int_U \left(\int_Z f^3 \, \mathrm{d}m_Z - \delta/3 - \delta/3 \right) m(\mathrm{d}u)$$
$$= \delta m(U)/3 > 0.$$

□

4.6.1 Generalization: The Furstenberg Tower and Host–Kra–Ziegler Factors

Furstenberg's original proof of Theorem 4.7 in [Fur77] relies on a generalization of Proposition 4.48 for larger values of k. Formulating this requires a few definitions.

Definition 4.51 (Isometric Extension) Let (Y, S) be a $\mathbb{Z}$-space, let (Z, d) be a compact metric space, and let $\sigma : Y \longrightarrow \mathrm{Isom}(Z, d)$ be a measurable function, where the isometry group is given the same compact group metric as previously. Then the **concretely isometric extension** of (Y, S) with **fibre** Z and **cocycle** σ is the $\mathbb{Z}$-space $(Y \times Z, S_\sigma)$ with transformation defined by

$$S_\sigma(y, z) = (Sy, \sigma(y)(z)).$$

Now let $\mathbf{Y} = (Y, \nu, S)$ be a $\mathbb{Z}$-system. Then a **concretely isometric extension** of $\mathbf{Y}$ is a $\mathbb{Z}$-system extension $\pi : (X, \mu, T) \longrightarrow \mathbf{Y}$ such that (X, T) is a concretely isometric extension of (Y, S), π is the obvious coordinate projection, and μ is any T-invariant lift of ν.

In general, an extension $\pi : \mathbf{X} \longrightarrow \mathbf{Y}$ is **isometric** if it is isomorphic to some concretely isometric extension of $\mathbf{Y}$. To be precise, this means that there are a concretely isometric extension $\pi' : \mathbf{X}' \longrightarrow \mathbf{Y}$ and a commutative diagram

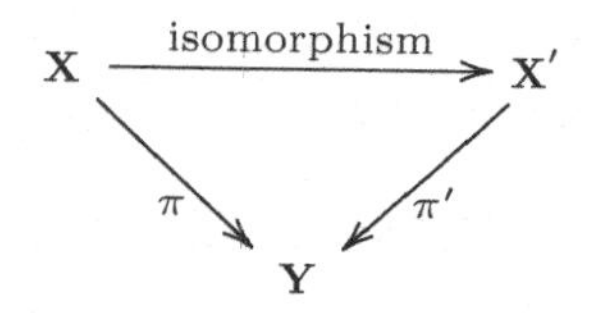

In this definition, rather than think of $Y \times Z$ as a product, one should visualize it as a 'measurable bundle' of copies of Z indexed by the base space Y. The transformation S_σ acts on the fibre $\{y\} \times Z$ by (i) moving the base-point y to Sy and (ii) acting on the fibre above that base-point by the transformation $\sigma(y)$. Clearly one can extend this definition to allow other kinds of fibre-transformation than isometries, but the definition above meets the needs of this subsection.

An isometric system is an isometric extension of the trivial one-point system. Much of our work from Section 4.5 has a generalization to extensions of systems, rather than single systems, and isometric extensions play the rôle of isometric systems in that generalization.

In the first place, an extension $\pi : \mathbf{X} \longrightarrow \mathbf{Y}$ of ergodic G-systems is called **relatively weakly mixing** if it has the following property: for any other extension $\xi : \mathbf{Z} = (Z, \theta, R) \longrightarrow \mathbf{Y}$ with $\mathbf{Z}$ ergodic, the relative product (see Subsection 4.3.4) of $\mathbf{X}$ and $\mathbf{Z}$ over the diagonal joining of two copies of $\mathbf{Y}$ is still ergodic. An important generalization of Theorem 4.42 asserts that an extension π fails to have this property if and only if it has a factorization

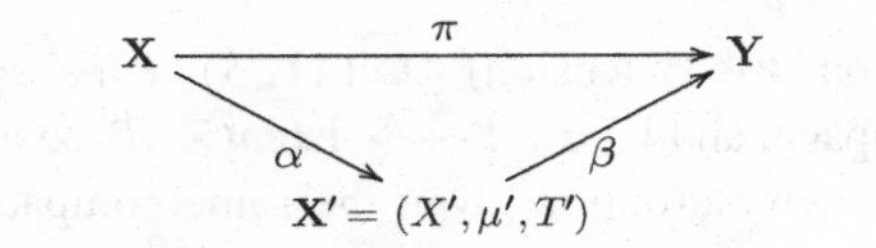

in which β is a nontrivial isometric extension. Moreover, in this case any $(T \times R)$-invariant measurable function on $X \times Z$ agrees a.e. (for the relative product measure) with a function lifted from $X' \times Z$. This result is a part of 'Furstenberg–Zimmer theory', which first appeared independently in [Fur77] and [Zim76b, Zim76a]; see also [Gla03, Chapter 9] for a textbook treatment.

Secondly, there is a description of the invariant measures on an isometric extension which generalizes Proposition 4.46. It is essentially due to Mackey [Mac66]; see also [Fur77, Theorem 8.1] or [Gla03, Theorem 3.25].

Using this theory, Furstenberg gave a generalization of Proposition 4.48 to larger values of k in terms of a generalization of Kronecker systems.

Definition 4.52 (Distal Class; Distal Systems) The k**-step distal class**, $\mathfrak{D}^k$, is the class of $\mathbb{Z}$-systems $\mathbf{X}$ which admit height-k towers of factor maps

$$\mathbf{X} = \mathbf{X}_k \xrightarrow{\pi_{k-1}} \mathbf{X}_{k-1} \xrightarrow{\pi_{k-2}} \cdots \xrightarrow{\pi_1} \mathbf{X}_1$$

in which $\mathbf{X}_1$ is an isometric system and each extension $\mathbf{X}_{i+1} \xrightarrow{\pi_i} \mathbf{X}_i$ is isometric.

A member of $\mathfrak{D}^k$ is called a k**-step distal system**.

Exercise Prove that $\mathfrak{D}^k$ is idempotent. (This is similar to the proof of Lemma 4.38, with a few extra technicalities.) ◁

The technical heart of Furstenberg's original paper [Fur77], although he does not use the language of idempotent classes, is the following.

Theorem 4.53 *If* $\mathbf{X} = (X, \mu, T)$ *is a system,*

$$\lambda_N := \frac{1}{N}\sum_{n=1}^{N}(T \times \cdots \times T^k)^n_* \mu^{\Delta(k)} \quad \text{for each } N,$$

and $\lambda_\infty \in \mathrm{J}^{(k)}(\mathbf{X})$ *is any subsequential limit of this sequence of joinings, then* λ_∞ *is relatively independent over* $(\mathfrak{D}^{k-2}_{\mathbf{X}}, \ldots, \mathfrak{D}^{k-2}_{\mathbf{X}})$. □

This neatly extends the structure given by Proposition 4.48 in case $k = 3$, since $\mathfrak{D}^1$ is the Kronecker class.

Theorem 4.53 tells one a great deal about a subsequential limit joining λ_∞, but it does not obviously imply its uniqueness, and Furstenberg does not prove that in his paper. However, Theorem 4.53 does imply that, for proving Theorem 4.7, it suffices to study the special case of $(k-2)$-step distal systems. For these, a more hands-on analysis shows multiple recurrence, even without knowing convergence of the joinings λ_N.

Following Furstenberg's work in [Fur77], he and Katznelson proved the generalization Theorem 4.8 in [FK78]. That paper adopts the same ergodic-theoretic point of view, but its use of the structure theory is different. First, they say that an extension $\pi : \mathbf{X} \longrightarrow \mathbf{Y}$ of $\mathbb{Z}^d$-systems is **primitive** if there is a group-theoretic splitting $\mathbb{Z}^d \cong \Gamma \oplus \Gamma'$, for some subgroups $\Gamma, \Gamma' \leq \mathbb{Z}^d$, such that π is isometric when regarded as an extension of Γ-systems, and relatively weakly mixing as an extension of Γ'-systems. By repeatedly applying Furstenberg–Zimmer theory to different subgroups of $\mathbb{Z}^d$, Furstenberg and Katznelson show that any $\mathbb{Z}^d$-system admits a tower of factors in which each single step is a primitive extension, possibly with different splittings of $\mathbb{Z}^d$ at every step. Importantly, one must allow this tower of factors to be 'transfinite', in that it includes taking several inverse limits during the ascent. Having shown this, they carry out a proof of Theorem 4.8 by 'transfinite induction', starting at the base of that tower and then obtaining the conclusion of multiple recurrence for every system appearing in the tower.

The success of this approach in the multidimensional setting made it popular in later accounts: it reappeared in [Fur81] and [FKO82]. However, it gives little information about the limit joinings, and so lies further from the present notes.

Much more recently than Furstenberg's work, the case of Theorem 4.31 for the powers of a single transformation was proved by Host and Kra in [HK05b] and by Ziegler in [Zie07]. Although their arguments are different, both of those works refine Furstenberg's by narrowing the class of factors needed to understand the behaviour of these limits. Once again, we sketch the story as it applies to limiting joinings. It requires some familiarity with nilpotent Lie groups, but these will not reappear later in this course.

Definition 4.54 A ***k*-step nilspace** is a $\mathbb{Z}$-space of the form

$$(G/\Gamma, R_g),$$

where

- G is a k-step nilpotent Lie group;
- $\Gamma < G$ is a co-compact lattice; and
- R_g is the left-rotation action on G/Γ of an element $g \in G$.

A ***k*-step nilsystem** is a k-step nilspace equipped with an invariant Borel probability measure.

Old results of Parry [Par69, Par70] show that any ergodic invariant probability measure on a nilspace $(G/\Gamma, R_g)$ must be the Haar measure corresponding to some closed subgroup $H \leq G$ which contains g. By restricting to the sub-nilmanifold $H/(H \cap \Gamma)$, one may then assume that the invariant measure is simply the full Haar measure and that R_g is acting ergodically. Often one restricts attention to ergodic nilsystems, and includes this choice of invariant measure in the definition. We do not because it is less convenient for the formalism of idempotent classes.

Using the basic structure theory of nilpotent Lie groups, one can show that a k-step nilsystem is always k-step distal. More specifically, it has a tower of k factor maps which starts with a rotation of a finite-dimensional compact Abelian Lie group, and then extends by an action on a bundle of tori at each higher step. See, for instance, those papers of Parry, the classic monograph [AGH63], or the introduction to [GT07].

As previously, we will also need to allow for inverse limits in our applications.

Definition 4.55 (Pro-Nil Classes; Pro-Nil Systems) Given $k \geq 1$, the ***k*-step pro-nil class** of $\mathbb{Z}$-systems is the smallest class that contains all k-step nilsystems and is closed under isomorphisms and inverse limits. It is denoted $\mathfrak{Z}^k$.

A member of $\mathfrak{Z}^k$ is called a ***k*-step pro-nil system**.

More concretely, $\mathbf{Z}$ is a k-step pro-nilsystem if it is an inverse limit of some tower of k-step nilsystems

$$\mathbf{Z} \longrightarrow \cdots \longrightarrow (G_2/\Gamma_2, \nu_2, R_{g_2}) \longrightarrow (G_1/\Gamma_1, \nu_1, R_{g_1}).$$

One can now prove that each class $\mathfrak{Z}^k$ is idempotent. The proof of this is mostly an application of the classical theory of dynamics on nilspaces, and we do not give it here. Note, however, that it is necessary to introduce inverse limits explicitly, as we have above. For instance, an ergodic rotation on the infinite-dimensional torus $\mathbb{T}^{\mathbb{N}}$ with its Haar measure is not a 1-step nilsystem, since the underlying space is not a finite-dimensional manifold; but it is a 1-step pro-nilsystem, because it is an inverse limit for the tower of coordinate-projection factor maps $\mathbb{T}^{\mathbb{N}} \longrightarrow \mathbb{T}^n$.

Since a k-step nilsystem is k-step distal, it follows that $\mathfrak{Z}^k \subseteq \mathfrak{D}^k$. In case $k = 1$ this is an equality.

Exercise Prove that $\mathfrak{Z}^1$ is the Kronecker class, using Proposition 4.46 and the fact that characters separate points on any compact metric Abelian group. $\lhd$

The main structural result of [HK05b] and [Zie07] can be translated to apply to joining averages (as in Theorem 4.30) rather than functional averages (as in Theorem 4.31). It becomes the following.

Theorem 4.56 (Derived from [HK05b], [Zie07]) *Let (X, μ, T) be an ergodic system, and let*

$$\lambda_N := \frac{1}{N}\sum_{n=1}^{N}(T \times \cdots \times T^k)^n_*\mu^{\Delta(k)} \quad \textit{for each } N.$$

Then:

i) These joinings λ_N tend to a limit $\lambda_\infty \in \mathrm{J}^{(k)}(\mathbf{X})$ as $N \longrightarrow \infty$.
ii) This λ_∞ is relatively independent over $(\mathfrak{Z}^{k-2}_{\mathbf{X}}, \mathfrak{Z}^{k-2}_{\mathbf{X}}, \ldots, \mathfrak{Z}^{k-2}_{\mathbf{X}})$.
iii) For each of the nilsystem-factors

$$\mathbf{X} \longrightarrow (G/\Gamma, \nu, R_g)$$

which generate $\mathfrak{Z}^{k-2}_{\mathbf{X}}$, the joining limit of the off-diagonal averages

$$\frac{1}{N}\sum_{n=1}^{N}(R_g \times \cdots \times R_{g^k})^n_*\nu^{\Delta(k)}$$

is the Haar probability measure on the nilmanifold $H/(H \cap \Gamma^k)$ for some intermediate nilpotent Lie group

$$\{(g, g, \ldots, g) \in G^k : g \in G\} \leq H \leq G^k.$$

Moreover, the factor $\mathfrak{Z}^{k-2}_{\mathbf{X}}$ is the smallest for which (ii) holds. □

Since $\mathfrak{Z}^k \subseteq \mathfrak{D}^k$, Theorem 4.56 refines Theorem 4.53. Actually, nilsystems are a *much* more constrained class than general distal systems, and so this is a very great refinement. This is what makes the precise 'algebraic' description of the limit joining in part (iii) possible, and it is why this structure, unlike Theorem 4.53, is also enough to prove convergence. Part (iii) follows from those older works of Parry; see also Leibman [Lei05]. In the proof of the above, the structure that appears in part (iii) is actually used in the deduction of part (i).

4.7 Towards Convergence in General

The last ambition for this course is a proof of Theorem 4.30. The general multidimensional case requires some new ideas. Perhaps surprisingly, the current proofs of convergence still leave the actual structure of the limit joining rather mysterious, in contrast with the explicit picture sketched in Section 4.6.

As before, we should like to follow the strategy outlined in Section 4.4: using the fact that any subsequential limit joining λ_∞ has the off-diagonal invariance given by Lemma 4.32, deduce enough about the structure of λ_∞ to conclude that it is unique. The extra structure that we will seek for λ_∞ is a tuple of 'special' factors $\mathscr{S}_i \leq \mathscr{B}_X$, $i = 1, 2, \ldots, d$, such that λ_∞ must be relatively independent over $(\mathscr{S}_1, \ldots, \mathscr{S}_d)$.

However, in the multidimensional setting one cannot expect these factors $\mathscr{S}_i$ to be too simple.

Example Let (X, μ, T) be a $\mathbb{Z}^d$-system, and suppose $A \in \mathscr{B}_X$ is $T^{\mathbf{e}_1 - \mathbf{e}_2}$-invariant. Then

$$\lambda_N(A \times A \times X^{d-2}) = \frac{1}{N} \sum_{n=1}^{N} \mu(T^{-n\mathbf{e}_1} A \cap T^{-n\mathbf{e}_2} A) = \mu(A),$$

so any subsequential limit also gives $\lambda_\infty(A \times A \times X^{d-2}) = \mu(A)$. Therefore, if λ_∞ is relatively independent over $(\mathscr{S}_1, \ldots, \mathscr{S}_d)$, then the Cauchy–Bunyakowski–Schwarz inequality gives

$$\|1_A\|_2^2 = \mu(A) = \int \mathsf{E}(1_A \mid \mathscr{S}_1) \otimes \mathsf{E}(1_A \mid \mathscr{S}_2) \otimes 1_X \otimes \cdots \otimes 1_X \, \mathrm{d}\lambda_\infty$$
$$\leq \|\mathsf{E}(1_A \mid \mathscr{S}_1)\|_2 \|\mathsf{E}(1_A \mid \mathscr{S}_2)\|_2.$$

This is possible only if $1_A = \mathsf{E}(1_A \mid \mathscr{S}_1) = \mathsf{E}(1_A \mid \mathscr{S}_2)$, and hence if A actually lies in both $\mathscr{S}_1$ and $\mathscr{S}_2$ modulo μ.

Generalizing this argument, one finds that any such tuple of factors $(\mathscr{S}_1, \ldots, \mathscr{S}_d)$ must satisfy

$$\mathscr{S}_i \supseteq \bigvee_{j \in \{1,\ldots,d\} \setminus \{i\}} \mathscr{B}_X^{T^{\mathbf{e}_i - \mathbf{e}_j}}. \tag{4.7}$$

◁

This example seems intimidating: each of the factors $\mathscr{B}_X^{T^{\mathbf{e}_i - \mathbf{e}_j}}$ appearing in this lower bound for $\mathscr{S}_i$ could still involve a completely arbitrary action of the other transformations $T^{\mathbf{e}_k}$, $k \neq i, j$. Note that one cannot use an ergodic decomposition to assume that the individual transformations $T^{\mathbf{e}_i - \mathbf{e}_j}$ are ergodic, since the disintegrands μ_y in the ergodic decomposition for this transformation need not be invariant under the other $T^{\mathbf{e}_k}$s. Ergodic decomposition allows one to assume only that the $\mathbb{Z}^d$-action T is ergodic as a whole.

However, with a little more thought, one sees that the factors on the right-hand side of (4.7) do not pose a great problem for proving convergence by themselves. This will be explained shortly, but first let us give a name to systems with the property that we may take $\mathscr{S}_i$ *equal* to the corresponding right-hand side in (4.7).

Definition 4.57 (Pleasant System) Let

$$\mathfrak{C} = \mathfrak{C}^{\mathbf{e}_1 - \mathbf{e}_2} \vee \cdots \vee \mathfrak{C}^{\mathbf{e}_1 - \mathbf{e}_d},$$

the example idempotent class considered following Lemma 4.25.

A $\mathbb{Z}^d$-system is **pleasant** if every self-joining $\lambda \in \mathrm{J}^{(d)}(\mathbf{X})$ which is off-diagonal-invariant is relatively independent over $(\mathfrak{C}_{\mathbf{X}}, \mathscr{B}_X, \ldots, \mathscr{B}_X)$.

Not all systems are pleasant.

Example Let $d \geq 3$, and let $(G/\Gamma, \nu, R_g)$ be a nontrivial $(d-2)$-step nilsystem for which R_{g^j} is ergodic for all $j \neq 0$ (such examples are plentiful among nilsystems). Define a $\mathbb{Z}^d$-system on $(X, \mu) := (G/\Gamma, \nu)$ by setting $T^{\mathbf{e}_j} := R_{g^j}$ for $j = 1, 2, \ldots, d$. Then each of the factors $\mathscr{B}_X^{T^{\mathbf{e}_i - \mathbf{e}_j}}$ for $i \neq j$ is trivial, because $T^{\mathbf{e}_i - \mathbf{e}_j} = R_{g^{i-j}}$ is ergodic. However, the last part of Theorem 4.56 shows that the relevant limit joinings for these systems are not product measures, and a slightly more careful argument shows that the first coordinate alone is not independent from the others under this limit joining. Therefore this system (X, μ, T) is not pleasant. $\lhd$

The previous example is obtained by using different powers of the single transformation R_g as the generators of a $\mathbb{Z}^d$-action. This feels a bit like a pretence, but less artificial examples are also available. In fact, characteristic factors for general $\mathbb{Z}^d$-systems can be described in detail when $d = 3$, and that description includes other examples for which one cannot have equality in (4.7): see [Aus10b, Subsection 7.4]. For $\mathbb{Z}^d$-actions with $d \geq 4$ no corresponding general analysis is known.

Nevertheless, pleasant systems are plentiful enough that they can be used to prove convergence in general. This is a consequence of the following.

Theorem 4.58 (Pleasant Extensions Theorem) *Every $\mathbb{Z}^d$-system has an extension which is pleasant.*

Proof of Theorem 4.30 Given Pleasant Extensions Let $\lambda_N \in \mathrm{J}^{(d)}(\mathbf{X})$ be the off-diagonal averages of the diagonal joining. We will prove that $(\lambda_N)_N$ converges to some limit joining by induction on d.

When $d = 1$, $\lambda_N = \mu$ for all N, so the result is trivial.

Now suppose it is known for $\mathbb{Z}^{d-1}$-systems, and let $\mathbf{X}$ be a $\mathbb{Z}^d$-system. Let $\pi : \widetilde{\mathbf{X}} \longrightarrow \mathbf{X}$ be a pleasant extension, as provided by Theorem 4.58, and let $\widetilde{\lambda}_N$ be the corresponding sequence of self-joinings for $\widetilde{\mathbf{X}}$. Then a simple calculation gives $\lambda_N = (\pi \times \cdots \times \pi)_* \widetilde{\lambda}_N$, and the pushforward map $(\pi \times \cdots \times \pi)_*$ is continuous for the joining topology (Lemma 4.19), so it suffices to prove that $\widetilde{\lambda}_N$ converges. Equivalently, replacing $\mathbf{X}$ with $\widetilde{\mathbf{X}}$, we may simply assume that $\mathbf{X}$ itself is pleasant.

By compactness of the joining topology, the sequence $(\lambda_N)_N$ has subsequential limits, and convergence will follow if we prove that they are all equal. By the pleasantness of $\mathbf{X}$, any subsequential limit joining is relatively independent over $(\mathfrak{C}_{\mathbf{X}}, \mathscr{B}_X, \ldots, \mathscr{B}_X)$. It therefore suffices to prove that the sequence

$$\int f_1 \otimes \cdots \otimes f_d \, \mathrm{d}\lambda_N = \frac{1}{N} \sum_{n=1}^{N} \int (f_1 \circ T^{n\mathbf{e}_1}) \cdots (f_d \circ T^{n\mathbf{e}_d}) \, \mathrm{d}\mu$$

converges to a single limit whenever $f_1, f_2, \ldots, f_d \in L^\infty(\mu)$ and f_1 is $\mathfrak{C}_{\mathbf{X}}$-measurable. Fix $f_2, \ldots, f_d$. By multilinearity and a simple approximation argument, it suffices to prove this convergence for all f_1 drawn from a subset of such $\mathfrak{C}_{\mathbf{X}}$-measurable functions whose linear span is dense in the norm $\|\cdot\|_2$.

By the definition of $\mathfrak{C}$ and some routine measure theory (exercise!), such a subset is provided by the product functions of the form $g_2 \cdot g_3 \cdots g_d$, where each $g_j \in L^\infty(\mu)$ is $T^{\mathbf{e}_1 - \mathbf{e}_j}$-invariant. Substituting such a product into our averages now gives the sequence

$$\frac{1}{N}\sum_{n=1}^{N} \int ((g_2 \dots g_d) \circ T^{n\mathbf{e}_1})(f_2 \circ T^{n\mathbf{e}_2}) \cdots (f_d \circ T^{n\mathbf{e}_d})\, \mathrm{d}\mu.$$

Using the fact that $g_j \circ T^{n\mathbf{e}_1} = g_j \circ T^{n\mathbf{e}_j}$ for each j and n, this can be rearranged to

$$\frac{1}{N}\sum_{n=1}^{N} \int ((g_2 \cdot f_2) \circ T^{n\mathbf{e}_2}) \cdots ((g_d \cdot f_d) \circ T^{n\mathbf{e}_d})\, \mathrm{d}\mu.$$

Finally, this an analogous sequence of joining averages for the $\mathbb{Z}^{d-1}$-system generated by $T^{\mathbf{e}_2}, \dots, T^{\mathbf{e}_d}$. Our inductive hypothesis therefore gives convergence for all choices of f_j and g_j, $2 \le j \le d$, so the proof is complete. □

Exercise

1. With $\mathfrak{C}$ the class in Definition 4.57, prove the following: for any $\mathbb{Z}^d$-system (X, μ, T) with ergodic decomposition

$$\mu = \int_Y \mu_y \, \nu(\mathrm{d}y),$$

if $\mathscr{S}$ is a choice of maximal $\mathfrak{C}$-factor for (X, μ, T), then it is also a choice of maximal $\mathfrak{C}$-factor for (X, μ_y, T) for ν-a.e. y. [Hints: First prove this for each class $\mathfrak{C}^{\mathbf{e}_1 - \mathbf{e}_j}$ separately; adapt the proof of Proposition 4.13.]
2. Prove that a general $\mathbb{Z}^d$-system is pleasant if and only if almost all the measures in its ergodic decomposition define pleasant systems.
3. Starting from Theorem 4.58, prove that every ergodic $\mathbb{Z}^d$-system has an extension which is pleasant and ergodic.
4. Now let (X, μ, T) be a $\mathbb{Z}$-system with the property that T^n is ergodic for every $n \neq 0$ (such as system is called **totally ergodic**), and define a $\mathbb{Z}^d$-system $\widetilde{\mathbf{X}} := (X, \mu, \widetilde{T})$ on the same probability space by setting $\widetilde{T}^{\mathbf{e}_j} := T^j$ for $j = 1, 2, \dots, d$. Show that in this case

$$\mathfrak{C}_{\widetilde{\mathbf{X}}} = \mathscr{B}_X^T,$$

so this is trivial if T is ergodic. In light of this, why is there no contradiction between part 3. of this exercise and the example preceding Theorem 4.58? ◁

Most of the remaining work will go into proving the Pleasant Extensions Theorem. One can also give a proof of Theorem 4.8 using that theorem, but this is rather more work, and we leave it aside: see [Aus10a] or [Aus10c].

4.8 Sated Systems and Pleasant Extensions

4.8.1 Satedness

We now make a brief return to the abstract study of idempotent classes. Let $\mathfrak{C}$ be an idempotent class of G-systems for some countable group G.

Definition 4.59 (Adjoining) A **$\mathfrak{C}$-adjoining** of $\mathbf{X} = (X, \mu, T)$ is an extension of $\mathbf{X}$ of the form

$$(X \times Y, \lambda, T \times S) \xrightarrow{\pi} \mathbf{X},$$

where

- $\mathbf{Y} = (Y, \nu, S) \in \mathfrak{C}$;
- $\lambda \in \mathrm{J}(\mathbf{X}, \mathbf{Y})$;
- and π is the first coordinate projection,

or any extension isomorphic to such a π.

Lemma 4.60 *A $\mathfrak{C}$-adjoining of a $\mathfrak{C}$-adjoining is a $\mathfrak{C}$-adjoining.*

Proof Given $\mathbf{X}$, any $\mathfrak{C}$-adjoining of a $\mathfrak{C}$-adjoining of $\mathbf{X}$ may be written in the form

$$(X \times Y_1 \times Y_2, \lambda, T \times S_1 \times S_2) \xrightarrow{\pi} \mathbf{X},$$

where $\mathbf{Y}_i = (Y_i, \nu_i, S_i) \in \mathfrak{C}$ for $i = 1, 2$, $\lambda \in \mathrm{J}(\mathbf{X}, \mathbf{Y}_1, \mathbf{Y}_2)$, and π is the first coordinate projection.

Letting λ_{12} be the projection of λ onto the coordinates in $Y_1 \times Y_2$, it follows that $\mathbf{Y} := (Y_1 \times Y_2, \lambda_{12}, S_1 \times S_2) \in \mathfrak{C}$, because $\mathfrak{C}$ is closed under joinings. Since the above triple-joining may be written as a joining of $\mathbf{X}$ and this $\mathbf{Y}$, it is a $\mathfrak{C}$-adjoining of $\mathbf{X}$. □

Lemma 4.61 *Let*

$$\cdots \xrightarrow{\pi_2} \mathbf{X}_2 \xrightarrow{\pi_1} \mathbf{X}_1 \xrightarrow{\pi_0} \mathbf{X}_0 = \mathbf{X}$$

be a tower in which each π_i is a $\mathfrak{C}$-adjoining of $\mathbf{X}_i$. Then its inverse limit is a $\mathfrak{C}$-adjoining of $\mathbf{X}$.

Proof By a simple induction on i, there are systems $\mathbf{Y}_i = (Y_i, \nu_i, S_i) \in \mathfrak{C}$ and joinings $\lambda_i \in \mathrm{J}(\mathbf{X}, \mathbf{Y}_1, \ldots, , \mathbf{Y}_i)$ for $i \geq 1$ such that

$$\mathbf{X}_i = (X \times Y_1 \times \cdots \times Y_i, \lambda_i, T \times S_1 \times \cdots \times S_i)$$

for each i, and π_{i-1} is the coordinate projection that omits the factor Y_i.

Given this, the proof of Proposition 4.22 constructs an inverse limit for the original sequence of the form

$$\Big(X \times \prod_{i\geq 1} Y_i, \lambda, T \times \prod_{i\geq 1} S_i\Big),$$

where $\lambda \in \mathrm{J}(\mathbf{X}, \mathbf{Y}_1, \mathbf{Y}_2, \ldots)$. Projecting out the coordinate-copy of $\mathbf{X}$ gives some $\lambda' \in \mathrm{J}(\mathbf{Y}_1, \mathbf{Y}_2, \ldots)$, and the corresponding system

$$\mathbf{Y} := \Big(\prod_{i\geq 1} Y_i, \lambda', \prod_{i\geq 1} S_i\Big).$$

This is an inverse limit of finite joinings among the systems $\mathbf{Y}_i$, so it defines an inverse limit of joinings of members of $\mathfrak{C}$, hence is itself a member of $\mathfrak{C}$.

Therefore our original inverse limit is a joining of $\mathbf{X}$ and $\mathbf{Y} \in \mathfrak{C}$, so is a $\mathfrak{C}$-adjoining of $\mathbf{X}$. □

With the above preliminaries in hand, the key new definition is the following.

Definition 4.62 (Sated System) A G-system $\mathbf{X}$ is **$\mathfrak{C}$-sated** if the following holds: if $\mathbf{Y} \in \mathfrak{C}$, then any $\lambda \in \mathrm{J}(\mathbf{X}, \mathbf{Y})$ is relatively independent over $(\mathfrak{C}_{\mathbf{X}}, \mathscr{B}_Y)$: that is,

$$\int_{X\times Y} f \otimes g \, \mathrm{d}\lambda = \int_{X\times Y} \mathsf{E}(f \mid \mathfrak{C}_{\mathbf{X}}) \otimes g \, \mathrm{d}\lambda \tag{4.8}$$

for all $f \in L^\infty(\mu)$ and $g \in L^\infty(\nu)$.

The intuition here is that a $\mathfrak{C}$-sated system admits only very restricted joinings with members of the class $\mathfrak{C}$: they must be relatively independent over some member of $\mathfrak{C}$ that is already a factor of $\mathbf{X}$. The following alternative characterization will be useful shortly.

Proposition 4.63 *For a G-system $\mathbf{X}$, the following are equivalent:*

i) $\mathbf{X}$ is $\mathfrak{C}$-sated;
ii) for every extension $\pi : \widetilde{\mathbf{X}} \longrightarrow \mathbf{X}$ and every $f \in L^\infty(\mu)$, one has

$$\mathsf{E}_{\tilde{\mu}}(f \circ \pi \mid \mathfrak{C}_{\widetilde{\mathbf{X}}}) = \mathsf{E}_\mu(f \mid \mathfrak{C}_{\mathbf{X}}) \circ \pi; \tag{4.9}$$

Proof *(i)* $\Longrightarrow$ *(ii).* Let $\pi : \widetilde{\mathbf{X}} \longrightarrow \mathbf{X}$ be an extension, and let $\xi : \widetilde{\mathbf{X}} \longrightarrow \mathbf{Y}$ be another factor map which generates the factor $\mathfrak{C}_{\widetilde{\mathbf{X}}}$ modulo $\widetilde{\mu}$. Let $\lambda := (\pi, \xi)_* \widetilde{\mu} \in \mathrm{J}(\mathbf{X}, \mathbf{Y})$ be their joint distribution in $\widetilde{\mathbf{X}}$ (see Subsection 4.3.3). Applying Definition 4.62 to this joining gives

$$\int (f \circ \pi)(g \circ \xi)\, \mathrm{d}\widetilde{\mu} = \int f \otimes g \, \mathrm{d}\lambda \quad \text{(by the definition of } \lambda)$$

$$= \int \mathsf{E}(f \mid \mathfrak{C}_{\mathbf{X}}) \otimes g \, \mathrm{d}\lambda = \int (\mathsf{E}(f \mid \mathfrak{C}_{\mathbf{X}}) \circ \pi)(g \circ \xi)\, \mathrm{d}\widetilde{\mu}$$

for all $f \in L^\infty(\mu)$ and $g \in L^\infty(\nu)$. Since the functions of the form $g \circ \xi$ for $g \in L^\infty(\nu)$ are precisely the $\mathfrak{C}_{\widetilde{\mathbf{X}}}$-measurable members of $L^\infty(\widetilde{\mu})$, this equality and the definition of conditional expectation prove (4.9).

(ii) $\Longrightarrow$ *(i).* Suppose that $\mathbf{Y} = (Y, \nu, S) \in \mathfrak{C}$ and $\lambda \in \mathrm{J}(\mathbf{X}, \mathbf{Y})$, and let $\pi : \widetilde{\mathbf{X}} \longrightarrow \mathbf{X}$ be the corresponding $\mathfrak{C}$-adjoining. Let $f \in L^\infty(\mu)$ and $g \in L^\infty(\nu)$, and let $\xi : \widetilde{X} = X \times Y \longrightarrow Y$ be the second coordinate projection. Then $\mathfrak{C}_{\widetilde{\mathbf{X}}} \supseteq \xi^{-1}(\mathfrak{C}_{\mathbf{Y}})$ modulo $\widetilde{\mu}$, by the maximality of $\mathfrak{C}_{\widetilde{\mathbf{X}}}$ among $\mathfrak{C}$-factors of $\widetilde{\mathbf{X}}$. Therefore $g \circ \xi$ is $\mathfrak{C}_{\widetilde{\mathbf{X}}}$-measurable, and so applying (4.9) gives

$$\int f \otimes g \, \mathrm{d}\lambda = \int (f \circ \pi)(g \circ \xi)\, \mathrm{d}\widetilde{\mu} = \int \mathsf{E}_{\widetilde{\mu}}(f \circ \pi \mid \mathfrak{C}_{\widetilde{\mathbf{X}}})(g \circ \xi)\, \mathrm{d}\widetilde{\mu}$$

$$= \int (\mathsf{E}_\mu(f \mid \mathfrak{C}_{\mathbf{X}}) \circ \pi)(g \circ \xi)\, \mathrm{d}\widetilde{\mu} = \int \mathsf{E}_\mu(f \mid \mathfrak{C}_{\mathbf{X}}) \otimes g \, \mathrm{d}\lambda.$$

□

Sometimes condition (ii) of the preceding proposition is easier to verify than condition (i), often because of the following useful equivalence.

Lemma 4.64 *If* $\pi : \mathbf{X} = (X, \mu, T) \longrightarrow \mathbf{Y} = (Y, \nu, S)$ *is an extension of* G*-systems and* $f \in L^2(\nu)$*, then the following are equivalent:*

i) $\mathsf{E}_\mu(f \circ \pi \mid \mathfrak{C}_{\mathbf{X}}) = \mathsf{E}_\nu(f \mid \mathfrak{C}_{\mathbf{Y}}) \circ \pi$;
ii) $\|\mathsf{E}_\mu(f \circ \pi \mid \mathfrak{C}_{\mathbf{X}})\|_{L^2(\mu)} = \|\mathsf{E}_\nu(f \mid \mathfrak{C}_{\mathbf{Y}})\|_{L^2(\nu)}$;
iii) $\|\mathsf{E}_\mu(f \circ \pi \mid \mathfrak{C}_{\mathbf{X}})\|_{L^2(\mu)} \leq \|\mathsf{E}_\nu(f \mid \mathfrak{C}_{\mathbf{Y}})\|_{L^2(\nu)}$.

Proof *(i)* $\Longrightarrow$ *(ii)* This is immediate, since $\|F \circ \pi\|_{L^2(\mu)} = \|F\|_{L^2(\nu)}$ for any $F \in L^2(\nu)$.

(ii) $\Longrightarrow$ *(iii)* Trivial.

(iii) $\Longrightarrow$ *(i)* The definition of conditional expectation gives

$$\mathsf{E}_\nu(f \mid \mathfrak{C}_{\mathbf{Y}}) \circ \pi = \mathsf{E}_\mu(f \circ \pi \mid \pi^{-1}(\mathfrak{C}_{\mathbf{Y}})),$$

so condition (iii) is equivalent to

$$\|\mathsf{E}_\mu(f \circ \pi \mid \mathfrak{C}_{\mathbf{X}})\|_{L^2(\mu)} \leq \|\mathsf{E}_\mu(f \circ \pi \mid \pi^{-1}(\mathfrak{C}_{\mathbf{Y}}))\|_{L^2(\mu)}. \tag{4.10}$$

Clearly $\pi^{-1}(\mathfrak{C}_{\mathbf{Y}})$ is a $\mathfrak{C}$-factor of $\mathbf{X}$; since $\mathfrak{C}_{\mathbf{X}}$ is the maximal $\mathfrak{C}$-factor, it follows that $\mathfrak{C}_{\mathbf{X}} \supseteq \pi^{-1}(\mathfrak{C}_{\mathbf{Y}})$ modulo μ. Therefore the law of iterated conditional expectation gives

$$\mathsf{E}_\mu(f \circ \pi \mid \pi^{-1}(\mathfrak{C}_{\mathbf{Y}})) = \mathsf{E}_\mu\big(\mathsf{E}_\mu(f \circ \pi \mid \mathfrak{C}_{\mathbf{X}}) \bigm| \pi^{-1}(\mathfrak{C}_{\mathbf{Y}})\big). \tag{4.11}$$

Now, the conditional expectation operator $\mathsf{E}(\cdot \mid \pi^{-1}(\mathfrak{C}_{\mathbf{Y}}))$ is an orthogonal projection on $L^2(\mu)$ (recall the discussion following Theorem 4.12). Combining (4.11) and (4.10), we see that this orthogonal projection does not reduce the norm of $\mathsf{E}_\mu(f \circ \pi \mid \mathfrak{C}_{\mathbf{X}})$, and this is possible only if it actually leaves that function fixed: that is, only if

$$\mathsf{E}_\mu(f \circ \pi \mid \pi^{-1}(\mathfrak{C}_{\mathbf{Y}})) = \mathsf{E}_\mu(f \circ \pi \mid \mathfrak{C}_{\mathbf{X}}).$$

□

Theorem 4.65 (Sated Extensions Theorem) *Let $\mathfrak{C}$ be an idempotent class of G-systems. Every G-system $\mathbf{X}$ has a $\mathfrak{C}$-sated extension.*

Proof *Step 1.* Let S be a countable subset of the unit ball of $L^\infty(\mu)$ which is dense for the norm $\|\cdot\|_2$: this is possible because this ball is contained in $L^2(\mu)$, which is separable. Now let $(f_r)_{r\geq 1}$ be a sequence in S with the property that every $f \in S$ is equal to f_r for infinitely many values of r.

Having chosen these, we construct a tower

$$\cdots \xrightarrow{\pi_2} \mathbf{X}_2 \xrightarrow{\pi_1} \mathbf{X}_1 \xrightarrow{\pi_0} \mathbf{X}_0 = \mathbf{X}$$

of $\mathfrak{C}$-adjoinings by the following recursion. Assuming $\mathbf{X}_i$ has already been constructed for some $i \geq 0$, let $\pi_0^i := \pi_0 \circ \cdots \circ \pi_{i-1} : \mathbf{X}_i \longrightarrow \mathbf{X}$, and consider the real number

$$\alpha_i := \sup\big\{\|\mathsf{E}_\nu(f_i \circ \pi_0^i \circ \xi \mid \mathfrak{C}_{\mathbf{Y}})\|_{L^2(\nu)} : \\ \mathbf{Y} = (Y, \nu, S) \xrightarrow{\xi} \mathbf{X}_i \text{ is a } \mathfrak{C}\text{-adjoining of } \mathbf{X}_i\big\}.$$

Let $\pi_i : \mathbf{X}_{i+1} \longrightarrow \mathbf{X}_i$ be some choice of $\mathfrak{C}$-adjoining with the property that

$$\|\mathsf{E}_{\mu_{i+1}}(f_i \circ \pi_0^i \circ \pi_i \mid \mathfrak{C}_{\mathbf{X}_{i+1}})\|_{L^2(\mu_{i+1})} \geq \alpha_i - 2^{-i}.$$

This continues the recursion.

Let $\widetilde{\mathbf{X}}$ be an inverse limit of this tower, and let $\widetilde{\pi} : \widetilde{\mathbf{X}} \longrightarrow \mathbf{X}$ be the resulting factor map. By Lemma 4.61, this is still a $\mathfrak{C}$-adjoining of $\mathbf{X}$, so we may identify it with

$$(X \times Y, \lambda, T \times S) \overset{\text{coord. proj.}}{\longrightarrow} \mathbf{X}$$

for some $\mathbf{Y} \in \mathfrak{C}$ and some $\lambda \in \mathrm{J}(\mathbf{X}, \mathbf{Y})$.

Step 2. We will prove that $\widetilde{\mathbf{X}}$ satisfies condition (ii) in Proposition 4.63. Thus, suppose that $\psi : \mathbf{Z} = (Z, \rho, R) \longrightarrow \widetilde{\mathbf{X}}$ is a further extension. We must show that

$$\mathsf{E}_\rho(F \circ \psi \mid \mathfrak{C}_\mathbf{Z}) = \mathsf{E}_{\widetilde{\mu}}(F \mid \mathfrak{C}_{\widetilde{\mathbf{X}}}) \circ \psi$$

for all $F \in L^\infty(\widetilde{\mu})$.

Both sides of the desired equation are linear in F and also continuous in F for the norm $\|\cdot\|_{L^1(\widetilde{\mu})}$. It therefore suffices to prove the equation for F drawn from a subset of $L^\infty(\widetilde{\mu})$ whose linear span is dense in that space for the norm $\|\cdot\|_{L^1(\widetilde{\mu})}$. In particular, it suffices to do so for functions of the form $f \otimes h$ for some $f \in S$ and $h \in L^\infty(\nu)$.

However, if

$$F = f \otimes h = (f \otimes 1_Y) \cdot (1_X \otimes h)$$

for some $f \in S$ and $g \in L^\infty(\nu)$, then the second factor here is lifted from the system $\mathbf{Y} \in \mathfrak{C}$, hence must be $\mathfrak{C}_{\widetilde{\mathbf{X}}}$-measurable. Therefore $(1_X \otimes h) \circ \psi$ is $\mathfrak{C}_\mathbf{Z}$-measurable. Using this, the desired equality becomes

$$\mathsf{E}_\rho((f \otimes 1_Y) \circ \psi \mid \mathfrak{C}_\mathbf{Z}) \cdot ((1_X \otimes h) \circ \psi) = (\mathsf{E}_{\widetilde{\mu}}(f \otimes 1_Y \mid \mathfrak{C}_{\widetilde{\mathbf{X}}}) \circ \psi) \cdot ((1_X \otimes h) \circ \psi).$$

This will now follow if we prove it without the presence of h.

To do this, observe that for any i, the definition of α_i and the way we chose $\mathbf{X}_{i+1}$ give that

$$\begin{aligned}\|\mathsf{E}_{\widetilde{\mu}}(f_i \otimes 1_Y \mid \mathfrak{C}_{\widetilde{\mathbf{X}}})\|_{L^2(\widetilde{\mu})} &\geq \|\mathsf{E}_{\mu_{i+1}}(f_i \circ \pi_0^{i+1} \mid \mathfrak{C}_{\mathbf{X}_{i+1}})\|_{L^2(\mu_{i+1})} \\ &\geq \alpha_i - 2^{-i} \geq \|\mathsf{E}_\rho((f_i \otimes 1_Y) \circ \psi \mid \mathfrak{C}_\mathbf{Z})\|_{L^2(\rho)} - 2^{-i}.\end{aligned}$$

The last inequality here holds because $\mathbf{Z}$ is a $\mathfrak{C}$-adjoining of each $\mathbf{X}_i$, by Lemmas 4.60 and 4.61, and hence falls within the scope of the supremum that defined α_i. Since $f_i = f$ for infinitely many i, it follows that

$$\|\mathsf{E}_{\widetilde{\mu}}(f \otimes 1_Y \mid \mathfrak{C}_{\widetilde{\mathbf{X}}})\|_{L^2(\widetilde{\mu})} \geq \|\mathsf{E}_\rho((f \otimes 1_Y) \circ \psi \mid \mathfrak{C}_\mathbf{Z})\|_{L^2(\rho)}.$$

This verifies condition (iii) of Lemma 4.64, and hence proves the desired equality. □

4.8.2 Pleasant Extensions from Satedness

Theorem 4.66 *If $\mathfrak{C}$ is as in Theorem 4.58, then a $\mathfrak{C}$-sated system is pleasant.*

Proof Suppose $\mathbf{X}$ is $\mathfrak{C}$-sated, and let $\lambda \in \mathrm{J}^{(d)}(\mathbf{X})$ be an off-diagonal-invariant self-joining. Let $Y := X^{d-1}$, and let $\nu \in \Pr Y$ be the marginal of λ on the coordinates indexed by $2, 3, \ldots, d$. We may consider λ as a coupling of μ and ν on $X \times X^{d-1}$.

Now define a $\mathbb{Z}^d$-action S on (Y, ν) as follows:

$$S^{\mathbf{e}_i} := \begin{cases} T^{\mathbf{e}_2} \times \cdots \times T^{\mathbf{e}_d} & \text{if } i = 1 \\ T^{\mathbf{e}_i} \times \cdots \times T^{\mathbf{e}_i} & \text{if } i = 2, 3, \ldots, d. \end{cases}$$

Crucially, this is *not* the diagonal action: the first generator, $S^{\mathbf{e}_1}$, is part of the *off*-diagonal transformation.

This gives a well-defined $\mathbb{Z}^d$-system $\mathbf{Y} = (Y, \nu, S)$ owing to the off-diagonal-invariance of λ, which implies that ν is $S^{\mathbf{e}_1}$-invariant. Moreover, λ is invariant under both

$$T^{\mathbf{e}_i} \times S^{\mathbf{e}_i} \quad \text{for each } i = 2, 3, \ldots, d$$

and

$$T^{\mathbf{e}_1} \times S^{\mathbf{e}_1} = T^{\mathbf{e}_1} \times T^{\mathbf{e}_2} \times \cdots \times T^{\mathbf{e}_d},$$

so $\lambda \in \mathrm{J}(\mathbf{X}, \mathbf{Y})$.

Finally, observe that $\mathbf{Y}$ is itself a joining of the $\mathbb{Z}^d$-systems $\mathbf{Y}_j := (X, \mu, S_j)$, $j = 2, 3, \ldots, d$, where

$$S_j^{\mathbf{e}_1} := T^{\mathbf{e}_j} \quad \text{and} \quad S_j^{\mathbf{e}_i} := T^{\mathbf{e}_i} \text{ for } i = 2, 3, \ldots, d.$$

These systems satisfy $\mathbf{Y}_j \in \mathfrak{C}^{\mathbf{e}_1 - \mathbf{e}_j}$, and so $\mathbf{Y} \in \mathfrak{C}$. Therefore λ defines a $\mathfrak{C}$-adjoining of $\mathbf{X}$, and so for any functions $f_1, f_2, \ldots, f_d \in L^\infty(\mu)$ the $\mathfrak{C}$-satedness of $\mathbf{X}$ gives

$$\int_{X^d} f_1 \otimes (f_2 \otimes \cdots \otimes f_d)\,\mathrm{d}\lambda = \int_{X^d} \mathsf{E}_\mu(f_1 \,|\, \mathfrak{C}_{\mathbf{X}}) \otimes (f_2 \otimes \cdots \otimes f_d)\,\mathrm{d}\lambda.$$

This is equivalent to pleasantness. □

Proof of Theorem 4.58 This follows immediately from the conjunction of Theorems 4.65 and 4.66. □

Exercise This exercise will deduce Theorem 4.31 from an analysis of self-joinings, much as Theorem 4.35 was deduced from Corollary 4.34. Let $\mathbf{X} = (X, \mu, T)$ be a $\mathbb{Z}^d$-system, let $f_1, \ldots, f_d \in L^\infty(\mu)$, and let

$$A_N(f_1, \ldots, f_d) := \frac{1}{N} \sum_{n=1}^{N} (f_1 \circ T^{n\mathbf{e}_1}) \cdots (f_d \circ T^{n\mathbf{e}_d}) \quad \text{for } N \geq 1.$$

1. For each $N \geq 1$, let

$$\theta_N := \frac{1}{N^2} \sum_{m,n=1}^{N} (T^{m\mathbf{e}_1} \times \cdots \times T^{m\mathbf{e}_d} \times T^{n\mathbf{e}_1} \times \cdots \times T^{n\mathbf{e}_d})_* \mu^{\Delta},$$

where μ^{Δ} is the diagonal $(2d)$-fold self-joining of $\mathbf{X}$ on $X^d \times X^d$. Prove that each θ_N is a self-joining of $\mathbf{X}$, and that

$$\|A_N(f_1, \ldots, f_d)\|_2^2 = \int_{X^d \times X^d} f_1 \otimes \cdots \otimes f_d \otimes \overline{f_1} \otimes \cdots \otimes \overline{f_d} \, \mathrm{d}\theta_N.$$

2. Let θ be a subsequential joining-topology limit of $(\theta_N)_{N \geq 1}$. Show that θ is also invariant under

$$\vec{T}_1 := T^{\mathbf{e}_1} \times \cdots \times T^{\mathbf{e}_d} \times \mathrm{id} \times \cdots \times \mathrm{id}$$

and

$$\vec{T}_2 := \mathrm{id} \times \cdots \times \mathrm{id} \times T^{\mathbf{e}_1} \times \cdots \times T^{\mathbf{e}_d}.$$

3. Let $\mathfrak{C}$ be the idempotent class

$$\mathfrak{C}^{\mathbf{e}_1 - \mathbf{e}_2} \vee \cdots \vee \mathfrak{C}^{\mathbf{e}_1 - \mathbf{e}_d} \vee \mathfrak{C}^{\mathbf{e}_1}.$$

Prove that if $\mathbf{X}$ is $\mathfrak{C}$-sated, then θ is relatively independent over $(\mathfrak{C}_{\mathbf{X}}, \mathscr{B}_X, \mathscr{B}_X, \ldots, \mathscr{B}_X)$. [This is the tricky part: as in the proof of Theorem 4.66, the key is to choose a $\mathbb{Z}^d$-action on $(X^d \times X^d, \theta)$, using some combination of diagonal transformations and $\vec{T}_1$ and $\vec{T}_2$, so that it may be interpreted as a $\mathfrak{C}$-adjoining of $\mathbf{X}$.]
4. Deduce that if $\mathbf{X}$ is $\mathfrak{C}$-sated, then

$$\|A_N(f_1, \ldots, f_d) - A_N(\mathsf{E}_\mu(f_1 \,|\, \mathfrak{C}_{\mathbf{X}}), f_2, \ldots, f_d)\|_2 \longrightarrow 0 \text{ as } N \longrightarrow \infty.$$

Use this, together with an approximation of $\mathfrak{C}_{\mathbf{X}}$-measurable functions similar to that in the proof of Theorem 4.30, to complete the proof of Theorem 4.31 by induction on d. ◁

4.9 Further Reading

I have been quite sparing with references in this course. But it gives only a very narrow view of quite a large area of ergodic theory. The following references, though still incomplete, offer several directions for further study.

General Ergodic Theory

There are many good ergodic theory textbooks. Among these books, [Pet83] and [Gla03] include factors, joinings, and their rôles in other parts of ergodic theory; [Gla03] is thorough, but [Pet83] is somewhat gentler. The recent book [EW11] also covers this machinery, as well as giving its own treatment of ergodic Ramsey theory. That presentation is close to [Fur81] and omits some of the more modern ideas presented in this course.

Idempotent classes are a more recent definition than the others, having appeared first in preprint versions of [Aus15]. They have not yet been used outside the study of multiple recurrence, so do not appear in other accounts of general ergodic theory.

Finer Analysis in the One-Dimensional Setting

For $\mathbb{Z}$-systems, Theorem 4.56 has enabled several much finer results on nonconventional averages and multiple recurrence, both in ergodic theory and in the neighbouring field of topological dynamics. For some examples, see [BHK05, HKM10, HKM14, SY12, DDM+13].

The first proof of Theorem 4.56 is due to Host and Kra [HK05b]. Their approach introduces a very useful family of seminorms on $L^\infty(\mu)$ for a general system (X, μ, T), defined in terms of certain self-joinings and which turn out to detect the presence of nilsystem-factors. These 'Host–Kra' seminorms appear again in several of the other papers listed above, and have been subjected to a more detailed analysis of their own in [HK08, HK09]. They are related to the 'Gowers uniformity' norms in additive combinatorics, mentioned again below.

The paper [FK05] shows that Theorem 4.56 can also be brought to bear on nonconventional averages of $\mathbb{Z}^d$-systems under the assumption of total ergodicity. That assumption makes all the factors on the right-hand side of (4.7) trivial, and prevents one from needing to construct any extensions.

Another intriguing relation between one and higher dimensions is Frantzikinakis' recent result from [Fra]. It characterizes the sequences

$$\int (f_1 \circ T^{n\mathbf{e}_1}) \cdot \dots \cdot (f_d \circ T^{n\mathbf{e}_d})\, \mathrm{d}\mu, \quad n \in \mathbb{N},$$

that one can obtain from an arbitrary $\mathbb{Z}^d$-system (X, μ, T) and functions $f_1, \dots, f_d \in L^\infty(\mu)$. He shows that such a sequence can always be decomposed into a nilsequence (roughly, this is a sequence generated by sampling some function along the orbit of a nilsystem) and a sequence that tends to zero in uniform density. Thus, these sequences of integrals for $\mathbb{Z}^d$-systems can be described in terms of nilsystems, a class of 'one-dimensional' objects. It would be very interesting to know whether this characterization can be turned into any further information about limit joinings.

Polynomial Sequences, Nilpotent Groups, and Prime Numbers

Several generalizations of Theorem 4.7 or 4.8 are obtained by restricting the possible values of n. One usually obtains a counterpart generalization of Theorem 4.3 or 4.4.

One class of these generalizations is obtained by demanding that n be a value taken by a certain polynomial. More generally, one can consider tuples in $\mathbb{Z}^d$ produced by several different polynomials. One of the main results of [BL96] is a sufficient condition on a tuple $\mathbf{p}_1, \dots, \mathbf{p}_k$ of polynomials $\mathbb{Z} \longrightarrow \mathbb{Z}^d$ for the following to hold: if $E \subseteq \mathbb{Z}^d$ has $\overline{\mathrm{d}}(E) > 0$, then there is some $n \in \mathbb{Z}$ such that

$$\{\mathbf{p}_1(n), \dots, \mathbf{p}_k(n)\} \subseteq E.$$

Leibman has given a further generalization to subsets and polynomial mappings of discrete nilpotent groups in [Lei98]. Other more recent developments involving polynomials or nilpotent groups, some based on Theorem 4.56, can be found in [HK05a, FK06, BLL08, Fra08, Ausd, CFH11]. If one studies polynomials that are 'different enough' from one another, then there are also results for tuples of transformations that do not commute, and hence do not define a $\mathbb{Z}^d$-action: see [CF12, FZK15].

In this polynomial setting, the problem of norm convergence for the associated functional nonconventional ergodic averages was open for a long time. It was recently solved by Walsh [Wal12], using an approach similar to [Tao08] together with some clever new ideas. Zorin-Kranich has generalized Walsh's argument further in [ZK].

A more recent trend is to study multiple recurrence along the sequence of prime numbers or other sequences related to them. Results in this direction are proved by adapting the ergodic-theoretic machinery and combining it with estimates from analytic number theory on the distribution of the primes. For examples, see [BLZ11, FHK13, WZ12].

Pointwise Convergence

Among ergodic theorists, perhaps the best-known open problem in this area is whether the functional averages of Theorem 4.31 converge pointwise a.e., as well as in norm. For general $\mathbb{Z}$-systems, one such sequence of averages was shown to converge pointwise in [Bou90]. Aside from this, results have been obtained for some related but 'smoother' sequences of averages, such as 'cubical averages', or under some extra assumptions on the system: see [Les87, Ass05, Ass07, Ass10, HSY, DS]. The general case still seems far out of reach.

Extensions and Satedness

Theorem 4.58 was first proved in [Aus09]. Shortly after that paper appeared, Host gave an alternative construction in [Hos09] which he called 'magic' extensions. The rather more abstract approach via satedness originated in [Aus15], which seeks to construct extensions that retain some extra algebraic structure in the original group action. See also [Aus10c] for an overview of this story.

Host's construction in [Hos09] introduces the analogues of the Host–Kra seminorms for a general $\mathbb{Z}^d$-system. These offer a good way to organize the discussion of extensions and characteristic factors for any of the different extension methods, although I have not followed that route in the present course.

Machinery for constructing extensions with improved behaviour has now found other uses in this area, and has been adapted to other settings, such as systems defined by several commuting actions of a general amenable group. See [Aus13, Chu11, CZK, Ausa, DS].

Even after one has ascended to a pleasant extension, the limit joining of Theorem 4.30 for a $\mathbb{Z}^d$-system has not been described completely. As far as I know, the most detailed results are still those in [Aus10a].

Finitary Approaches to Additive Combinatorics

In addition to the ergodic-theoretic and hypergraph approaches to Szemerédi's theorem, Gowers introduced an approach based on Fourier analysis in [Gow98, Gow01]. This major breakthrough gives by far the best-known bounds for quantitative versions of Szemerédi's theorem. It builds on Roth's proof of his theorem in [Rot53], which is Fourier-theoretic; but to extend Roth's ideas it is necessary to replace traditional Fourier analysis with a new theory, now sometimes referred to as 'higher-order Fourier analysis'.

An important part of this new theory is a family of norms on functions on cyclic groups, now often called 'Gowers uniformity norms'. These are closely analogous to the Host–Kra seminorms in ergodic theory.

Gowers uniformity norms and higher-order Fourier analysis have since been the subject of intense development, at least in the one-dimensional setting of Szemerédi's theorem. Most famously, Green and Tao proved in [GT08] that the set of prime numbers contain arithmetic progressions. This fact is not covered by Szemerédi's theorem since the primes have upper Banach density equal to zero, but a proof can be given by building on some of Gowers' ideas and incorporating results about the distribution of the primes. More recently, Green, Tao and Ziegler have proved an important structural result for the uniformity norms which can be viewed as a finitary analog of Theorem 4.56:

see the papers [GTZ11, GTZ12] and their predecessors listed there, and also Szegedy's alternative approach to this result in [Szea, Szeb]. This result can be turned into precise asymptotics for the density of arithmetic progressions appearing in the prime numbers [GT10]. The body of works in additive combinatorics that derive other consequences from this kind of machinery is growing fast, but I will not try to list it further.

Once again, the story in higher dimensions is far less complete. Some important first steps were taken by Shkredov in [Shk05, Shk06]. Also, the papers [Ausb, Ausc] study an essentially algebraic problem which arises from certain questions about Gowers-like uniformity norms in higher dimensions.

Tim Austin
Courant Institute, NYU
251 Mercer St, New York, NY 10012, USA
email: tim@cims.nyu.edu
cims.nyu.edu/~tim

References

[AGH63] L. Auslander, L. Green, and F. Hahn. *Flows on Homogeneous Spaces*. With the assistance of L. Markus and W. Massey, and an appendix by L. Greenberg. Annals of Mathematics Studies, No. 53. Princeton University Press, Princeton, NJ, 1963.

[Ass05] I. Assani. Pointwise convergence of nonconventional averages. *Colloq. Math.*, 102(2):245–262, 2005.

[Ass07] I. Assani. Averages along cubes for not necessarily commuting m.p.t. In *Ergodic theory and related fields*, volume 430 of *Contemp. Math.*, pages 1–19. Amer. Math. Soc., Providence, RI, 2007.

[Ass10] I. Assani. Pointwise convergence of ergodic averages along cubes. *J. Anal. Math.*, 110:241–269, 2010.

[Ausa] T. Austin. Non-conventional ergodic averages for several commuting actions of an amenable group. To appear, *J. Anal. Math.*. Preprint at arXiv.org: 1309.4315.

[Ausb] T. Austin. Partial difference equations over compact Abelian groups, I: modules of solutions. Preprint, available online at arXiv.org: 1305.7269.

[Ausc] T. Austin. Partial difference equations over compact Abelian groups, II: step-polynomial solutions. Preprint, available online at arXiv.org: 1309.3577.

[Ausd] T. Austin. Pleasant extensions retaining algebraic structure, II. *J. Anal. Math.* 126 (2015), 1–111.

[Aus09] T. Austin. On the norm convergence of nonconventional ergodic averages. *Ergodic Theory Dynam. Systems*, 30(2):321–338, 2009.

[Aus10a] T. Austin. Deducing the multidimensional Szemerédi theorem from an infinitary removal lemma. *J. d'Analyse Math.*, 111:131–150, 2010.

[Aus10b] T. Austin. Extensions of probability-preserving systems by measurably-varying homogeneous spaces and applications. *Fund. Math.*, 210(2):133–206, 2010.

[Aus10c] T. Austin. *Multiple recurrence and the structure of probability-preserving systems*. ProQuest LLC, Ann Arbor, MI, 2010. Thesis (Ph.D.)–University of California, Los Angeles; preprint online at `arXiv.org`: 1006.0491.

[Aus13] T. Austin. Equidistribution of joinings under off-diagonal polynomial flows of nilpotent Lie groups. *Ergodic Theory Dynam. Systems*, 33(6):1667–1708, 2013.

[Aus15] T. Austin. Pleasant extensions retaining algebraic structure, I. *J. Anal. Math.*, 125(1):1–36, 2015.

[Ber87] V. Bergelson. Weakly mixing PET. *Ergodic Theory Dynam. Systems*, 7(3):337–349, 1987.

[BHK05] V. Bergelson, B. Host, and B. Kra. Multiple recurrence and nilsequences. *Invent. Math.*, 160(2):261–303, 2005. With an appendix by I. Ruzsa.

[Bil95] P. Billingsley. *Probability and Measure*. Wiley Series in Probability and Mathematical Statistics. John Wiley & Sons, Inc., New York, third edition, 1995. A Wiley-Interscience Publication.

[BL96] V. Bergelson and A. Leibman. Polynomial extensions of van der Waerden's and Szemerédi's theorems. *J. Amer. Math. Soc.*, 9(3):725–753, 1996.

[BLL08] V. Bergelson, A. Leibman, and E. Lesigne. Intersective polynomials and the polynomial Szemerédi theorem. *Adv. Math.*, 219(1):369–388, 2008.

[BLZ11] V. Bergelson, A. Leibman, and T. Ziegler. The shifted primes and the multidimensional Szemerédi and polynomial van der Waerden theorems. *C. R. Math. Acad. Sci. Paris*, 349(3–4):123–125, 2011.

[Bou90] J. Bourgain. Double recurrence and almost sure convergence. *J. Reine Angew. Math.*, 404:140–161, 1990.

[Bre68] L. Breiman. *Probability*. Addison-Wesley Publishing Company, Reading, MA – London – Don Mills, ON, 1968.

[CF12] Q. Chu and N. Frantzikinakis. Pointwise convergence for cubic and polynomial multiple ergodic averages of non-commuting transformations. *Ergodic Theory Dynam. Systems*, 32(3):877–897, 2012.

[CFH11] Q. Chu, N. Frantzikinakis, and B. Host. Ergodic averages of commuting transformations with distinct degree polynomial iterates. *Proc. Lond. Math. Soc. (3)*, 102(5):801–842, 2011.

[Chu11] Q. Chu. Multiple recurrence for two commuting transformations. *Ergodic Theory Dynam. Systems*, 31(3):771–792, 2011.

[CL84] J.-P. Conze and E. Lesigne. Théorèmes ergodiques pour des mesures diagonales. *Bull. Soc. Math. France*, 112(2):143–175, 1984.

[CL88a] J.-P. Conze and E. Lesigne. Sur un théorème ergodique pour des mesures diagonales. In *Probabilités*, volume 1987 of *Publ. Inst. Rech. Math. Rennes*, pages 1–31. Univ. Rennes I, Rennes, 1988.

[CL88b] J.-P. Conze and E. Lesigne. Sur un théorème ergodique pour des mesures diagonales. *C. R. Acad. Sci. Paris Sér. I Math.*, 306(12):491–493, 1988.

[CZK] Q. Chu and P. Zorin-Kranich. Lower bound in the Roth Theorem for amenable groups. To appear, *Ergodic Theory Dynam. Systems.*

[DDM+13] P. Dong, S. Donoso, A. Maass, Song Shao, and Xiangdong Ye. Infinite-step nilsystems, independence and complexity. *Ergodic Theory Dynam. Systems*, 33(1):118–143, 2013.

[DS] S. Donoso and W. Sun. A pointwise cubic average for two commuting transformations. Preprint, available online at `arXiv.org`: 1410.4887.

[ET36] P. Erdős and P. Turán. On some sequences of integers. *J. London Math. Soc.*, 11:261–264, 1936.

[EW11] M. Einsiedler and T. Ward. *Ergodic Theory with a View Towards Number Theory*, volume 259 of *Graduate Texts in Mathematics*. Springer-Verlag London, Ltd., London, 2011.

[FHK13] N. Frantzikinakis, B. Host, and B. Kra. The polynomial multidimensional Szemerédi theorem along shifted primes. *Israel J. Math.*, 194(1):331–348, 2013.

[FK78] H. Furstenberg and Y. Katznelson. An ergodic Szemerédi Theorem for commuting transformations. *J. d'Analyse Math.*, 34:275–291, 1978.

[FK85] H. Furstenberg and Y. Katznelson. An ergodic Szemerédi theorem for IP-systems and combinatorial theory. *J. d'Analyse Math.*, 45:117–168, 1985.

[FK91] H. Furstenberg and Y. Katznelson. A Density Version of the Hales-Jewett Theorem. *J. d'Analyse Math.*, 57:64–119, 1991.

[FK05] N. Frantzikinakis and B. Frantzikinakis. Convergence of multiple ergodic averages for some commuting transformations. *Ergodic Theory Dynam. Systems*, 25(3):799–809, 2005.

[FK06] N. Frantzikinakis and B. Kra. Ergodic averages for independent polynomials and applications. *J. London Math. Soc. (2)*, 74(1):131–142, 2006.

[FKO82] H. Furstenberg, Y. Katznelson, and D. Ornstein. The ergodic theoretical proof of Szemerédi's theorem. *Bull. Amer. Math. Soc. (N.S.)*, 7(3):527–552, 1982.

[Fol99] G. B. Folland. *Real Analysis*. Pure and Applied Mathematics (New York). John Wiley & Sons Inc., New York, second edition, 1999. Modern techniques and their applications, A Wiley-Interscience Publication.

[Fra] N. Frantzikinakis. Multiple correlation sequences and nilsequences. To appear in *Invent. Math.*. Preprint online at `arXiv.org`: 1407.0631.

[Fra08] N. Frantzikinakis. Multiple ergodic averages for three polynomials and applications. *Trans. Amer. Math. Soc.*, 360(10):5435–5475, 2008.

[Fur67] H. Furstenberg. Disjointness in ergodic theory, minimal sets, and a problem in Diophantine approximation. *Math. Systems Theory*, 1:1–49, 1967.

[Fur77] H. Furstenberg. Ergodic behaviour of diagonal measures and a theorem of Szemerédi on arithmetic progressions. *J. d'Analyse Math.*, 31:204–256, 1977.

[Fur81] H. Furstenberg. *Recurrence in Ergodic Theory and Combinatorial Number Theory*. Princeton University Press, Princeton, 1981.

[FW96] H. Furstenberg and B. Weiss. A mean ergodic theorem for $\frac{1}{N}\sum_{n=1}^{N} f(T^n x)g(T^{n^2} x)$. In V. Bergelson, A. B. C March, and

J. Rosenblatt, editors, *Convergence in Ergodic Theory and Probability*, pages 193–227. De Gruyter, Berlin, 1996.

[FZK15] N. Frantzikinakis and P. Zorin-Kranich. Multiple recurrence for non-commuting transformations along rationally independent polynomials. *Ergodic Theory Dynam. Systems*, 35(2):403–411, 2015.

[Gla03] E. Glasner. *Ergodic Theory via Joinings*. American Mathematical Society, Providence, 2003.

[Gow98] W. T. Gowers. A new proof of Szemerédi's theorem for arithmetic progressions of length four. *Geom. Funct. Anal.*, 8(3):529–551, 1998.

[Gow01] W. T. Gowers. A new proof of Szemerédi's theorem. *Geom. Funct. Anal.*, 11(3):465–588, 2001.

[Gow06] W. T. Gowers. Quasirandomness, counting and regularity for 3-uniform hypergraphs. *Combin. Probab. Comput.*, 15(1-2):143–184, 2006.

[GRS90] R. L. Graham, B. L. Rothschild, and J. H. Spencer. *Ramsey Theory*. John Wiley & Sons, New York, 1990.

[GT07] B. J. Green and T. Tao. The quantitative behaviour of polynomial orbits on nilmanifolds. Preprint, available online at `arXiv.org`: 0709.3562, 2007.

[GT08] B. J. Green and T. Tao. The primes contains arbitrarily long arithmetic progressions. *Ann. Math.*, 167:481–547, 2008.

[GT10] B. Green and T. Tao. Linear equations in primes. *Ann. of Math. (2)*, 171(3):1753–1850, 2010.

[GTZ11] B. Green, T. Tao, and T. Ziegler. An inverse theorem for the Gowers U^4-norm. *Glasg. Math. J.*, 53(1):1–50, 2011.

[GTZ12] B. Green, T. Tao, and T. Ziegler. An inverse theorem for the Gowers $U^{s+1}[N]$-norm. *Ann. of Math. (2)*, 176(2):1231–1372, 2012.

[HK01] B. Host and B. Kra. Convergence of Conze-Lesigne averages. *Ergodic Theory Dynam. Systems*, 21(2):493–509, 2001.

[HK05a] B. Host and B. Kra. Convergence of polynomial ergodic averages. *Israel J. Math.*, 149:1–19, 2005. Probability in mathematics.

[HK05b] B. Host and B. Kra. Nonconventional ergodic averages and nilmanifolds. *Ann. Math.*, 161(1):397–488, 2005.

[HK08] B. Host and B. Kra. Parallelepipeds, nilpotent groups and Gowers norms. *Bull. Soc. Math. France*, 136(3):405–437, 2008.

[HK09] B. Host and B. Kra. Uniformity seminorms on ℓ^∞ and applications. *J. Anal. Math.*, 108:219–276, 2009.

[HKM10] B. Host, B. Kra and A. Maass. Nilsequences and a structure theorem for topological dynamical systems. *Adv. Math.*, 224(1):103–129, 2010.

[HKM14] B. Host, B. Kra, and A. Maass. Complexity of nilsystems and systems lacking nilfactors. *J. Anal. Math.*, 124:261–295, 2014.

[Hos09] B. Host. Ergodic seminorms for commuting transformations and applications. *Studia Math.*, 195(1):31–49, 2009.

[HSY] W. Huang, S. Shao, and X. Ye. Pointwise convergence of multiple ergodic averages and strictly ergodic models. Preprint, available online at `arXiv.org`: 1406.5930.

[Lei98] A. Leibman. Multiple recurrence theorem for measure preserving actions of a nilpotent group. *Geom. Funct. Anal.*, 8(5):853–931, 1998.

[Lei05] A. Leibman. Pointwise convergence of ergodic averages for polynomial sequences of translations on a nilmanifold. *Ergodic Theory Dynam. Systems*, 25(1):201–213, 2005.

[Les87] E. Lesigne. Théorèmes ergodiques ponctuels pour des mesures diagonales. Cas des systèmes distaux. *Ann. Inst. H. Poincaré Probab. Statist.*, 23(4):593–612, 1987.

[Mac66] G. W. Mackey. Ergodic theory and virtual groups. *Math. Ann.*, 166:187–207, 1966.

[NRS06] B. Nagle, V. Rödl, and M. Schacht. The counting lemma for regular k-uniform hypergraphs. *Random Structures Algorithms*, 28(2):113–179, 2006.

[Par69] W. Parry. Ergodic properties of affine transformations and flows on nilmanifolds. *Amer. J. Math.*, 91:757–771, 1969.

[Par70] W. Parry. Dynamical systems on nilmanifolds. *Bull. London Math. Soc.*, 2:37–40, 1970.

[Pet83] K. E. Petersen. *Ergodic Theory*. Cambridge University Press, Cambridge, 1983.

[Pol09] D. H. J. Polymath. A new proof of the density Hales-Jewett theorem. Preprint, available online at `arXiv.org`: 0910.3926, 2009.

[Rot53] K. F. Roth. On certain sets of integers. *J. London Math. Soc.*, 28:104–109, 1953.

[Roy88] H. L. Royden. *Real Analysis*. Macmillan Publishing Company, New York, third edition, 1988.

[Rud95] D. J. Rudolph. Eigenfunctions of $T \times S$ and the Conze-Lesigne algebra. In *Ergodic theory and its connections with harmonic analysis (Alexandria, 1993)*, volume 205 of *London Math. Soc. Lecture Note Ser.*, pages 369–432. Cambridge University Press, Cambridge, 1995.

[Shk05] I. D. Shkredov. On a problem of Gowers. *Dokl. Akad. Nauk*, 400(2):169–172, 2005. (Russian).

[Shk06] I. D. Shkredov. On a problem of Gowers. *Izv. Ross. Akad. Nauk Ser. Mat.*, 70(2):179–221, 2006. (Russian).

[Sta00] A. N. Starkov. *Dynamical Systems on Homogeneous Spaces*, volume 190 of *Translations of Mathematical Monographs*. American Mathematical Society, Providence, RI, 2000. Translated from the 1999 Russian original by the author.

[SY12] S. Shao and X. Ye. Regionally proximal relation of order d is an equivalence one for minimal systems and a combinatorial consequence. *Adv. Math.*, 231(3–4):1786–1817, 2012.

[Szea] B. Szegedy. Gowers norms, regularization and limits of functions on Abelian groups. Preprint, available online at `arXiv.org`: 1010.6211.

[Szeb] B. Szegedy. On higher-order Fourier analysis. Preprint, available online at `arXiv.org`: 1203.2260.

[Sze75] E. Szemerédi. On sets of integers containing no k elements in arithmetic progression. *Acta Arith.*, 27:199–245, 1975.

[Tao06a] T. Tao. A quantitative ergodic theory proof of Szemerédi's theorem. *Electron. J. Combin.*, 13(1):Research Paper 99, 49 pp. (electronic), 2006.

[Tao06b] T. Tao. Szemerédi's regularity lemma revisited. *Contrib. Discrete Math.*, 1(1):8–28, 2006.

[Tao08] T. Tao. Norm convergence of multiple ergodic averages for commuting transformations. *Ergodic Theory and Dynamical Systems*, 28:657–688, 2008.

[vdW27] B. L. van der Waerden. Beweis einer baudetschen vermutung. *Nieuw. Arch. Wisk.*, 15:212–216, 1927.

[Wal12] M. N. Walsh. Norm convergence of nilpotent ergodic averages. *Ann. of Math. (2)*, 175(3):1667–1688, 2012.

[WZ12] T. D. Wooley and T. D. Ziegler. Multiple recurrence and convergence along the primes. *Amer. J. Math.*, 134(6):1705–1732, 2012.

[Zha96] Q. Zhang. On convergence of the averages $(1/N)\sum_{n=1}^{N} f_1(R^n x) f_2(S^n x) f_3(T^n x)$. *Monatsh. Math.*, 122(3):275–300, 1996.

[Zie07] T. Ziegler. Universal characteristic factors and Furstenberg averages. *J. Amer. Math. Soc.*, 20(1):53–97 (electronic), 2007.

[Zim76a] R. J. Zimmer. Ergodic actions with generalized discrete spectrum. *Illinois J. Math.*, 20(4):555–588, 1976.

[Zim76b] R. J. Zimmer. Extensions of ergodic group actions. *Illinois J. Math.*, 20(3):373–409, 1976.

[ZK] P. Zorin-Kranich. Norm convergence of multiple ergodic averages on amenable groups. To appear, *J. d'Analyse Math.*

5

Diophantine Problems and Homogeneous Dynamics

Manfred Einsiedler and Tom Ward

Abstract

We will discuss some classical questions that have their origins in the work of Gauss from 1863, along with some more recent developments due primarily to Duke, Rudnick and Sarnak (Duke Math. J. 71 (1993), no. 1, 143–179), and to Eskin and McMullen (ibid., 181–209). Rather than aiming for maximal generality, we will try to expose the striking connection between certain equidistribution problems in dynamics and some asymptotic counting problems. In the second part we will discuss the connection between the theory of Diophantine approximation and homogeneous dynamics.

5.1 Equidistribution and the Gauss Circle Problem

We start[1] with a connection between an equidistribution result and a lattice-point counting problem in the classical setting of the Gauss circle problem. This problem asks for estimates of the number of integral points in the disk of radius R with centre at the origin. It is clear that the first estimate is given by the area of the disk, so the emphasis is on the error term. We write $\|\cdot\|$ for the Euclidean norm on $\mathbb{R}^2$.

Proposition 5.1 *For any $R > 0$ let*

$$N(R) = \left|\{n \in \mathbb{Z}^2 \mid \|n\| \leqslant R\}\right|. \tag{5.1}$$

[1] In these notes we will make use of several concepts from ergodic theory and dynamical systems. For most of these a suitable source is [13], and for some of the material on the space of lattices particularly a suitable source is [14].

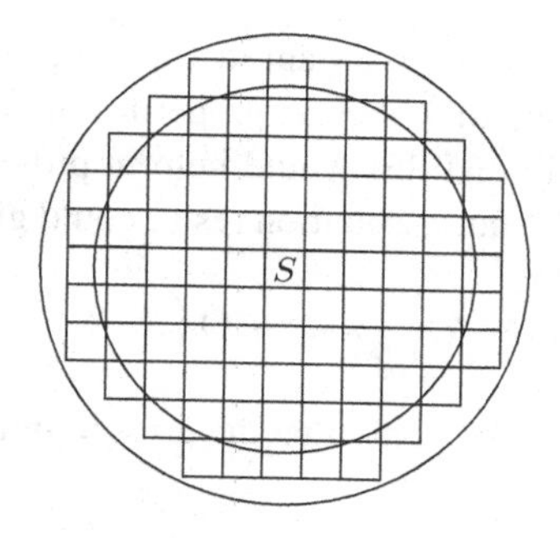

Figure 5.1 Containing the error term for $N(R)$ inside an annulus.

Then

$$N(R) = \pi R^2 + \mathrm{O}(R).$$

The proof of this initial estimate is geometric, and has no connection to dynamics. The main term πR^2 is the area of the two-dimensional ball of radius R, and the error term is controlled by the area of the annulus between the largest circle inside the grid of unit squares lying inside the circle of radius R and the smallest circle containing all the unit squares that intersect the circle of radius R, as indicated in Figure 5.1.

PROOF. Consider the unit square $S = [-\frac{1}{2}, \frac{1}{2}) \times [-\frac{1}{2}, \frac{1}{2})$, which is a fundamental domain for $\mathbb{Z}^2 < \mathbb{R}^2$. Then, as indicated in Figure 5.1, we have

$$B_{R-\frac{1}{\sqrt{2}}}(0) \subseteq S + \{n \in \mathbb{Z}^2 \mid \|n\| \leqslant R\} \subseteq B_{R+\frac{1}{\sqrt{2}}}(0).$$

By taking areas, we conclude that

$$\left(R - \frac{1}{\sqrt{2}}\right)^2 \pi \leqslant N(R) \leqslant \left(R + \frac{1}{\sqrt{2}}\right)^2 \pi$$

so $N(R) = \pi R^2 + \mathrm{O}(R)$ as required. □

It is conjectured that

$$N(R) = \pi R^2 + \mathrm{O}_\varepsilon\left(R^{\frac{1}{2}+\varepsilon}\right) \tag{5.2}$$

for all $\varepsilon > 0$. We refer to the paper of Ivić, Krätzel, Kühleitner and Nowak [19] for a survey of the many partial results towards this conjecture[2].

[2] The error term $N(R) - \pi R^2$ was shown to be bounded above by $2\sqrt{2}\pi R$ by Gauss [16, pp. 269–291]. Hardy [17] and Landau [23] found a *lower* bound for the error by showing that the error is not $\mathrm{o}(R^{1/2}(\log R)^{1/4})$, so the power of R must be at least $\frac{1}{2}$. Huxley showed the estimate $O(R^{\frac{131}{208}})$ in [18].

To motivate later arguments, we want to explain a connection between the error term in $N(R)$ and equidistribution properties of large circles in $\mathbb{R}^2$ modulo $\mathbb{Z}^2$. It seems unlikely that this would help in proving the conjecture, but with some effort such an equidistribution result could give

$$N(R) = \pi R^2 + \mathrm{O}\left(R^{1-\delta}\right)$$

for some $\delta > 0$. We will prove only that the error term is $\mathrm{o}(R)$.

Let

$$\mathrm{T}^1\mathbb{R}^2 = \{(x, v) \mid x \in \mathbb{R}^2, v \in \mathbb{R}^2, \|v\| = 1\} = \mathbb{R}^2 \times \mathbb{S}^1$$

be the unit tangent bundle of $\mathbb{R}^2$, and let

$$\mathrm{T}^1\mathbb{T}^2 = \{(x, v) \mid x \in \mathbb{T}^2, v \in \mathbb{R}^2, \|v\| = 1\} = \mathbb{T}^2 \times \mathbb{S}^1$$

be the unit tangent bundle of $\mathbb{T}^2$. Also write

$$(x, v) \pmod{\mathbb{Z}}^2 = (x \pmod{\mathbb{Z}}^2, v)$$

for the canonical map from $\mathrm{T}^1\mathbb{R}^2$ to $\mathrm{T}^1\mathbb{T}^2$ and $\mathrm{d}(x, v)$ for the canonical volume form in $\mathrm{T}^1\mathbb{T}^2$.

Proposition 5.2 *Let*

$$\begin{aligned}\gamma_R : [0, 1] &\longrightarrow \mathrm{T}^1\mathbb{R}^2\\ t &\longmapsto (R\mathrm{e}^{2\pi i t}, \mathrm{e}^{2\pi i t})\end{aligned}$$

be the constant speed parametrization of the outward tangent vectors on the circle of radius R. Then

$$\int_0^1 f(\gamma_R(t))\,\mathrm{d}t \longrightarrow \int_{\mathrm{T}^1\mathbb{T}^2} f(x, v)\,\mathrm{d}(x, v)$$

as $R \to \infty$, for every $f \in C\left(\mathrm{T}^1\mathbb{T}^2\right)$.

SKETCH PROOF. As t varies in $[0, 1]$, $\gamma_R(t) = (x_R(t), v(t))$ goes through all directions with unit speed, so it is enough to show that the positional part $x_R(t)$ restricted to any interval $[\alpha, \beta] \subseteq [0, 1]$ equidistributes in $\mathbb{T}^2$. Indeed, if this equidistribution is known then we can split $[0, 1]$ into subintervals

$$[0, \tfrac{1}{n}] \cup [\tfrac{1}{n}, \tfrac{2}{n}] \cup \cdots \cup [\tfrac{n-1}{n}, 1]$$

and use the continuity of $f \in C\left(\mathrm{T}^1\mathbb{T}^2\right)$ together with the equidistribution of $x_R(t)$ to see that for large $n = n(f)$ and $R = R(n)$,

$$\begin{aligned}\int_0^1 f(\gamma(t))\,\mathrm{d}t &\approx \sum_{j=0}^{n-1}\int_{j/n}^{(j+1)/n} f(x_R(t), \mathrm{e}^{2\pi \mathrm{i} j/n})\,\mathrm{d}t \\ &\approx \frac{1}{n}\sum_{j=0}^{n-1}\int_{\mathbb{T}^2} f(x, \mathrm{e}^{2\pi \mathrm{i} j/n})\,\mathrm{d}x \\ &\approx \int_{\mathrm{T}^1\mathbb{T}^2} f(x, v)\,\mathrm{d}(x, v)\end{aligned}$$

where the suggested errors are $\mathrm{o}(1)$ as $n \to \infty$ (and also $R \to \infty$).

Thus it remains to show that for any $\alpha, \beta \in [0, 1]$ with $\alpha < \beta$, we have

$$\frac{1}{\beta - \alpha}\int_\alpha^\beta f\left(x_R(t)\right)\mathrm{d}t \longrightarrow \int_{\mathbb{T}^2} f(x)\,\mathrm{d}x \tag{5.3}$$

as $R \to \infty$, for any $f \in C(\mathbb{T}^2)$. Since the linear hull of the space of characters

$$\mathrm{e}_n(x) = \mathrm{e}^{2\pi \mathrm{i}(n_1 x_1 + n_2 x_2)}$$

for $n = (n_1, n_2) \in \mathbb{Z}^2$ is dense in $C(\mathbb{T}^2)$, it is enough to show (5.3) for characters e_n with $n \neq (0, 0)$ (the case $n = (0, 0)$ is clear).

We now fix $n \neq (0, 0)$ and claim that we may assume that

$$\begin{pmatrix} -\sin 2\pi t \\ \cos 2\pi t \end{pmatrix}$$

is never orthogonal to n for $t \in [\alpha, \beta]$. To see this, notice that if it fails for some $t_0 \in [\alpha, \beta]$, then we may split the interval, writing

$$[\alpha, \beta] = [\alpha, t_0] \cup [t_0, \beta].$$

Moreover, for a sufficiently small $\varepsilon > 0$ we may replace these with the intervals $[\alpha, t_0 - \varepsilon]$ and $[t_0 + \varepsilon, \beta]$. Repeating this if necessary one more time (with t_0 replaced by the fractional part of $t_0 + \frac{1}{2}$), we obtain finitely many disjoint intervals satisfying the non-orthogonality condition whose union differs from the original interval $[\alpha, \beta]$ in a set of small measure. The equidistribution in (5.3) for each of the small intervals implies that (5.3) also holds for $[\alpha, \beta]$ up to a small error. Thus we may assume that

$$|-n_1 \sin 2\pi t + n_2 \cos 2\pi t| \geqslant \kappa$$

for some $\kappa > 0$ and all $t \in [\alpha, \beta]$.

Now split the interval $[\alpha, \beta]$ into $m = \lfloor R^{3/4} \rfloor$ subintervals

$$[\alpha_1, \alpha_2] \cup [\alpha_2, \alpha_3] \cup \cdots \cup [\alpha_m, \alpha_{m+1}]$$

with $\alpha_j = \alpha + (j-1)\frac{\beta-\alpha}{m}$. By the Taylor expansion we have

$$\begin{aligned} x_R(t) &= R\left[\begin{pmatrix} \cos 2\pi\alpha_j \\ \sin 2\pi\alpha_j \end{pmatrix} + (t-\alpha_j)2\pi \begin{pmatrix} -\sin 2\pi\alpha_j \\ \cos 2\pi\alpha_j \end{pmatrix} + \mathrm{O}\left(\tfrac{1}{m^2}\right)\right] \\ &= R\begin{pmatrix} \cos 2\pi\alpha_j \\ \sin 2\pi\alpha_j \end{pmatrix} + R(t-\alpha_j)2\pi \begin{pmatrix} -\sin 2\pi\alpha_j \\ \cos 2\pi\alpha_j \end{pmatrix} + \mathrm{O}\left(\tfrac{1}{R^{1/2}}\right) \end{aligned}$$

for all $t \in [\alpha_j, \alpha_{j+1}]$ and $j = 1, \dots, m$. As $n \neq 0$ is fixed and $R \to \infty$, it is clear that this gives

$$\begin{aligned} \left|\frac{1}{\beta-\alpha}\int_\alpha^\beta \mathrm{e}_n(x_R(t))\,\mathrm{d}t\right| &= \left|\frac{1}{\beta-\alpha}\sum_{j=1}^m \int_{\alpha_j}^{\alpha_{j+1}} \mathrm{e}_n\left(R\begin{pmatrix} \cos 2\pi\alpha_j \\ \sin 2\pi\alpha_j \end{pmatrix}\right.\right. \\ &\qquad \left.\left. +R(t-\alpha_j)2\pi\begin{pmatrix} -\sin 2\pi\alpha_j \\ \cos 2\pi\alpha_j \end{pmatrix}\right)\mathrm{d}t\right| + \mathrm{o}(1) \\ &\leqslant \frac{1}{\beta-\alpha}\sum_{j=1}^m \left|\int_{\alpha_j}^{\alpha_{j+1}} \mathrm{e}^{4\pi^2 \mathrm{i} Rt(-n_1\sin 2\pi\alpha_j + n_2\cos 2\pi\alpha_j)}\,\mathrm{d}t\right| + \mathrm{o}(1) \\ &\leqslant \frac{1}{\beta-\alpha}\sum_{j=1}^m \frac{2}{4\pi^2 R|-n_1\sin 2\pi\alpha_j + n_2\cos 2\pi\alpha_j|} + \mathrm{o}(1). \end{aligned}$$

However, since we assume that $|-n_1\sin 2\pi\alpha_j + n_2\cos 2\pi\alpha_j| \geqslant \kappa$ we deduce that

$$\frac{1}{\beta-\alpha}\int_\alpha^\beta \mathrm{e}_n(x_R(t))\,\mathrm{d}t \longrightarrow 0$$

as $R \to \infty$, as required. □

We are now ready to present a modest improvement to the error term in Proposition 5.1 using the equidistribution from Proposition 5.2. As before we write $N(R) = \left|\{\mathbf{n} \in \mathbb{Z}^2 \mid \|\mathbf{n}\| \leqslant R\}\right|$.

Theorem 5.3 $N(R) = \pi R^2 + \mathrm{o}(R)$.

PROOF. In order to take advantage of the equidistribution result, we need to find a function defined on the unit tangent bundle with the property that the

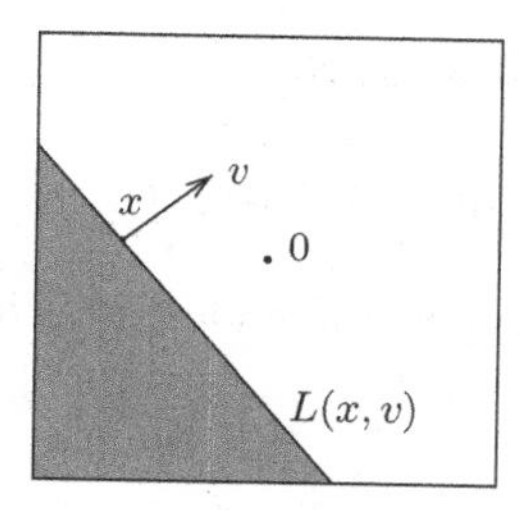

Figure 5.2 The value $h(x, v)$ is the area of the polygon determined by S and the half-space with x on its boundary and v as outward normal, minus 1 if and only if 0 lies in the polygon.

integral along a line segment relates to the difference between an area calculation and a lattice point count. To this end we define a function h on the fundamental domain

$$\mathrm{T}^1 S = [-\tfrac{1}{2}, \tfrac{1}{2}) \times [-\tfrac{1}{2}, \tfrac{1}{2}) \times \mathbb{S}^1$$

using Figure 5.2 (recall that S is the unit square $[-\frac{1}{2}, \frac{1}{2}) \times [-\frac{1}{2}, \frac{1}{2})$).

In other words, we define $h(x, v)$ to be the difference between an area calculation and the simple lattice count of whether or not 0 belongs to the polygon. In anticipation of a later line integral calculation, we also define a function $f : S \times \mathbb{S}^1 \to \mathbb{R}$ by

$$f(x, v) = \frac{h(x, v)}{\text{length of } L(x, v)},$$

where $L(x, v)$ is the line segment in S going through x and normal to v, as illustrated in Figure 5.2.

We claim that f is Riemann integrable. First notice that it is bounded, since the length of $L(x, v)$ is small if and only if the whole line segment is close to one of the corners of S, in which case we have

$$h(x, v) = \mathrm{O}\left((\text{length of } L(x, v))^2\right).$$

Moreover, the set of discontinuities of f is contained in the null set

$$(\partial S) \times \mathbb{S}^1 \cup \{(x, v) \mid 0 \in L(x, v)\}.$$

This implies that f is Riemann integrable. By the symmetries of the construction we have $\int_{\mathrm{T}^1\mathbb{T}^2} f \,\mathrm{d}(x, v) = 0$. Moreover, the equidistribution claim in Proposition 5.2 extends to the function f by the following argument. For any $\varepsilon > 0$, there are continuous functions $f_-, f_+ \in C(\mathrm{T}^1\mathbb{T}^2)$ with the properties that $f_- \leqslant f \leqslant f_+$ and

$$\int_{\mathrm{T}^1\mathbb{T}^2} (f_+ - f_-)\,\mathrm{d}(x, v) \leqslant \varepsilon.$$

Since $\int_{\mathrm{T}^1\mathbb{T}^2} f \, \mathrm{d}(x, v) = 0$ we also get

$$-\varepsilon \leqslant \int_{\mathrm{T}^1\mathbb{T}^2} f_- \, \mathrm{d}(x, v) \leqslant 0 \leqslant \int_{\mathrm{T}^1\mathbb{T}^2} f_+ \, \mathrm{d}(x, v) \leqslant \varepsilon.$$

Applying Proposition 5.2 to the continuous functions f_- and f_+ shows that if R is large enough then

$$\begin{aligned} -2\varepsilon \leqslant \int_{\mathrm{T}^1\mathbb{T}^2} f_- \, \mathrm{d}(x, v) - \varepsilon \leqslant \int_0^1 f_-(\gamma_R(t)) \, \mathrm{d}t \leqslant \int_0^1 f(\gamma_R(t)) \, \mathrm{d}t \\ \leqslant \int_0^1 f_+(\gamma_R(t)) \, \mathrm{d}t \leqslant \int_{\mathrm{T}^1\mathbb{T}^2} f_+ \, \mathrm{d}(x, v) + \varepsilon \leqslant 2\varepsilon, \quad (5.4) \end{aligned}$$

which shows that Proposition 5.2 also holds for f. Nonetheless, it will be convenient to continue to work with the continuous functions f_- and f_+ instead of f a little longer.

We now split the path γ_R into segments by splitting $[0, 1]$ into a disjoint union of adjacent intervals I_k for $k = 1, \ldots, K$ in such a way that for each k there is a unique $n_k \in \mathbb{Z}^2$ with $R\mathrm{e}^{2\pi \mathrm{i} t} - n_k \in S$ for $t \in I_k$. We also require, as we may, that $n_k \neq n_\ell$ for $k \neq \ell$, as illustrated in Figure 5.3.

Now define a new path $\overline{\gamma}_R$ that is affine on each of the intervals I_k and coincides with the path γ_R at the boundaries ∂I_k for $1 \leqslant k \leqslant K$. In other words, the path $\overline{\gamma}_R$ is the piecewise linear path forming a straight line from one intersection on the circle of radius R with the grid $\partial S + \mathbb{Z}^2$ to the next one (shown by the dots in Figure 5.3). Since γ_R and $\overline{\gamma}_R$ are uniformly $\mathrm{O}(R^{-1})$-close, we may use (5.4) for large R to see that

$$-3\varepsilon \leqslant \int_0^1 f_-(\overline{\gamma}_R(t)) \, \mathrm{d}t \leqslant \int_0^1 f(\overline{\gamma}_R(t)) \, \mathrm{d}t \leqslant \int_0^1 f_+(\overline{\gamma}_R(t)) \, \mathrm{d}t \leqslant 3\varepsilon,$$

in other words we have shown that

$$\int_0^1 f(\overline{\gamma}_R(t)) \, \mathrm{d}t = \mathrm{o}(1)$$

as $R \to \infty$.

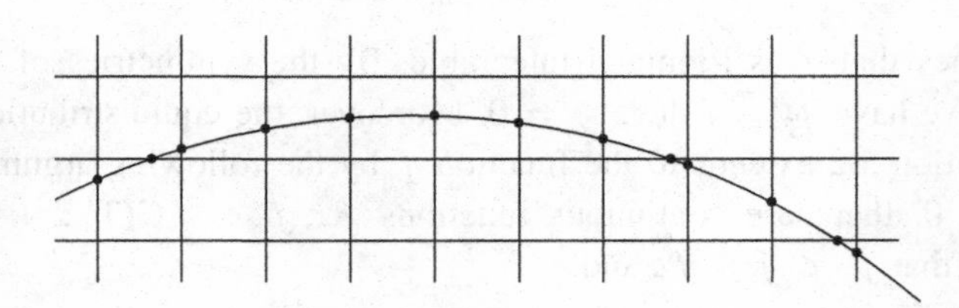

Figure 5.3 The dots indicate the boundaries of the intervals I_k.

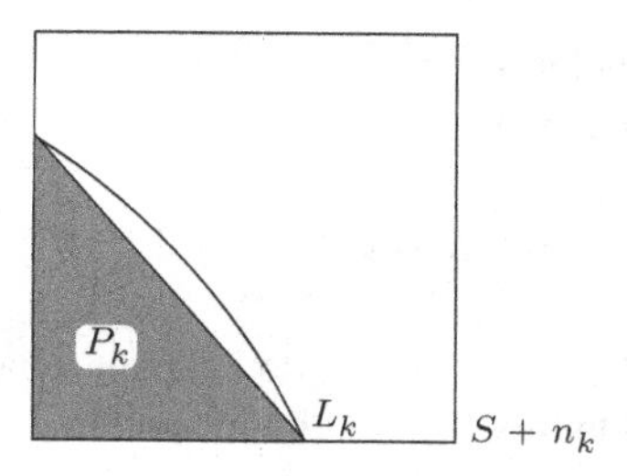

Figure 5.4 The polygon P_k.

Recalling the definition of f, the integral over one subinterval is given by

$$\int_{I_k} f(\overline{\gamma}_R(t))\,\mathrm{d}t = (\text{area of } P_k - \mathbb{1}_{P_k}(n_k))\frac{|I_k|}{\text{length of } L_k},$$

where P_k is the polygon inside $S + n_k$ determined by a portion of $\partial S + n_k$ and the line segment $L_k = \overline{\gamma}_R(I_k)$, and $|I_k|$ is the length of the interval I_k (see Figure 5.4).

By construction, the length $|I_k|$ of I_k is $\frac{\phi_k}{2\pi}$ where ϕ_k is the angle of the arc on the circle of radius R corresponding to I_k, so $\phi_k = \mathrm{O}(R^{-1})$. On the other hand,

$$\text{length of } L_k = 2R\sin\frac{\phi_k}{2} = 2R\left(\frac{\phi_k}{2} + \mathrm{O}\left(\phi_k^3\right)\right).$$

Therefore

$$\begin{aligned}\int_0^1 f\left(\overline{\gamma}_R(t)\right)\,\mathrm{d}t &= \sum_{k=1}^{K}\left(\text{area of } P_k - \mathbb{1}_{P_k}(n_k)\right)\frac{|I_k|}{\text{length of } L_k}\\ &= \frac{1}{2\pi R}\left(\sum_{k=1}^{K}\left(\text{area of } P_k - \mathbb{1}_{P_k}(n_k)\right) + \sum_{k=1}^{K}\mathrm{O}\left(\phi_k^3\right)\right)\\ &= \frac{1}{2\pi R}\big(\text{area of polygon enclosed by } \overline{\gamma}_R\\ &\qquad -\text{no. of lattice points inside}\big) + \mathrm{O}(R^{-3}).\end{aligned}$$

To summarize, we have shown that the difference between the area enclosed by $\overline{\gamma}_R$ and the number of lattice points within this polygon is $\mathrm{o}(R)$.

We now need to analyse how the area and lattice point count for the polygon defined above differ from the same quantities for the circle. The area of the polygon enclosed by $\overline{\gamma}_R$ differs from the area of the circle by $\mathrm{O}(1)$. We claim that the number of lattice points inside the circle but outside the polygon is $\mathrm{o}(R)$. These claims together prove the theorem.

In order to prove the claim, notice that any lattice point n inside the circle but outside the polygon satisfies

$$0 \leqslant R - \|n\| = \mathrm{O}(R^{-1}).$$

Fix $\delta > 0$ and let $g(x, v) = \mathbb{1}_{B_{2\delta}(0)}(x)$ be the characteristic function of the 2δ-ball around the identity $0 \in \mathbb{T}^2$ (but considered as a function on $\mathrm{T}^1\mathbb{T}^2$). If now $\|n\|$ lies in $[R - \delta, R]$, then there is a corresponding subinterval of length $\gg \frac{\delta}{R}$ such that $g(\gamma_R(t)) = 1$ for all t in that interval. Therefore

$$N_{R,\delta} = \left|\left\{n \in \mathbb{Z}^2 \mid R - \delta \leqslant \|n\| \leqslant R\right\}\right| \ll \frac{R}{\delta} \int_0^1 g(\gamma_R(t))\,\mathrm{d}t.$$

By construction, g is Riemann integrable and so, for sufficiently large R,

$$N_{R,\delta} \ll \frac{2R}{\delta} \int_{\mathrm{T}^1\mathbb{T}^2} g(x, v)\,\mathrm{d}(x, v) \ll \frac{R}{\delta}\delta^2 = R\delta.$$

This proves the claim, and hence the theorem. □

Remark Similar methods may be used to prove various extensions and refinements of these results, including the following.

1. For $d \geqslant 2$ let $N^*(R) = |\{n \in \mathbb{Z}^d \mid n \text{ is primitive and } \|n\| \leqslant R\}|$ (an integer vector is said to be primitive if the greatest common divisor of its entries is 1). Then $N^*(R) = (\zeta(d)^{-1}V_d + o(1))R^2$ as $R \to \infty$. Here V_d is the volume of the unit ball in $\mathbb{R}^d$ and $\zeta(s) = \sum_{n=1}^{\infty} n^{-s}$ denotes the Riemann zeta function.
2. Proposition 5.1 may be improved to give an effective statement of the form

$$\left| \int_0^1 f\left(\gamma_R(t) \pmod{\mathbb{Z}}^2\right) \mathrm{d}t - \int_{\mathrm{T}^1\mathbb{T}^2} f(x, v)\,\mathrm{d}(x, v) \right| \ll R^{-\delta_1}\,\mathrm{S}(f),$$

   for some fixed $\delta_1 > 0$, where $\mathrm{S}(f)$ denotes some Sobolev norm, and f is a function in $C^\infty\left(\mathrm{T}^1\mathbb{T}^2\right)$.
3. Inserting this into the proof of Theorem 5.3 gives

$$N(R) = \pi R^2 + \mathrm{O}\left(R^{1-\delta_2}\right)$$

   for some $\delta_2 > 0$.

5.2 Counting Points in $\mathrm{SL}_2(\mathbb{Z}) \cdot \mathrm{i} \subseteq \mathbb{H}$

The Upper Half-Plane Model of the Hyperbolic Plane

In order to formulate the next counting problem we briefly recall the basic properties of the hyperbolic plane, starting with the upper half-plane

$$\mathbb{H} = \{x + \mathrm{i}y \in \mathbb{C} \mid y > 0\}$$

and its tangent bundle $\mathrm{T}\mathbb{H} = \mathbb{H} \times \mathbb{C}$. If $\phi : [0, 1] \to \mathbb{H}$ is differentiable at $t \in [0, 1]$ with $\phi(t) = z$, then the derivative of ϕ at t is defined to be

$$\mathrm{D}\phi(t) = (\phi(t), \phi'(t)) \in \mathrm{T}_z\mathbb{H},$$

where $\mathrm{T}_z\mathbb{H} = \{z\} \times \mathbb{C}$ is the tangent space at z, which may be viewed as a complex vector space simply by ignoring the first component. This structure allows us to define the hyperbolic Riemannian metric using the collection of inner products

$$\langle v, w\rangle_z = \frac{1}{y^2}(v \cdot w)$$

for $z = x + \mathrm{i}y \in \mathbb{H}$ and $v, w \in \mathrm{T}_z\mathbb{H}$, where $(v \cdot w)$ is the inner product in $\mathbb{C}$ under the identification of $\mathbb{C}$ with $\mathbb{R}^2$ as real vector spaces. If

$$\phi : [0, 1] \to \mathbb{H}$$

is a path (that is, a continuous piecewise differentiable curve), then its length is

$$\mathrm{L}(\phi) = \int_0^1 \| \mathrm{D}\phi(t) \|_{\phi(t)} \, \mathrm{d}t$$

where $\| \mathrm{D}\phi(t) \|_{\phi(t)}$ denotes the length of the tangent vector

$$\mathrm{D}\phi(t) = (\phi(t), \phi'(t)) \in \mathrm{T}_{\phi(t)}\mathbb{H}$$

with respect to the norm derived from $\langle \cdot, \cdot \rangle_{\phi(t)}$. In other words the speed of the path at time t is $\| \mathrm{D}\phi(t) \|_{\phi(t)}$ and we obtain the length of the curve by integration. The hyperbolic metric is

$$\mathsf{d}(z_0, z_1) = \inf_{\phi} \mathrm{L}(\phi),$$

where the infimum is taken over all continuous piecewise differentiable curves ϕ with $\phi(0) = z_0$ and $\phi(1) = z_1$. The real line $\mathbb{R} \subseteq \mathbb{C}$ together with a single point ∞ is the *boundary* $\partial\mathbb{H}$ of $\mathbb{H}$.

The geometry of the upper half-plane model is connected to dynamics via the natural action of $\mathrm{SL}_2(\mathbb{R})$. Recall that an action of a group G on a set X is said to be transitive if for any $x_1, x_2 \in X$ there is a $g \in G$ with $g \cdot x_1 = x_2$, and is simply transitive if there is a unique $g \in G$ with $g \bullet x_1 = x_2$. The group $\mathrm{SL}_2(\mathbb{R})$ acts on $\mathbb{H}$ by the Möbius transformations

$$g = \begin{pmatrix} a & b \\ c & d \end{pmatrix} : z \mapsto \frac{az+b}{cz+d}. \tag{5.5}$$

The matrix $-I_2$ acts trivially on $\mathbb{H}$, so (5.5) actually defines an action of the projective special linear group

$$\mathrm{PSL}_2(\mathbb{R}) = \mathrm{SL}_2(\mathbb{R})/\{\pm I_2\},$$

but we will write $\begin{pmatrix} a & b \\ c & d \end{pmatrix}$ for the element $\pm \begin{pmatrix} a & b \\ c & d \end{pmatrix}$ of $\mathrm{PSL}_2(\mathbb{R})$.

We assemble the main properties arising from this action (see [13, Ch. 9] for the details).

1. The action of $\mathrm{PSL}_2(\mathbb{R})$ on $\mathbb{H}$ in (5.5) is isometric, meaning that

$$\mathsf{d}\left(g(z_0), g(z_1)\right) = \mathsf{d}(z_0, z_1)$$

   for any $z_0, z_1 \in \mathbb{H}$ and $g \in \mathrm{PSL}_2(\mathbb{R})$, and is transitive.
2. The action of $\mathrm{PSL}_2(\mathbb{R})$ on $\mathrm{T}\mathbb{H}$ defined by the derivative $\mathrm{D}\,g$ of the action of $g \in \mathrm{PSL}_2(\mathbb{R})$ on $\mathbb{H}$ preserves the Riemannian metric. This also implies that the volume measure defined by the formula

$$\mathrm{vol}(B) = \int_B \frac{\mathrm{d}x\,\mathrm{d}y}{y^2}$$

   is preserved by the action of $\mathrm{PSL}_2(\mathbb{R})$ (which can also be checked rigorously by a direct calculation without knowing what a Riemannian metric is).
3. The stabilizer

$$\mathrm{Stab}_{\mathrm{PSL}_2(\mathbb{R})}(\mathrm{i}) = \{g \in \mathrm{PSL}_2(\mathbb{R}) \mid g(\mathrm{i}) = \mathrm{i}\}$$

   of $\mathrm{i} \in \mathbb{H}$ is the projective special orthogonal group

$$\mathrm{PSO}(2) = \mathrm{SO}(2)/\{\pm I_2\}$$

   where

$$\mathrm{SO}(2) = \left\{ \begin{pmatrix} \cos\theta & -\sin\theta \\ \sin\theta & \cos\theta \end{pmatrix} \mid \theta \in \mathbb{R} \right\}$$

   is the compact group of rotations of the plane.
4. In particular, the action gives an identification

$$\mathbb{H} \cong \mathrm{PSL}_2(\mathbb{R})/\,\mathrm{PSO}(2),$$

   and under this identification the coset $g\,\mathrm{PSO}(2)$ corresponds to $g(\mathrm{i})$.

Write $\mathrm{T}^1\mathbb{H} = \{(z, v) \in \mathrm{T}\mathbb{H} \mid \|v\|_z = 1\}$ for the unit tangent bundle of $\mathbb{H}$ consisting of all unit vectors v attached to all possible points $z \in \mathbb{H}$. Since the action of $\mathrm{D}\,g$ preserves the length of tangent vectors, the restriction of $\mathrm{D}\,g$ defines an action of $\mathrm{PSL}_2(\mathbb{R})$ on $\mathrm{T}^1\mathbb{H}$ (naturally extending the action on $\mathbb{H}$ itself).

5. The action of $\mathrm{PSL}_2(\mathbb{R})$ on $\mathrm{T}^1\mathbb{H}$ is simply transitive, giving the identification

$$\mathrm{T}^1\mathbb{H} \cong \mathrm{PSL}_2(\mathbb{R}) \tag{5.6}$$

by choosing an arbitrary reference vector (z_0, v_0) in $\mathrm{T}^1\mathbb{H}$ corresponding to $I_2 \in \mathrm{PSL}_2(\mathbb{R})$; the identification is then given by sending g to $\mathrm{D}\,g(z_0, v_0)$. We will make the convenient choice $z_0 = \mathrm{i}$ and $v_0 = \mathrm{i}$.

The isometric action of elements of $\mathrm{PSL}_2(\mathbb{R})$ can be used to fold $\mathbb{H}$ into a smaller surface (whose unit tangent bundle can be obtained by folding $\mathrm{T}^1\mathbb{H}$). We note that this is similar to how one uses the isometric translations on $\mathbb{R}^d$ by vectors in $\mathbb{Z}^d$ to obtain the quotient space $\mathbb{T}^d = \mathbb{R}^d/\mathbb{Z}^d$. The prototypical example of a discrete subgroup of $\mathrm{PSL}_2(\mathbb{R})$ that we can use in this way is $\mathrm{PSL}_2(\mathbb{Z})$. The quotient space $\mathrm{PSL}_2(\mathbb{Z})\backslash\mathbb{H}$ actually has finite volume (see Figure 5.5, where the identification of $z \in \mathbb{H}$ with $-1/z$ and with $1 + z$ are used), and is called the modular surface. A discrete subgroup $\Gamma < \mathrm{PSL}_2(\mathbb{R})$ is called a lattice if $\Gamma\backslash\mathbb{H}$ (or equivalently $\Gamma\backslash\mathrm{PSL}_2(\mathbb{R})$) has finite volume, once again we refer to [13] for the details. Note that after taking the quotient by the lattice on the left there is no natural action on the left, but there is still a natural action of $\mathrm{PSL}_2(\mathbb{R})$ on $\Gamma\backslash\mathrm{PSL}_2(\mathbb{R})$ on the right. Here the diagonal subgroup has the natural interpretation as the geodesic flow on the unit tangent bundle of the surface $\Gamma\backslash\mathbb{H}$.

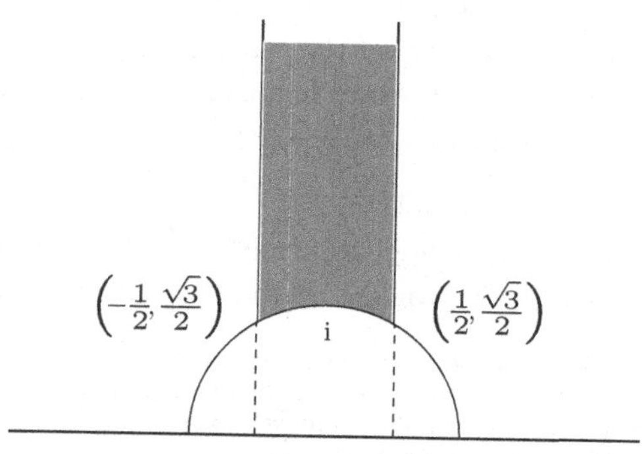

Figure 5.5 A fundamental domain for $\mathrm{PSL}_2(\mathbb{R})\backslash\mathbb{H}$ obtained by using the identifications $z \mapsto -1/z$ and $z \mapsto 1 + z$.

The above provides the first example of homogeneous dynamics, that is the study of actions of subgroups of a Lie group G on quotients $\Gamma\backslash G$, where Γ is a lattice subgroup (a discrete subgroup whose associated quotient space has finite volume with respect to the measure inherited from the Haar measure on G). We write $g\centerdot x = xg^{-1}$ for the action of an element $g \in G$ on $x = \Gamma h$ in $\Gamma\backslash G$. We will also make use of several results from homogeneous dynamics. The first of these will be a rather general mixing[3] principle: for any lattice Γ in $\mathrm{SL}_2(\mathbb{R})$ the action of $\mathrm{SL}_2(\mathbb{R})$ on $\Gamma\backslash\mathrm{SL}_2(\mathbb{R})$ is mixing (and so, in particular, the action of any infinite subgroup will also be mixing). We refer to [13, Sec. 11.4] for the proof.

5.2.1 The Counting Problem

For $\Gamma = \mathrm{PSL}_2(\mathbb{Z})$ (or any other lattice in $\mathrm{PSL}_2(\mathbb{R})$), we define

$$N(R) = |\{\gamma(\mathrm{i}) \mid \mathsf{d}(\gamma(\mathrm{i}), \mathrm{i}) < R, \gamma \in \Gamma\}|.$$

Theorem 5.4 (Selberg)

$$N(R) = \frac{\mathrm{vol}\left(B_R^{\mathbb{H}}(\mathrm{i})\right)}{\mathrm{vol}\left(\Gamma\backslash\mathbb{H}\right) |\,\mathrm{Stab}_\Gamma(\mathrm{i})|} + \mathrm{o}\Big(\mathrm{vol}\left(B_R^{\mathbb{H}}(\mathrm{i})\right)\Big)$$

as $R \to \infty$.

Selberg [29] used spectral methods to prove this[4], and also obtained more information about the error term. We present an approach following Eskin and McMullen [15] that uses the ergodic-theoretic property of mixing, following the approach of Duke, Rudnick and Sarnak [12]. In Section 5.1, the main term for the lattice point counting problem came from the simple geometric approach of tiling a ball of large radius with translates of a chosen fundamental domain for $\mathbb{Z}^2$ in $\mathbb{R}^2$, and this might be expected to give a heuristic rationale for the main term once again. However in the new setting this thinking cannot carry us further for two reasons connected to the geometry of the space.

1. We will see shortly that $\mathrm{vol}\left(B_R^{\mathbb{H}}(\mathrm{i})\right) \sim \pi \mathrm{e}^R$, so the volume of an annulus $B_{R+c}^{\mathbb{H}}(\mathrm{i}) \smallsetminus B_R^{\mathbb{H}}(\mathrm{i})$ is approximately $\pi \mathrm{e}^R(\mathrm{e}^c - 1)$, and is therefore comparable in size to the volume of the ball. In other words, the error

[3] The action of G is called *mixing* if for any Borel measurable sets $A, B \subseteq X$ we have $m(A \cap g_n\centerdot B) \to m(A)m(B)$ as $g_n \to \infty$ in G, where m is the measure inherited from Haar measure on G and $g_n \to \infty$ means that for any compact subset $K \subseteq \Gamma$ there is some $N = N(K)$ such that $g_n \notin K$ for $n \geqslant N$.

[4] For the history and primary references of these developments we refer to the paper of Phillips and Rudnick [27].

term produced by the annulus has the same order of magnitude as the main term. In Section 5.1 the analogous volume grew quadratically, giving the corresponding annulus linear growth.
2. The tiles that arise in the case $\Gamma = \mathrm{SL}_2(\mathbb{R})$ are unbounded with respect to the metric, so in order to use an annulus to capture all of them we should really use $c = \infty$ (or at least some large value to capture most of the translates of the fundamental domain).

These phenomena – manifestations of the hyperbolic geometry of the space $\mathrm{SL}_2(\mathbb{Z})\backslash \mathrm{SL}_2(\mathbb{R})$ – make the lattice point counting problem considerably more subtle. Nonetheless, the volume of the ball does give us a starting point to discuss the problem.

Lemma 5.5 $\mathrm{vol}\left(B_R^{\mathbb{H}}(\mathrm{i})\right) = 2\pi\,(\cosh(R) - 1)$ *for all* $R > 0$.

An immediate consequence is that $\mathrm{vol}\left(B_R^{\mathbb{H}}(\mathrm{i})\right)$ is asymptotic to $\pi \mathrm{e}^R$ as R goes to infinity, as claimed above.

OUTLINE PROOF OF LEMMA 5.5. The volume calculation may be carried out using the disc model $\mathbb{D} = \{w \in \mathbb{C} \mid |w| < 1\}$ of the hyperbolic plane, which carries the Riemannian metric

$$4\frac{\mathrm{d}x^2 + \mathrm{d}y^2}{(1 - r^2)^2}$$

at the point $w = x + \mathrm{i}y$, where $r^2 = x^2 + y^2$. A calculation shows that the maps

$$\mathbb{D} \ni w \longmapsto z(w) = \mathrm{i}\frac{1 + w}{1 - w} \in \mathbb{H};$$

$$\mathbb{H} \ni z \longmapsto w(z) = \frac{z - \mathrm{i}}{z + \mathrm{i}} \in \mathbb{D}$$

are holomorphic and are inverses to each other. Moreover,

$$\Im\,(z(w)) = \frac{1 - |w|^2}{|1 - w|^2}$$

and

$$\frac{\mathrm{d}z}{\mathrm{d}w} = \frac{2\mathrm{i}}{(1 - w)^2}.$$

From this, one can check that the hyperbolic Riemannian metric

$$\frac{\mathrm{d}x^2 + \mathrm{d}y^2}{y^2}$$

is mapped to

$$4\frac{\mathrm{d}x^2+\mathrm{d}y^2}{(1-r^2)^2}.$$

So it is sufficient to take $w=0$ and calculate the volume of the ball of radius R around w, which by symmetry is a disc of some radius ρ around w in the Euclidean metric as well. The Euclidean radius ρ may be calculated using the relation

$$R=2\int_0^\rho \frac{\mathrm{d}r}{1-r^2}=\ln\left(\frac{1+r}{1-r}\right)\Big|_0^\rho=\ln\left(\frac{1+\rho}{1-\rho}\right),$$

or equivalently

$$\rho=\frac{\mathrm{e}^R-1}{\mathrm{e}^R+1}.$$

Thus the volume of the hyperbolic ball of radius R around w is given by

$$\begin{aligned}
\operatorname{vol}\left(B_R^{\mathbb{H}}(\mathrm{i})\right)&=4\int_0^\rho\int_0^{2\pi}\frac{r\,\mathrm{d}r\,\mathrm{d}\phi}{(1-r^2)^2}\\
&=4\pi\int_0^{\rho^2}\frac{\mathrm{d}u}{(1-u)^2}\\
&=4\pi\frac{1}{1-u}\Big|_0^{\rho^2}\\
&=4\pi\left(\frac{1}{1-\rho^2}-1\right)\\
&=2\pi\left(\frac{2(\mathrm{e}^R+1)^2}{(\mathrm{e}^R+1)^2-(\mathrm{e}^R-1)^2}-2\right)\\
&=2\pi\left(\cosh(R)-1\right).
\end{aligned}$$

□

We will also need the following equidistribution result concerning large circles as illustrated in Figure 5.6 (which will be seen as a consequence of mixing). Below we will work with $\mathrm{PSL}_2(\mathbb{R})$ but will still use the matrix notation for the elements of $\mathrm{PSL}_2(\mathbb{R})$. Thus, for example, we will write

$$k_\phi=\begin{pmatrix}\cos\phi & -\sin\phi\\ \sin\phi & \cos\phi\end{pmatrix}\in K=\mathrm{SO}(2)/\{\pm1\}=\{k_\phi\mid\phi\in[0,\pi)\}.$$

Theorem 5.6 (Equidistribution of Large Circles) *For any point z in $\mathbb{H}$, the circles obtained by following geodesics from z in all directions for time t equidistribute in $\mathrm{PSL}_2(\mathbb{Z})\backslash\mathrm{T}^1\mathbb{H}$. Indeed, for any finite volume quotient*

$$X=\Gamma\backslash\mathrm{PSL}_2(\mathbb{R})$$

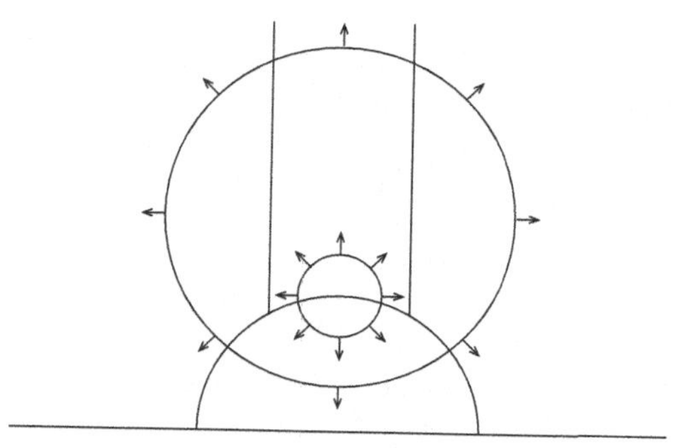

Figure 5.6 Equidistribution of large circles in the modular surface becomes visible after the circle is moved into the fundamental domain using the isometries in Γ.

we have[5]

$$\frac{1}{\pi}\int_0^{\pi} f\left(gk_\phi \cdot x\right)\, \mathrm{d}\phi \longrightarrow \frac{1}{m_X(X)}\int_X f\, \mathrm{d}m_X \tag{5.7}$$

as $g \to \infty$ *in* $\mathrm{PSL}_2(\mathbb{R})$.

Proof of Theorem 5.6. Recall that any element g can be written

$$g = k_1 a k_2$$

with $k_1, k_2 \in K$, $a = \begin{pmatrix} \lambda & \\ & \lambda^{-1} \end{pmatrix}$, and $\lambda > 0$. Here k_2 simply represents a rotation inside the integral on the left-hand side of (5.7), while the effect of k_1 may be thought of as replacing the function f by the function f^{k_1}, defined by $f^{k_1}(y) = f(k_1 \cdot y)$. Along a sequence $g_n = k_1^{(n)} a^{(n)} k_2^{(n)} \to \infty$, the choices we make for $k_1^{(n)}$ lie in the compact group K, so we may choose a converging subsequence. So assume that $k_1^{(n)}$ converges to k_0 say. Uniform continuity of $f \in C_c(X)$ then shows that $f^{k_1^{(n)}}$ converges uniformly to f^{k_0}. This reduces the problem to the case of a sequence

$$g_n = a_n = \begin{pmatrix} \lambda_n & \\ & \lambda_n^{-1} \end{pmatrix}$$

with $\lambda_n \to \infty$ as $n \to \infty$ (also since the case $\lambda_n \to 0$ as $n \to \infty$ may be dealt with similarly using the other unipotent). Given $f \in C_c(X)$ and any $\varepsilon > 0$ there is some $\delta_0 > 0$ with

[5] We will not normalize the Haar measure on $\Gamma\backslash \mathrm{PSL}_2(\mathbb{R})$ to be a probability measure, and instead will assume a canonical compatibility of the various Haar measures involved.

$$\mathsf{d}(e,h) < \delta_0 \implies |f(y) - f(h{\boldsymbol\cdot}y)| < \varepsilon$$

for all $y \in X$. Write

$$B = \left\{ \begin{pmatrix} \lambda & \\ s & \lambda^{-1} \end{pmatrix} \right\}$$

and let

$$B_\delta^B = \{h \in B \mid \mathsf{d}(h, I) < \delta\},$$

(where the metric is inherited from the metric on $\mathbb{H}$) so that

$$a_n B_\delta^B a_n^{-1} \subseteq B_{\delta_0}^B$$

for δ sufficiently small and all $n \geqslant 1$. Thus

$$\left| f(a_n k_\phi{\boldsymbol\cdot}x) - f(a_n h a_n^{-1} a_n k_\phi{\boldsymbol\cdot}x) \right| < \varepsilon,$$

and so

$$\left| \frac{1}{\pi}\int_0^\pi f(a_n k_\phi{\boldsymbol\cdot}x)\,\mathrm{d}\phi - \frac{1}{m_B\left(B_\delta^B\right)} \int_{B_\delta^B} \frac{1}{\pi}\int_0^\pi f(a_n h k_\phi{\boldsymbol\cdot}x)\,\mathrm{d}\phi\,\mathrm{d}m_B(h) \right| < \varepsilon.$$

Now notice that $B_\delta^B K \subseteq BK$ is a neighbourhood of the identity in the group $\mathrm{PSL}_2(\mathbb{R})$, and the Haar measure is locally the product of the two Haar measures, so the second integral in the above estimate equals

$$\frac{1}{m_G\left(B_\delta^B K\right)} \int_{B_\delta^B K} f(a_n g{\boldsymbol\cdot}x)\,\mathrm{d}m_G(g) = \frac{1}{m_X\left(B_\delta^B K{\boldsymbol\cdot}x\right)} \left\langle f, \mathbb{1}_{a_n B_\delta^B K{\boldsymbol\cdot}x} \right\rangle,$$

which converges to

$$\frac{1}{m_X(X)} \int f\,\mathrm{d}m_X$$

by the mixing property as $a_n \to \infty$. □

Proof of Theorem 5.4. We define for every $g \in G$ the counting function for the modified counting problem by

$$F_R(g) = \frac{1}{\mathrm{vol}(B_R^{\mathbb{H}})} \left| \{g\gamma(\mathrm{i}) \mid \gamma \in \Gamma\} \cap B_R^{\mathbb{H}}(\mathrm{i}) \right|,$$

where we write $\mathrm{vol}(B_R^{\mathbb{H}})$ for the volume of the hyperbolic R-ball around i (or any other point). It is clear that $F_R(g) = F_R(g\gamma)$ for any $\gamma \in \Gamma$ so that we can think of $F_R(g\Gamma) = F_R(g)$ as a function of[6]

$$x = g\Gamma \in X = G/\Gamma.$$

[6] This shows that in the connection of equidistribution problems to counting problems one is automatically led to the quotient $X = G/\Gamma$ (where of course the same theorems hold as on $\Gamma\backslash G$).

We want to show the convergence

$$F_R(\Gamma) \longrightarrow \frac{1}{\mathrm{vol}\,(\Gamma\backslash\mathbb{H})\,|\,\mathrm{Stab}_\Gamma(\mathrm{i})|} = c,$$

as $R \to \infty$. The more general definition allows us to first prove the weaker claim that

$$F_R(x)\,\mathrm{d}m_X \longrightarrow c\,\mathrm{d}m_X$$

in the weak* topology as $R \to \infty$. This will be deduced from Theorem 5.6 and a technique called *unfolding*, which means using different ways to express the integrals on the spaces shown below:

$$\begin{array}{ccc} & G & \\ \swarrow & & \searrow \\ X = G/\Gamma & & G/\mathrm{SO}(2) \cong \mathbb{H}. \end{array}$$

We use the hyperbolic area measure on $\mathbb{H}$ and the isomorphism

$$B = \left\{ \begin{pmatrix} 1 & x \\ & 1 \end{pmatrix} \begin{pmatrix} \sqrt{y} & \\ & \frac{1}{\sqrt{y}} \end{pmatrix} \mid x \in \mathbb{R}, y > 0 \right\} \cong \mathbb{H} = \{x + yi \mid x \in \mathbb{R}, y > 0\}$$

to normalize the left Haar measure on B, normalize the Haar measure on

$$K = \mathrm{SO}(2)/\{\pm I\}$$

to satisfy $m_K(K) = \pi$, and use the Iwasawa decomposition $G = BK$ to normalize the Haar measure on G (by declaring the proportionality constant to be equal to one).

For $\alpha \in C_c(X)$ the claimed weak* convergence involves the integral[7]

$$\begin{aligned}
\int_X F_R \alpha \,\mathrm{d}m_X &= \frac{1}{\mathrm{vol}(B_R^{\mathbb{H}})} \int_{G/\Gamma} \sum_{\gamma \in \Gamma/\mathrm{Stab}_\Gamma(\mathrm{i})} \mathbb{1}_{B_R^{\mathbb{H}}(\mathrm{i})}(g\gamma(\mathrm{i}))\alpha(g\Gamma)\,\mathrm{d}m_X \\
&= \frac{1}{\mathrm{vol}(B_R^{\mathbb{H}})} \frac{1}{|\,\mathrm{Stab}_\Gamma(\mathrm{i})|} \int_{G/\Gamma} \sum_{\gamma \in \Gamma} \mathbb{1}_{B_R^{\mathbb{H}}(\mathrm{i})}(g\gamma(\mathrm{i}))\alpha(g\Gamma)\,\mathrm{d}m_X \\
&= \frac{1}{\mathrm{vol}(B_R^{\mathbb{H}})} \frac{1}{|\,\mathrm{Stab}_\Gamma(\mathrm{i})|} \int_G \mathbb{1}_{B_R^{\mathbb{H}}(\mathrm{i})}(g(\mathrm{i}))\alpha(g\Gamma)\,\mathrm{d}m_G \\
&= \frac{1}{\mathrm{vol}(B_R^{\mathbb{H}})} \frac{1}{|\,\mathrm{Stab}_\Gamma(\mathrm{i})|} \int_B \mathbb{1}_{B_R^{\mathbb{H}}(\mathrm{i})}(b(\mathrm{i})) \int_K \alpha(bk\Gamma)\,\mathrm{d}m_K(k)\,\mathrm{d}m_B(b) \\
&= \frac{\pi}{|\,\mathrm{Stab}_\Gamma(\mathrm{i})|} \frac{1}{\mathrm{vol}(B_R^{\mathbb{H}})} \int_{B_R^{\mathbb{H}}(\mathrm{i})} \frac{1}{\pi} \int_K \alpha(bk\Gamma)\,\mathrm{d}m_K(k)\,\mathrm{d}m_{\mathbb{H}}(b\cdot\mathrm{i}).
\end{aligned}$$

[7] The sum over $\gamma \in \Gamma/\mathrm{Stab}_\Gamma(\mathrm{i})$ is defined so that every coset contributes precisely one summand (by using an arbitrary representative of the coset).

However, by Theorem 5.6, the last expression is an integral version of a Cesàro average of a convergent sequence. It follows that

$$\int_K F_R \alpha \,\mathrm{d}m_X \longrightarrow \frac{\pi}{m_X(X)|\operatorname{Stab}_\Gamma(\mathrm{i})|}\int_X \alpha\,\mathrm{d}m_X$$

as $R \to \infty$. By our normalizations of the Haar measures we also have

$$m_X(X) = \operatorname{vol}(\Gamma\backslash\mathbb{H})\, m_K(K) = \operatorname{vol}(\Gamma\backslash\mathbb{H})\,\pi.$$

Hence we have shown that an averaged version of our orbit-point counting asymptotic exists. To arrive at the actual count from the averaged version we will localize the function α near $\Gamma \in X$.

Given $\varepsilon > 0$ there exists some $\delta > 0$ such that

$$\frac{\operatorname{vol}(B^{\mathbb{H}}_{R-\delta})}{\operatorname{vol}(B^{\mathbb{H}}_R)} > 1 - \varepsilon$$

for large enough R (since by Lemma 5.5 the former quantity is asymptotic to $\mathrm{e}^{-\delta}$). Given this δ there exists some $\rho > 0$ such that $g \in B^G_\rho$ implies $\mathsf{d}_{\mathbb{H}}(g(\mathrm{i}), \mathrm{i}) < \delta$.

Suppose now that $\alpha \in C_c(\Gamma\backslash G)$ has $\int \alpha\,\mathrm{d}m_X = 1$, $\alpha \geqslant 0$, and

$$\operatorname{supp}(\alpha) \subseteq B^G_\rho \Gamma.$$

Then for $x \in \operatorname{supp}(\alpha)$ we have

$$\begin{aligned} F_{R-\delta}(x) &= \frac{1}{\operatorname{vol}(B^{\mathbb{H}}_{R-\delta})}\left|\{g\gamma(\mathrm{i}) \mid \gamma \in \Gamma\} \cap B^{\mathbb{H}}_{R-\delta}(\mathrm{i})\right| \\ &= \frac{1}{\operatorname{vol}(B^{\mathbb{H}}_{R-\delta})}\left|\{\gamma(\mathrm{i}) \mid \gamma \in \Gamma\} \cap \underbrace{g^{-1}(B^{\mathbb{H}}_{R-\delta}(\mathrm{i}))}_{\subseteq B^{\mathbb{H}}_R(\mathrm{i})}\right| \\ &\leqslant \frac{\operatorname{vol}(B^{\mathbb{H}}_R)}{\operatorname{vol}(B^{\mathbb{H}}_{R-\delta})}\frac{1}{\operatorname{vol}(B^{\mathbb{H}}_R)}\left|\Gamma(\mathrm{i}) \cap B^{\mathbb{H}}_R\right|, \end{aligned}$$

where we used the fact that $\mathsf{d}_{\mathbb{H}}(z, \mathrm{i}) < R - \delta$ and $g \in B^G_\rho$ implies

$$\mathsf{d}_{\mathbb{H}}(g^{-1}(z), \mathrm{i}) \leqslant \mathsf{d}_{\mathbb{H}}(g^{-1}(z), g^{-1}(\mathrm{i})) + \mathsf{d}_{\mathbb{H}}(g^{-1}(\mathrm{i}), \mathrm{i}) < R.$$

We multiply by α and integrate to get

$$\frac{\operatorname{vol}(B^{\mathbb{H}}_{R-\delta})}{\operatorname{vol}(B^{\mathbb{H}}_R)} \underbrace{\int F_{R-\delta}(x)\alpha(x)\,\mathrm{d}m_X}_{\longrightarrow \frac{1}{\operatorname{vol}(\Gamma\backslash\mathbb{H})\operatorname{Stab}_\Gamma(\mathrm{i})}} \leqslant \frac{N(R)}{\operatorname{vol}(B^{\mathbb{H}}_R)}.$$

This gives

$$\liminf_{R\to\infty} \frac{N(R)}{\operatorname{vol}(B_R^{\mathbb{H}})} \geqslant (1-\varepsilon)\frac{1}{\operatorname{vol}(\Gamma\backslash\mathbb{H})|\operatorname{Stab}_\Gamma(\mathrm{i})|}.$$

The reverse inequality works in the same way. Choosing $\delta > 0$, $\rho > 0$, and $\alpha \in C_c(X)$ according to some $\varepsilon > 0$ we get

$$\begin{aligned} F_{R+\delta}(x) &= \frac{1}{\operatorname{vol}(B_{R+\delta}^{\mathbb{H}})}\Big|\Gamma(\mathrm{i}) \cap \underbrace{g^{-1}(B_{R+\delta}^{\mathbb{H}}(\mathrm{i}))}_{\supset B_R^{\mathbb{H}}(\mathrm{i})}\Big| \\ &\geqslant \frac{\operatorname{vol}(B_R^{\mathbb{H}})}{\operatorname{vol}(B_{R+\delta})}\frac{N(R)}{\operatorname{vol}(B_R^{\mathbb{H}})} \end{aligned}$$

whenever $x = g\Gamma$ with $g \in B_\rho^G$. As before, this leads to the bound

$$\limsup_{R\to\infty} \frac{N(R)}{\operatorname{vol}(B_R^{\mathbb{H}})} \leqslant (1+\varepsilon)\frac{1}{\operatorname{vol}(\Gamma\backslash\mathbb{H})|\operatorname{Stab}_\Gamma(\mathrm{i})|}.$$

□

These kinds of methods have been developed much further by Eskin and McMullen [15] (building on work of Duke, Rudnick and Sarnak [12], who were using a different argument to obtain the equidistribution) using mixing to establish asymptotic counting results in a more general context. This relates a counting problem for points in Γ-orbits on $V = G/H$ to the equidistribution problem for 'translated' H-orbits of the form

$$gH\Gamma \subseteq X = G/\Gamma.$$

In many cases (for example, in the context of affine symmetric spaces), these methods can be used to count integer points on varieties asymptotically. If G and H consist of the $\mathbb{R}$-points of algebraic groups defined over $\mathbb{Q}$, then the variety $V = G/H$ can be identified with a variety defined over $\mathbb{Q}$, and $V(\mathbb{Z})$ is non-empty. In this case $V(\mathbb{Z})$ is a disjoint union

$$V(\mathbb{Z}) = \bigsqcup_i G(\mathbb{Z})v_i$$

of different $\Gamma = G(\mathbb{Z})$-orbits. This is often a finite union, in which case the asymptotic for $|V(\mathbb{Z}) \cap B_t|$ can be obtained by gluing together the results for the individual counts $|G(\mathbb{Z})v_i \cap B_t|$. We refer to the original papers for these results.

5.3 Dirichlet's Theorem and Dani's Correspondence

We start with a classical result on simultaneous Diophantine approximation. In stating this we will write $\frac{p}{q}$ for the vector $(\frac{p_j}{q})_j$ if $p = (p_j)_j$ is a vector and $q \in \mathbb{N}$.

Theorem 5.7 (Dirichlet's Theorem [9]) *For any $v \in \mathbb{R}^d$ and any integer Q there exist an integer q with $1 \leqslant q \leqslant Q^d$, and an integer vector $p \in \mathbb{Z}^d$ with*

$$\left\| v - \tfrac{p}{q} \right\|_\infty \leqslant \tfrac{1}{qQ}.$$

THE CLASSICAL PROOF OF THEOREM 5.7. Consider the $(Q^d + 1)$ points

$$0, v, \dots, Q^d v \pmod{\mathbb{Z}^d} \tag{5.8}$$

as elements of $\mathbb{T}^d \cong [0, 1)^d$. Now partition $[0, 1)$ into the Q intervals

$$[0, \tfrac{1}{Q}), [\tfrac{1}{Q}, \tfrac{2}{Q}), \dots, [\tfrac{Q-1}{Q}, 1),$$

and correspondingly divide $[0, 1)^d$ into Q^d cubes with sides chosen from the partition of each of the d axes. By the Pigeonhole principle[8] there exist two integers k, ℓ with $0 \leqslant k < \ell \leqslant Q^d$ such that the points kv and ℓv considered modulo $\mathbb{Z}^d$ from (5.8) belong to the same subcube. Letting $q = \ell - k$ gives

$$\|qv - p\|_\infty \leqslant \tfrac{1}{Q}$$

for some $p \in \mathbb{Z}^d$ as required. □

We will describe the connection between the theory of Diophantine approximation and homogeneous dynamics by studying the following refined property.

Definition 5.8 Fix $\lambda \in (0, 1]$. A vector $v \in \mathbb{R}^d$ is called *λ-Dirichlet improvable* if for every large enough Q there exists an integer q satisfying

$$1 \leqslant q \leqslant \lambda Q^d \text{ and } \left\| v - \tfrac{p}{q} \right\|_\infty \leqslant \lambda \frac{1}{qQ} \tag{5.9}$$

for some $p \in \mathbb{Z}^d$. A vector is simply called *Dirichlet-improvable* if it is λ-improvable for some $\lambda < 1$.

[8] Dirichlet [10] called it the *Schubfachprinzip* (drawer or shelf principle), and it is sometimes called Dirichlet's principle because he used it in this setting.

In order to describe the correspondence between this notion and homogeneous dynamics, we write[9] $\Lambda_v = u_v\mathbb{Z}^{d+1}$ where

$$u_v = \begin{pmatrix} 1 & \\ v & I_d \end{pmatrix},$$

and

$$g_Q = \begin{pmatrix} Q^{-d} & \\ & QI_d \end{pmatrix}.$$

Proposition 5.9 (Dani Correspondence) *Let $v \in \mathbb{R}^d$, $Q > 1$, and λ be given with $0 < \lambda \leqslant 1$. Then there exists an integer q satisfying (5.9) if and only if the lattice in $\mathbb{R}^{d+1}$ corresponding to $g_Q\Lambda_v$ intersects $[-\lambda, \lambda]^{d+1}$ non-trivially.*

PROOF. Suppose that the integer q satisfies (5.9) for some $p \in \mathbb{Z}^d$. Then the vector

$$\begin{pmatrix} q \\ qv - p \end{pmatrix} = \begin{pmatrix} 1 & 0 \\ v & 1 \end{pmatrix} \begin{pmatrix} q \\ -p \end{pmatrix} \in \Lambda_v$$

belongs to the lattice corresponding to v, and

$$g_Q \begin{pmatrix} q \\ qv - p \end{pmatrix} \in g_Q\Lambda_v$$

satisfies

$$\left\| g_Q \begin{pmatrix} q \\ qv - p \end{pmatrix} \right\|_\infty = \max\left(|qQ^{-d}|, Q\|qv - p\|_\infty\right) \leqslant \lambda. \tag{5.10}$$

Now suppose on the other hand that there is a non-trivial vector

$$\begin{pmatrix} q \\ qv - p \end{pmatrix} \in \Lambda_v$$

satisfying (5.10). We claim that $q \neq 0$. Assuming this for the moment, we may also assume that q is positive[10], and then (5.10) is equivalent to (5.9).

To prove the claim, suppose that $q = 0$. However, in this case (5.10) becomes

$$\left\| g_Q \begin{pmatrix} 0 \\ p \end{pmatrix} \right\|_\infty = Q\|p\| \leqslant \lambda \leqslant 1,$$

[9] Most of the research papers concerning the interaction between homogeneous dynamics and Diophantine approximation use the description $X = G/\Gamma$ instead of $X = \Gamma\backslash G$, so we will adhere to this tradition here.

[10] For otherwise we may replace $\begin{pmatrix} q \\ qv - p \end{pmatrix}$ by $\begin{pmatrix} -q \\ -qv + p \end{pmatrix}$.

which forces $p = 0$ since $Q > 1$. This contradicts our assumption that

$$\begin{pmatrix} q \\ qv - p \end{pmatrix}$$

is non-trivial, which proves the claim and completes the proof. □

This correspondence allows Dirichlet's theorem to be proved using homogeneous dynamics. The most important locally homogeneous space for ergodic theory and its connections to number theory is the space

$$\mathsf{X}_d = \mathrm{SL}_d(\mathbb{R}) / \mathrm{SL}_d(\mathbb{Z}),$$

which may be identified with the space of unimodular lattices in $\mathbb{R}^d$. Just as in the case of the space X_2 from the last section, the space X_d has finite volume which enables us to use ergodic theoretic and dynamical properties of actions of subgroups of $\mathrm{SL}_d(\mathbb{R})$ on it. We refer to [14] for more details about the properties of this space and dynamical properties of various subgroups acting on it.

PROOF OF THEOREM 5.7 USING DYNAMICS. We set $\lambda = 1$ in Proposition 5.9. Then the theorem follows if we recall that any unimodular lattice Λ in $\mathbb{R}^{d+1}$ has to intersect $[-1, 1]^{d+1}$. In fact this is the content of Minkowski's theorem on convex bodies (and follows quite easily from the property that the volume of $[-\frac{1}{2} - \varepsilon, \frac{1}{2} + \varepsilon]^d$ is larger than one and so must contain points that are equivalent modulo Λ). □

Using ergodicity of the dynamics of

$$a_t = \begin{pmatrix} \mathrm{e}^{-dt} & \\ & \mathrm{e}^t I_d \end{pmatrix}$$

on $\mathsf{X}_{d+1} = \mathrm{SL}_{d+1}(\mathbb{R}) / \mathrm{SL}_{d+1}(\mathbb{Z})$ we can prove the following, recovering a result of Davenport and Schmidt [8].

Corollary 5.10 *Almost no vector $v \in \mathbb{R}^d$ is Dirichlet-improvable.*

PROOF. Let $\lambda \in (0, 1)$ and define the open neighbourhood

$$O_\lambda = \{\Lambda \in \mathsf{X}_{d+1} \mid \Lambda \cap [-\lambda, \lambda]^{d+1} = \{0\}\}$$

of $\mathbb{Z}^{d+1}$. Furthermore, let $O'_\lambda \subseteq O_\lambda$ be a non-trivial open subset with the property that $\overline{O'_\lambda} \subseteq O_\lambda$, so that in particular the Hausdorff distance

$$\varepsilon = \mathsf{d}(O'_\lambda, \mathsf{X}_{d+1} \smallsetminus O_\lambda)$$

from O'_λ to the complement of O_λ is positive.

As a consequence of ergodicity we obtain that almost every $v \in \mathbb{R}^d$ has the property that the a_t-orbit for $t \geqslant 0$ of $\Lambda_v = u_v \mathbb{Z}^{d+1}$ is dense. This should be surprising at first, since the space X_{d+1} is $(d^2 + 2d)$-dimensional and ergodicity concerns the uniform measure on this space, while v varies over only the d-dimensional plane. We sketch the argument: given some v_0, any lattice Λ near Λ_{v_0} can be written in the form

$$\Lambda = \begin{pmatrix} a & u \\ 0 & B \end{pmatrix} \Lambda_v$$

for some a near 1, some matrix $B \in \mathrm{GL}_d(\mathbb{R})$ near the identity, some row vector $u \in \mathbb{R}^d$, and some v near v_0 (as these elements of the various spaces just give a smooth coordinate system of a neighbourhood of the identity in $\mathrm{SL}_{d+1}(\mathbb{R})$ by the implicit function theorem). It is now easy to check that the vector u in the matrix $D = \begin{pmatrix} a & u \\ 0 & B \end{pmatrix}$ gets contracted by conjugation by a_t as $t \to \infty$. For the dynamics on X_{d+1} this implies that the orbit of $\Lambda = D\Lambda_v$ is dense if and only if the orbit of $\Lambda'_v = \begin{pmatrix} a & 0 \\ 0 & B \end{pmatrix} \Lambda_v$ is dense. Next we notice that the remaining matrix $D' = \begin{pmatrix} a & 0 \\ 0 & B \end{pmatrix}$ commutes with a_t. Using this we obtain that the orbit of $D'\Lambda_v$ is simply the orbit of Λ_v multiplied by D' on the left – so once more either both orbits are dense or neither of them is. Therefore the set of dense orbits is locally a product of some subset of $\mathbb{R}^d$ (corresponding to the parameter v) and the remaining directions, and the almost everywhere statement on X_{d+1} really just becomes an almost everywhere statement on $\mathbb{R}^d$.

Now let $t \geqslant 0$ be very large and chosen so that $a_t \bullet \Lambda_v \in O'_\lambda$. We set $Q = \lfloor \mathrm{e}^t \rfloor$ and deduce that

$$g_Q a_t^{-1} = \begin{pmatrix} Q^d \mathrm{e}^{-td} & \\ & Q^{-1} \mathrm{e}^t I_d \end{pmatrix}$$

is very close to I_{d+1}, in particular $\mathsf{d}(g_Q a_t^{-1}, I_{d+1}) < \varepsilon$ for sufficiently large t. From this we conclude that

$$g_Q \Lambda_v = g_Q a_t^{-1} a_t \Lambda_v \in O_\lambda.$$

By Proposition 5.9, there is no integer q satisfying (5.9) for Q. It follows that v is not λ-Dirichlet improvable.

Applying this to $\lambda = 1 - \frac{1}{n}$ gives the corollary. □

5.3.1 Nondivergence of Polynomial Trajectories and Diophantine Approximation

A general theme in the theory of Diophantine approximation is to try and show inheritance of Diophantine properties on $\mathbb{R}^d$ to submanifolds, or even more generally to fractals. We will not discuss the general framework concerning the inheritance of Diophantine properties to 'sufficiently curved smooth manifolds' but will explore some key developments by using a concrete polynomial curve.

As mentioned before, Davenport and Schmidt [8] showed that almost every point of $\mathbb{R}^d$ is not Dirichlet-improvable and later showed in [7] that almost every point on the curve (t, t^2) is not $(1/4)$-improvable. Baker [1] extended this to the same statement for almost every point on a sufficiently smooth curve in $\mathbb{R}^2$, and Dodson, Rynne, and Vickers [11] to almost every point on a sufficiently smooth curved manifold. Bugeaud [2] extended the result to the specific curve $(t, t^2, \ldots, t^d)$. Kleinbock and Weiss [21] used the correspondence introduced by Dani [5], [4] and the machinery of Kleinbock and Margulis [22] to formulate some of these questions in homogeneous dynamics, and the argument used for the proof of Corollary 5.12 is the argument used in [21]. We refer to a paper of Shah [30] for more details on the background and for another direction of similar results for curves that do not lie in translates of proper subspaces.

One of the driving forces for using dynamical theorems to prove results concerning Diophantine approximations is the following quantitative non-divergence result. Identifying an element $\Lambda = g\,\mathrm{SL}_d(\mathbb{Z}) \in \mathsf{X}_d$ with the corresponding unimodular lattice $g\mathbb{Z}^d$ in $\mathbb{R}^d$, we define

$$\mathsf{X}_d(\varepsilon) = \{\Lambda \in \mathsf{X}_d \mid \Lambda \cap B_\varepsilon(0) = \{0\}\},$$

and note that these are compact sets – this statement is usually referred to as Mahler's compactness criterion. Given a discrete subgroup $\Lambda < \mathbb{R}^d$ (which may be of lower rank than d) we write $\mathrm{covol}(\Lambda)$ for the volume of the quotient $\mathbb{R}\Lambda/\Lambda$, where $\mathbb{R}\Lambda$ is the $\mathbb{R}$-linear hull of Λ.

Theorem 5.11 (Quantitative Non-Divergence by Margulis, Dani and Kleinbock[11]) *Suppose that $p : \mathbb{R} \to \mathrm{SL}_d(\mathbb{R})$ is a polynomial and $T > 0$ is such that*

$$\sup_{t\in[0,T]} \mathrm{covol}(V, t) \geqslant \eta^{\dim V} \tag{5.11}$$

[11] This result has a long history; see Margulis [24], [25]; Dani [3], [6]; Kleinbock and Margulis [22]; Kleinbock [20]. For simplicity we present it here only for polynomial maps, but the actual result is more general.

(where $\mathrm{covol}(V, t) = \mathrm{covol}\left((V \cap \mathbb{Z}^d)p(t)\right)$*) for some* $\eta \in (0, 1]$ *and all rational subspaces* $V \subseteq \mathbb{R}^d$*. Then, for any choice of* $\varepsilon \in (0, \eta]$,

$$\frac{1}{T}\,|\{t \in [0, T] \mid \Gamma p(t) \notin \mathsf{X}_d(\varepsilon)\}| \ll_{d,D} \left(\frac{\varepsilon}{\eta}\right)^{1/D}, \tag{5.12}$$

where D depends only on the degree of the polynomial p.

Corollary 5.12 *For* $d \geqslant 2$ *there exists some* $\lambda_0 \in (0, 1)$ *such that almost no* $t \in \mathbb{R}$ *has the property that*

$$v(t) = \begin{pmatrix} t \\ t^2 \\ \vdots \\ t^d \end{pmatrix} \in \mathbb{R}^d$$

is λ_0*-Dirichlet improvable.*

For the proof we will need the following geometric input regarding the dynamics of g_Q on $\mathbb{R}^{d+1}$.

Lemma 5.13 *Let* $W \subseteq \mathbb{R}^{d+1}$ *be a k-dimensional subspace, and let*

$$c = \mathsf{d}(e_1, W) = \inf_{w \in W} \|e_1 - w\| \geqslant 0.$$

If $w_1, \ldots, w_k \in W$ *is an orthonormal basis of W, then*

$$\|g_Q w_1 \wedge \cdots \wedge g_Q w_k\| \geqslant c Q^k.$$

PROOF. Let

$$w_i = \varepsilon_i e_1 + w_i'$$

with $\varepsilon_i \in \mathbb{R}$ and $w_i' \in \langle e_2, \ldots, e_d \rangle$ for $i = 1, \ldots, d$. Clearly

$$w_1 \wedge \cdots \wedge w_k = w_1' \wedge \cdots \wedge w_k' + \sum_{i=1}^{k} \varepsilon_i \underbrace{w_1' \wedge \cdots \wedge e_1 \wedge \cdots \wedge w_k'}_{\text{with } e_1 \text{ in place of } w_i'}. \tag{5.13}$$

We calculate the norm using the fact that e_1 is normal to $w_1', \ldots, w_k'$, the identity (5.13), and the fact that $\|w_1 \wedge \cdots \wedge w_k\| = 1$ to obtain

$$\begin{aligned} \|w_1' \wedge \cdots \wedge w_k'\| &= \|e_1 \wedge w_1' \wedge \cdots \wedge w_k'\| \\ &= \|e_1 \wedge w_1 \wedge \cdots \wedge w_k\| = c \cdot 1 \end{aligned}$$

since the distance from e_1 to W is c. We now apply the map g_Q to get

$$\begin{aligned}\|g_Q w_1 \wedge \cdots \wedge g_Q w_k\|^2 &= \|g_Q w_1' \wedge \cdots \wedge g_Q w_k'\|^2 \\ &\quad + \Big\| \sum_{i=1}^{k} \varepsilon_i g_Q w_1' \wedge \cdots g_Q e_1 \wedge \cdots g_Q w_k' \Big\|^2 \\ &\geqslant (Q^k c)^2,\end{aligned}$$

where we have used the fact that

$$g_Q w_1' \wedge \cdots \wedge g_Q w_k' = Q^k w_1' \wedge \cdots \wedge w_k'$$

is orthogonal to the other sum. □

Proof of Corollary 5.12. We will use the notation O_ε from the proof of Corollary 5.10. We set $\eta = 1$ and wish to apply Theorem 5.11 on X_{d+1}. We will define the precise polynomial for the application of Theorem 5.11, but for now let us agree that this will be a modified version of the polynomial

$$p_0(t) = \begin{pmatrix} 1 & \\ v(t) & I_d \end{pmatrix} = \begin{pmatrix} 1 & & & \\ t & 1 & & \\ \vdots & & \ddots & \\ t^d & & & 1 \end{pmatrix},$$

so that the parameter D in Theorem 5.11 is already determined. By that theorem, and the equivalence of the norms $\|\cdot\|$ and $\|\cdot\|_\infty$ on $\mathbb{R}^{d+1}$, there exists some $\varepsilon > 0$ so that for any polynomial $p(t)$ (with the same D as p_0) satisfying

$$\sup_{t\in[0,T]} \operatorname{covol}(V, t) \geqslant 1 \tag{5.14}$$

for all rational subspaces $V \subseteq \mathbb{R}^d$ has

$$\frac{1}{T}\left|\left\{t \in [0, T] \mid p(t)\mathbb{Z}^{d+1} \notin O_\varepsilon\right\}\right| \leqslant \frac{1}{2}. \tag{5.15}$$

Now assume that the corollary is false for $\lambda_0 = \varepsilon$. Then

$$DT_\varepsilon = \{t \in \mathbb{R} \mid v(t) \text{ is } \varepsilon\text{-Dirichlet improvable}\}$$

must have a Lebesgue density point. In particular, there exists an interval $[\alpha, \beta] \subseteq \mathbb{R}$ such that

$$\frac{1}{\beta - \alpha} |\{t \in [\alpha, \beta] \mid v(t) \text{ is } \varepsilon\text{-Dirichlet improvable}\}| \geqslant \frac{9}{10}.$$

Using the definition of ε-Dirichlet improvable (and the basic property of measures), we find some Q_0 such that

$$\frac{1}{\beta-\alpha}\left|\{t\in[\alpha,\beta]\mid v(t)\text{ has (5.9) with }\lambda=\varepsilon\text{ for every }Q\geqslant Q_0\}\right|\geqslant\frac{3}{4}.$$

Using the matrix $p_0(t)=\begin{pmatrix}1&\\v(t)&I_d\end{pmatrix}$ and Dani's correspondence (Proposition 5.9), we can also phrase this as

$$\frac{1}{\beta-\alpha}\left|\left\{t\in[\alpha,\beta]\mid g_Q p_0(t)\mathbb{Z}^{d+1}\notin O_\varepsilon\text{ for every }Q\geqslant Q_0\right\}\right|\geqslant\frac{3}{4}. \quad (5.16)$$

To get a contradiction to (5.15), we have to show the assumption (5.14) for

$$p(t)=g_Q p_0(\alpha+t)$$

and $T=\beta-\alpha$. Assume therefore that (5.14) does not hold, meaning that for every $Q\geqslant Q_0$ there is a rational subspace $V_Q\subseteq\mathbb{R}^d$ with

$$\sup_{t\in[\alpha,\beta]}\operatorname{covol}(V_Q,t)<1,$$

where we are using $g_Q p(t)$ for the definition of $\operatorname{covol}(V_Q,t)$. Since the square of the covolume is a polynomial, this implies that

$$\sup_{t\in[0,1]}\operatorname{covol}(V_Q,t)\ll_{\alpha,\beta}1.$$

We set $t=\frac{i}{d}$, $W=p\left(\frac{i}{d}\right)V_Q$ and obtain

$$Q^{\dim V_Q}\mathsf{d}(e_1,W)\leqslant\|g_Q w_1\wedge\cdots\wedge g_Q w_k\|=\operatorname{covol}(V_Q,\tfrac{i}{d})\ll_{\alpha,\beta}1$$

from Lemma 5.13. Applying $p_0\left(\frac{i}{d}\right)^{-1}$, this gives

$$\mathsf{d}\left(p_0\left(\tfrac{i}{d}\right)^{-1}e_1,V_Q\right)\ll_{\alpha,\beta}Q^{-1} \quad (5.17)$$

for $i=1,\dots,d$. However, the vectors $p_0\left(\frac{i}{d}\right)^{-1}e_1$ for $i=0,1,\dots,d$ are easily checked to be linearly independent, and for large enough Q the condition (5.17) forces $V_Q=\mathbb{R}^{d+1}$. Since

$$\operatorname{covol}(\mathbb{R}^{d+1},t)=1$$

this contradicts our choice of V_Q, which proves (5.14) for large enough Q and gives a contradiction between (5.15) and (5.16). □

Using unipotent dynamics (Ratner's measure classification [28] – see also the monographs of Witte Morris [26] and the authors [14] for later treatments – and the full force of the linearization technique) Shah significantly strengthened Corollary 5.12, giving in particular the following result.

Theorem (Shah [31]) *Let $d \geqslant 2$. Then almost no $t \in \mathbb{R}$ has the property that*

$$v(t) = \begin{pmatrix} t \\ \vdots \\ t^d \end{pmatrix}$$

is Dirichlet improvable.

This result is a consequence of a more general equidistribution theorem, a special case of which is the following.

Theorem (Shah [31]) *Let $d \geqslant 2$, let $I \subseteq \mathbb{R}$ be a non-trivial compact interval, and let μ_I be the image of the Lebesgue measure under the map*

$$I \ni t \longmapsto \begin{pmatrix} 1 & \\ v(t) & I_d \end{pmatrix} \mathbb{Z}^{d+1}.$$

Then

$$(g_Q)_* \mu_I \longrightarrow m_{X_{d+1}}$$

*in the weak * topology as $Q \to \infty$.*

Manfred Einsiedler
Departement Mathematik, ETH Zürich
Rämistrasse 101, 8092 Zürich, Switzerland.
email: manfred.einsiedler@math.ethz.ch
www.math.ethz.ch/~einsiedl

Tom Ward
University Executive Group, University of Leeds, Leeds LS2 9JT, UK.
email: t.b.ward@leeds.ac.uk
https://www.maths.leeds.ac.uk

References

[1] R. C. Baker. Dirichlet's theorem on Diophantine approximation, *Math. Proc. Cambridge Philos. Soc.* **83** (1978), no. 1, 37–59.

[2] Y. Bugeaud. Approximation by algebraic integers and Hausdorff dimension, *J. London Math. Soc. (2)* **65** (2002), no. 3, 547–559.

[3] S. G. Dani. On orbits of unipotent flows on homogeneous spaces, *Ergodic Theory Dynam. Systems* **4** (1984), no. 1, 25–34.

[4] S. G. Dani. Divergent trajectories of flows on homogeneous spaces and Diophantine approximation, *J. Reine Angew. Math.* **359** (1985), 55–89.

[5] S. G. Dani. Correction to the paper: "Divergent trajectories of flows on homogeneous spaces and Diophantine approximation", *J. Reine Angew. Math.* **360** (1985), 214.

[6] S. G. Dani. On orbits of unipotent flows on homogeneous spaces. II, *Ergodic Theory Dynam. Systems* **6** (1986), no. 2, 167–182.

[7] H. Davenport and W. M. Schmidt. Dirichlet's theorem on diophantine approximation, in *Symposia Mathematica, Vol. IV (INDAM, Rome, 1968/69)*, pp. 113–132 (Academic Press, London, 1970).

[8] H. Davenport and W. M. Schmidt. Dirichlet's theorem on diophantine approximation. II, *Acta Arith.* **16** (1969/1970), 413–424.

[9] P. G. L. Dirichlet. Verallgemeinerung eines Satzes aus der Lehre von den Kettenbrüchen nebst einigen Anwendungen auf die Theorie der Zahlen, *S. B. Pruess. Akad. Wiss.* (1842), 93–95.

[10] P. G. L. Dirichlet. *Lectures on Number Theory.* Supplements by R. Dedekind. Transl. from the German by J. Stillwell. (American Mathematical Society, 1999).

[11] M. M. Dodson, B. P. Rynne, and J. A. G. Vickers. Dirichlet's theorem and Diophantine approximation on manifolds, *J. Number Theory* **36** (1990), no. 1, 85–88.

[12] W. Duke, Z. Rudnick, and P. Sarnak. Density of integer points on affine homogeneous varieties, *Duke Math. J.* **71** (1993), no. 1, 143–179.

[13] M. Einsiedler and T. Ward. Ergodic Theory with a View Towards Number Theory, in *Graduate Texts in Mathematics* **259** (Springer-Verlag London, Ltd., London, 2011).

[14] M. Einsiedler and T. Ward. *Homogeneous Dynamics and Applications*. To appear.

[15] A. Eskin and C. McMullen. Mixing, counting, and equidistribution in Lie groups, *Duke Math. J.* **71** (1993), no. 1, 181–209.

[16] C. F. Gauß. De nexu inter multitudinem classium, in quas formae binariae secundi gradus distribuuntur, earumque determinantem, in *Werke*, **2** (Cambridge University Press, 2011).

[17] G. H. Hardy. On the expression of a number as the sum of two squares, *Quart. J. Math.* **46** (1915), 263–283.

[18] M. N. Huxley. Integer points in plane regions and exponential sums, in *Number theory*, in *Trends Math.*, pp. 157–166 (Birkhäuser, Basel, 2000).

[19] A. Ivić, E. Krätzel, M. Kühleitner, and W. G. Nowak. Lattice points in large regions and related arithmetic functions: recent developments in a very classic topic, in *Elementare und analytische Zahlentheorie*, in *Schr. Wiss. Ges. Johann Wolfgang Goethe Univ. Frankfurt am Main, 20*, pp. 89–128 (Franz Steiner Verlag Stuttgart, Stuttgart, 2006).

[20] D. Kleinbock. An extension of quantitative nondivergence and applications to Diophantine exponents, *Trans. Amer. Math. Soc.* **360** (2008), no. 12, 6497–6523.

[21] D. Kleinbock and B. Weiss. Dirichlet's theorem on Diophantine approximation and homogeneous flows, *J. Mod. Dyn.* **2** (2008), no. 1, 43–62.

[22] D. Y. Kleinbock and G. A. Margulis. Flows on homogeneous spaces and Diophantine approximation on manifolds, *Ann. of Math. (2)* **148** (1998), no. 1, 339–360.

[23] E. Landau. Über dei Gitterpunkte in einem Kreise. II., *Göttingen Nachrichten* (1915), 161–171.
[24] G. A. Margulis. The action of unipotent groups in a lattice space, *Mat. Sb. (N.S.)* **86** (128) (1971), 552–556.
[25] G. A. Margulis. On the action of unipotent groups in the space of lattices, in *Lie groups and their representations (Proc. Summer School, Bolyai, János Math. Soc., Budapest, 1971)*, pp. 365–370 (Halsted, New York, 1975).
[26] D. Witte Morris. Ratner's Theorems on Unipotent Flows, in *Chicago Lectures in Mathematics* (University of Chicago Press, Chicago, IL, 2005).
[27] R. Phillips and Z. Rudnick. The circle problem in the hyperbolic plane, *J. Funct. Anal.* **121** (1994), no. 1, 78–116.
[28] M. Ratner. On Raghunathan's measure conjecture, *Ann. of Math. (2)* **134** (1991), no. 3, 545–607.
[29] A. Selberg. Recent developments in the theory of discontinuous groups of motions of symmetric spaces, in *Proceedings of the Fifteenth Scandinavian Congress (Oslo, 1968) Lecture Notes in Mathematics, Vol. 118* (Springer, Berlin, 1970), 99–120.
[30] N. A. Shah. Equidistribution of expanding translates of curves and Dirichlet's theorem on Diophantine approximation, *Invent. Math.* **177** (2009), no. 3, 509–532.
[31] N. A. Shah. Expanding translates of curves and Dirichlet-Minkowski theorem on linear forms, *J. Amer. Math. Soc.* **23** (2010), no. 2, 563–589.

6
Applications of Thin Orbits

Alex Kontorovich[1]

Abstract

This text is based on a series of three expository lectures on a variety of topics related to 'thin orbits', as delivered at Durham University's Easter School on 'Dynamics and Analytic Number Theory' in April 2014. The first lecture reviews closed geodesics on the modular surface and the reduction theory of binary quadratic forms before discussing Duke's equidistribution theorem (for indefinite classes). The second lecture exposits three quite different but (it turns out) not unrelated problems, due to Einsiedler–Lindenstrauss–Michel–Venkatesh, McMullen, and Zaremba. The third lecture reformulates these in terms of the aforementioned thin orbits, and shows how all three would follow from a single 'Local–Global' Conjecture of Bourgain and the author. We also describe some partial progress on the conjecture, which has led to some results on the original problems.

6.1 Lecture 1: Closed Geodesics, Binary Quadratic Forms, and Duke's Theorem

This first lecture has three parts. In §6.1.1, we review the geodesic flow on the hyperbolic plane to study closed geodesics on the modular surface. Then §6.1.2 discusses Gauss's reduction theory of binary quadratic forms. Finally, in §6.1.3, we combine the previous two discussions to connect indefinite classes to closed geodesics, and state Duke's equidistribution theorem.

[1] The author is partially supported by an NSF CAREER grant DMS-1455705, an NSF FRG grant DMS-1463940, an Alfred P. Sloan Research Fellowship, and a BSF grant.

6.1.1 Closed Geodesics

Let $\mathbb{H} = \{x + iy : y > 0\}$ denote the Poincaré (or, perhaps more precisely, Beltrami) upper half-plane, let $T\mathbb{H}$ be its tangent bundle, and for $(z, \zeta) \in T\mathbb{H}$, equip the tangent bundle with Riemannian metric $\|\zeta\|_z := |\zeta|/\Im z$. Here $z \in \mathbb{H}$ is the 'position' and $\zeta \in T_z\mathbb{H} \cong \mathbb{C}$ is the 'direction' vector. Let $T^1\mathbb{H}$ be the *unit* tangent bundle of all $(z, \zeta) \in T\mathbb{H}$ having $\|\zeta\|_z = 1$. The fractional linear action of the group $G = \mathrm{PSL}_2(\mathbb{R}) = \mathrm{SL}_2(\mathbb{R})/\{\pm I\}$ on $\mathbb{H}$ induces the following action on $T^1\mathbb{H}$:

$$G \ni \begin{pmatrix} a & b \\ c & d \end{pmatrix} : (z, \zeta) \in T^1\mathbb{H} \mapsto \left(\frac{az+b}{cz+d}, \frac{\zeta}{(cz+d)^2} \right), \tag{6.1}$$

with invariant measure

$$d\mu = \frac{dx\, dy\, d\theta}{y^2}, \tag{6.2}$$

in coordinates $(x + iy, \zeta)$, where $\arg \zeta = \theta$.

Exercise: This is indeed an action, which is moreover free, transitive, and invariant for the measure in (6.2).

The geodesics on $\mathbb{H}$ are vertical half-lines and semicircles orthogonal to the real line. Given $(z, \zeta) \in T^1\mathbb{H}$, the time-$t$ geodesic flow moves z along the geodesic determined by ζ to the point at distance t from z. The *visual point* from some $(z, \zeta) \in T^1\mathbb{H}$ is the point on the boundary $\partial\mathbb{H} \cong \mathbb{R} \cup \{\infty\}$ that one obtains by following the geodesic flow for infinite time.

People studied such flows on manifolds purely geometrically for some time before Gelfand championed the injection of algebraic and representation-theoretic ideas. With Fomin [GF51], he discovered that under the identification

$$G \cong T^1\mathbb{H}, \qquad g \leftrightsquigarrow g(i, \uparrow), \tag{6.3}$$

the geodesic flow on $T^1\mathbb{H}$ corresponds in G to right multiplication by the diagonal subgroup

$$A = \left\{ a_t := \begin{pmatrix} e^{t/2} & \\ & e^{-t/2} \end{pmatrix} \right\};$$

see e.g. [BM00, EW11].

To study (primitive, oriented) *closed* geodesics we move to the modular surface, defined as the quotient $\Gamma\backslash\mathbb{H}$, with $\Gamma = \mathrm{PSL}_2(\mathbb{Z})$. Its unit tangent bundle $\mathcal{X} := T^1(\Gamma\backslash\mathbb{H})$ is, as above, identified with $\Gamma\backslash G$, and the geodesic flow again corresponds to right multiplication by a_t. It is useful to think of this flow in two equivalent ways: (i) as a broken ray in a fundamental domain for $\mathcal{X}$ which is sent back inside when it tries to exit (see Figure 6.1a), or (ii) as a whole collection of Γ-translates of a single geodesic ray in the universal cover, $\mathbb{H}$, as

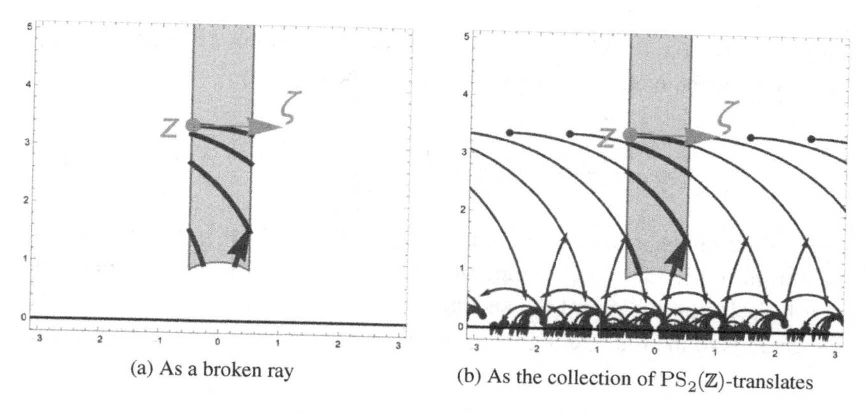

(a) As a broken ray

(b) As the collection of $PS_2(\mathbb{Z})$-translates

Figure 6.1 The geodesic flow on the modular surface

in Figure 6.1b. Thus, we will sometimes write $g \in \Gamma\backslash G$ for the first notion, and Γg for the second.

To obtain a closed geodesic on the modular surface, we start at some point Γg and come back to the same exact point (including the tangent vector) after a (least) time $\ell > 0$ (here ℓ is for *length*). That is,

$$\Gamma g a_\ell = \Gamma g, \qquad \text{or equivalently,} \qquad g a_\ell = M g,$$

for some matrix[2] $M \in \Gamma$. Then

$$M = g a_\ell g^{-1}, \tag{6.4}$$

so M has eigenvalues $e^{\pm \ell/2}$, and g is a matrix of eigenvectors. Note that M is hyperbolic with trace

$$\operatorname{tr} M = 2\cosh(\ell/2),$$

and the expanding eigenvalue, λ, say, is given by

$$\lambda = e^{\ell/2} = (\operatorname{tr} M + \sqrt{\operatorname{tr} M^2 - 4})/2. \tag{6.5}$$

Actually, since Γg is only determined up to left-Γ action, the matrix M is only determined up to Γ-conjugation (which of course leaves invariant its trace). In this way, primitive closed geodesics correspond to primitive (meaning: not of the form $[M_0^n]$ for some $M_0 \in \Gamma$, $n \geq 2$) hyperbolic conjugacy classes $[M]$ in Γ.

We note already from (6.5) that the lengths of closed geodesics are far from arbitrary; since $M \in \Gamma$, its eigenvalues are quadratic irrationals. It will also be

[2] Technically, as $\Gamma = SL_2(\mathbb{Z})/\{\pm I\}$, we should be using cosets $\pm M$ here. We will abuse notation and treat the elements of Γ as matrices, with the convention that their trace is positive.

useful later to note the visual point of g. **Exercise**: Writing M as $M = \left(\begin{smallmatrix} a & b \\ c & d \end{smallmatrix}\right)$, then if $c > 0$, the matrix g of eigenvectors can be given by

$$g = \frac{1}{(c^2(\mathrm{tr}^2 M - 4))^{1/4}} \begin{pmatrix} \lambda - d & 1/\lambda - d \\ c & c \end{pmatrix}. \tag{6.6}$$

The scaling factor is to ensure g has determinant 1. If $c < 0$, negate the first column in (6.6). Any other choice of g is obtained by rescaling the first column by a factor $\sigma e^{t/2}$ and the second by $\sigma e^{-t/2}$, $\sigma \in \{\pm 1\}$; this of course corresponds to the right action by a_t in $\mathrm{PSL}_2(\mathbb{R})$. The visual point α from g is determined by computing

$$\alpha = \lim_{t\to\infty} g a_t \cdot i = \frac{\lambda - d}{c} = \frac{a - d + \sqrt{\mathrm{tr}^2 M - 4}}{2c}. \tag{6.7}$$

Note again that this is a quadratic irrational, and its Galois conjugate $\overline{\alpha}$ is the visual point of the backwards geodesic flow. Note also that α is independent of the choice of g above. Finally, we record here that the fractional linear action of M on $\mathbb{R}$ fixes α; indeed, starting from $g a_\ell = Mg$, multiply both sides on the right by a_t, have that matrix act on the left by i, and take the limit as $t \to \infty$:

$$\alpha = \lim_{t\to\infty} g a_\ell a_t \cdot i = \lim_{t\to\infty} M g a_t \cdot i = M\alpha. \tag{6.8}$$

To see an explicit example, let us construct the geodesic corresponding to the hyperbolic matrix

$$M = \begin{pmatrix} 12 & 5 \\ -5 & -2 \end{pmatrix}. \tag{6.9}$$

From (6.5), (6.6), and (6.7), we compute

$$\lambda = 5 + 2\sqrt{6}, \quad g = \frac{1}{\sqrt[4]{2400}} \begin{pmatrix} -7 - 2\sqrt{6} & 7 - 2\sqrt{6} \\ 5 & -5 \end{pmatrix}, \quad \alpha = \frac{-7 - 2\sqrt{6}}{5}. \tag{6.10}$$

Using the identification (6.3) and action (6.1), the point $g \in G$ corresponds to the point $(z, \zeta) \in T^1\mathbb{H}$ where

$$z = -\frac{7}{5} + \frac{2i\sqrt{6}}{5}, \quad \zeta = -\frac{2\sqrt{6}}{5}. \tag{6.11}$$

The points z, ζ, and α are shown in Figure 6.2, as well as their images in the standard fundamental domain $\mathcal{F}$ for Γ. The resulting closed geodesic, also shown in $\mathcal{F}$, has length $\ell = 2\log\lambda \approx 4.58$. Had we started with M^2 instead of M, we would have obtained the same g, z, ζ, and α but the length would

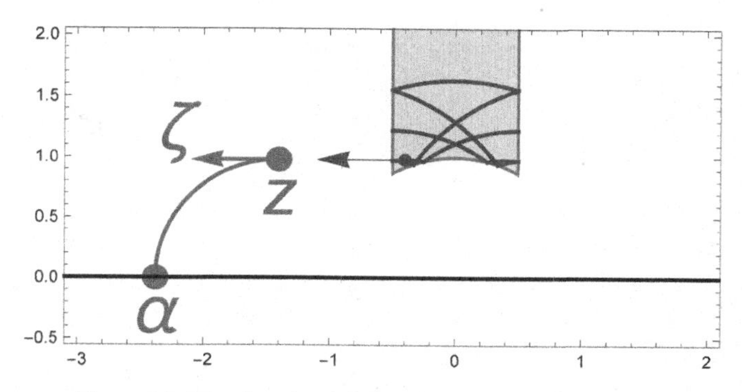

Figure 6.2 The closed geodesic corresponding to M in (6.9)

have doubled, corresponding to looping around the geodesic twice (hence our restriction to primitive geodesics and conjugacy classes). Replacing M by some conjugate, $\gamma M \gamma^{-1}$, with $\gamma \in \Gamma$, results in replacing g, z, and α by γg, γz, and $\gamma\alpha$, respectively (of course, the geodesic remains unchanged).

Next we wish to discuss the cutting sequence of the geodesic flow. Recall that Γ is generated by two elements $T = \left(\begin{smallmatrix} 1 & 1 \\ 0 & 1 \end{smallmatrix}\right)$ and $S = \left(\begin{smallmatrix} 0 & 1 \\ -1 & 0 \end{smallmatrix}\right)$, and that its standard fundamental domain $\mathfrak{F}$ is the intersection of the domains

$$\mathfrak{F}_T := \{\Re z > -1/2\}, \; \mathfrak{F}_{T^{-1}} := \{\Re z < 1/2\}, \; \text{and } \mathfrak{F}_S := \{|z| > 1\}.$$

As we follow the geodesic flow from $\mathfrak{F}$, thought of as a subset of the universal cover, $\mathbb{H}$, we pass through one of the boundary walls, leaving one of the domains $\mathfrak{F}_L$, $L \in \{T, T^{-1}, S\}$; here L is the 'letter' we must apply to return the flow to $\mathfrak{F}$. Given a starting point $(z, \zeta) \in T^1(\Gamma\backslash\mathbb{H})$, its cutting sequence is this sequence of letters L.

To illustrate this, consider again the example in Figure 6.2. The flow first hits the wall $\Re z = -1/2$, and must be translated by $L_1 = T$ back inside $\mathfrak{F}$. Next the flow encounters the wall $|z| = 1$, and is reflected using $L_2 = S$. Continuing in this way (see Figure 6.3), we find that the cutting sequence of (z, ζ) in (6.11) is:

$$T, S, T^{-1}, T^{-1}, S, T, S, T^{-1}, S, \dots, \tag{6.12}$$

repeating *ad infinitum*. It is easy to see from the geometry that such sequences are some number of Ts or T^{-1}s separated by single Ss. Computing these counts converts (6.12) into:

$$\underbrace{T}_{1}, S, \underbrace{T^{-1}, T^{-1}}_{2}, S, \underbrace{T}_{1}, S, \underbrace{T^{-1}}_{1}, S, \dots,$$

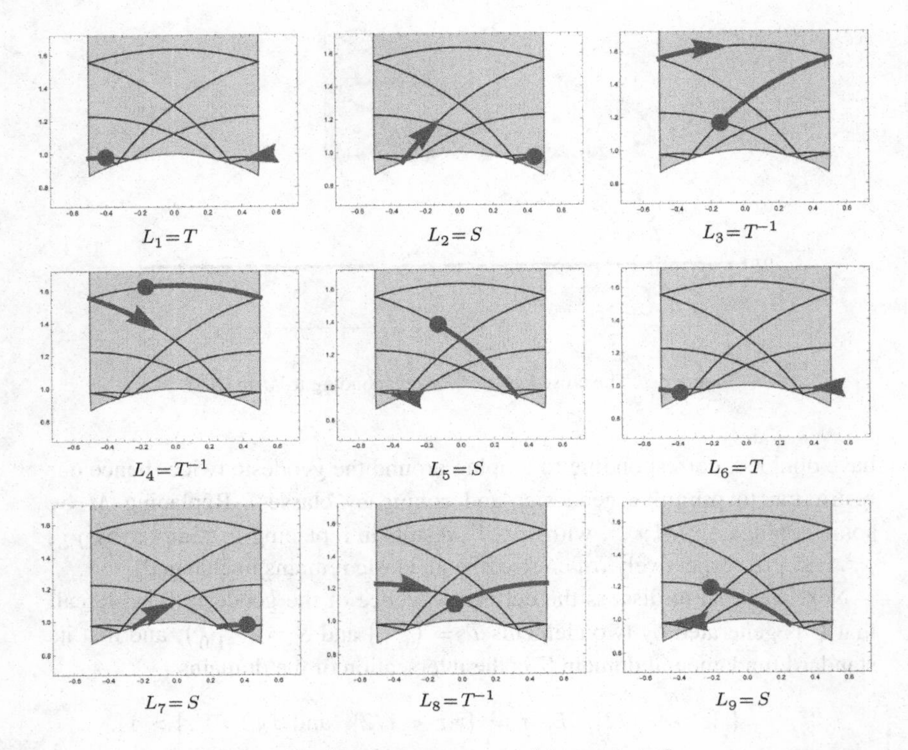

Figure 6.3 The cutting sequence of the geodesic flow

which corresponds to the sequence

$$1, 2, 1, 1, 1, 2, 1, 1, \ldots, \tag{6.13}$$

repeating.

It seems to have first been observed by Humbert [Hum16] that this sequence should be compared to the continued fraction expansion of the visual point α of (z, ζ). We write the continued fraction expansion of any $x \in \mathbb{R}$ as

$$x = a_0 + \cfrac{1}{a_1 + \cfrac{1}{a_2 + \ddots}} = [a_0, a_1, a_2, \ldots],$$

where $a_0 \in \mathbb{Z}$ and the other $a_j \in \mathbb{N}$ are positive. These numbers are called the *partial quotients* of x, and we will sometimes call them 'digits' or 'letters.' For the visual point α in (6.10), we compute:

$$\alpha = \frac{-7-2\sqrt{6}}{5} = [-3, 1, 1, \overline{1, 1, 1, 2}], \tag{6.14}$$

where the bar means 'repeating the last sequence of digits forever'.

Comparing (6.14) to (6.13), we see that the periodic parts match, up to cyclic permutation (since a closed geodesic has no canonical 'starting point' anyway). This leads us to the notion of a *reduced form* for α.

Definition *A quadratic irrational α is called 'reduced' if it and its Galois conjugate $\overline{\alpha}$ satisfy the inequalities:*

$$-1 < \overline{\alpha} < 0 < 1 < \alpha. \tag{6.15}$$

A representative M of a hyperbolic conjugacy class $[M]$ is also called reduced if its visual point α is reduced.

Exercise: A quadratic irrational α is reduced iff its continued fraction is *exactly* (as opposed to eventually) periodic.

How should we reduce the representative M in (6.9)? It's actually quite easy. Note that, in general, if α has continued fraction expansion

$$\alpha = [a_0, \ldots, a_h, \overline{a_{h+1}, \ldots, a_{h+\ell}}],$$

then

$$\begin{pmatrix} 0 & 1 \\ 1 & -a_0 \end{pmatrix} \cdot \alpha = \frac{1}{\alpha - a_0} = [a_1, \ldots, a_h, \overline{a_{h+1}, \ldots, a_{h+\ell}}]. \tag{6.16}$$

That is, such matrices eat away the first digit (this, of course, is the left-shift map from dynamics). For α in (6.14), we could try acting (on the left) by $\gamma_0 = \left(\begin{smallmatrix} 0 & 1 \\ 1 & -1 \end{smallmatrix}\right) \cdot \left(\begin{smallmatrix} 0 & 1 \\ 1 & -1 \end{smallmatrix}\right) \cdot \left(\begin{smallmatrix} 0 & 1 \\ 1 & 3 \end{smallmatrix}\right)$ to make an exactly periodic continued fraction $[\overline{1, 1, 1, 2}]$. But this matrix γ_0 has determinant -1, being an odd product of determinant -1 matrices. To act instead by an element of PSL_2, we eat away one more digit, using the matrix

$$\gamma = \begin{pmatrix} 0 & 1 \\ 1 & -1 \end{pmatrix} \cdot \begin{pmatrix} 0 & 1 \\ 1 & -1 \end{pmatrix} \cdot \begin{pmatrix} 0 & 1 \\ 1 & -1 \end{pmatrix} \cdot \begin{pmatrix} 0 & 1 \\ 1 & 3 \end{pmatrix}$$
$$= \begin{pmatrix} 2 & 5 \\ -3 & -7 \end{pmatrix}, \tag{6.17}$$

to obtain

$$\widetilde{\alpha} = \gamma\alpha = \frac{1+\sqrt{6}}{2} = [\overline{1, 1, 2, 1}]. \tag{6.18}$$

That is, we replace M by

$$\widetilde{M} = \gamma M \gamma^{-1} = \begin{pmatrix} 7 & 5 \\ 4 & 3 \end{pmatrix}, \tag{6.19}$$

with $\widetilde{M}$ now reduced.

In this way, the geodesic flow corresponds to the symbolic dynamics on the continued fraction expansion of the visual point of the flow. Actually, we have been quite sloppy; for example, it is not clear what one should do if the geodesic flow passes through an elliptic point of the orbifold $\Gamma\backslash\mathbb{H}$. There is in fact a better way of encoding the cutting sequence, as elucidated beautifully by Series [Ser85]. That said, the less precise (but more immediate) description given here will suffice for the purposes of our discussion below.

6.1.2 Binary Quadratic Forms

This will be a very quick introduction to an extremely well-studied and beautiful theory; see e.g. [Cas78] for a classical treatment. The theory is largely due to Gauss (building on Lagrange and Legendre), as developed in his 1801 magnum opus, *Disquisitiones Arithmeticae*.

For integers A, B, and C, let $Q = [A, B, C]$ denote the (integral) binary quadratic form $Q(x, y) = Ax^2 + Bxy + Cy^2$. The general problem being addressed was: Given Q, what numbers does it represent? That is, for which numbers $n \in \mathbb{Z}$ do there exist $x, y \in \mathbb{Z}$ so that $Q(x, y) = n$? The question is perhaps inspired by the famous resolution in the case $Q = x^2 + y^2$ by Fermat (and other special cases due to Euler and others). See [Cox13] for a beautiful exposition of this problem.

Some observations:

(i) If A, B, and C have a factor in common, then so do all numbers represented by Q, and by dividing out this factor, we may and will assume henceforth that Q is *primitive*, meaning $(A, B, C) = 1$.

(ii) The set of numbers represented by Q does not change if Q is replaced by $Q'(x, y) = Q(ax + by, cx + dy)$, with $ad - bc = \pm 1$; this of course is nothing but an invertible (over integers!) linear change of variables.

Gauss defined such a pair of forms Q, Q' to be equivalent but for us it will be more convenient to use *strict* (some authors call this *proper*, or *narrow*) equivalence, meaning we only allow 'orientation-preserving' transformations. That is, we will write $Q \sim Q'$ only when there is some $\gamma \in \mathrm{SL}_2(\mathbb{Z})$ (as opposed to GL_2) with $Q = Q' \circ \gamma$.

Exercise: This is indeed an *action*; that is, $(Q \circ \gamma_1) \circ \gamma_2 = Q \circ (\gamma_1\gamma_2)$, and hence $\sim$ is an equivalence relation.

For a given Q, the set of all $Q' \sim Q$ is called a *class* (or equivalence class) and denoted $[Q]$. Because we are considering strict equivalence here, this is often called the *narrow* class of Q.

Exercise: If $Q \sim Q'$ then $D_Q = D_{Q'}$, where

$$D_Q := \mathrm{discr}(Q) = B^2 - 4AC \tag{6.20}$$

is the discriminant. That is, the discriminant is a class function (invariant under equivalence). Observe that discriminants are quadratic residues $\pmod 4$, and hence $D_Q \equiv 0$ or $1 \pmod 4$.

Exercise: When $D < 0$, the form Q is *definite*; that is, it only takes either positive or negative values, but not both. When $D > 0$, the form is *indefinite*, representing both positive and negative numbers.

If $\mathrm{discr}(Q) = D = 0$, or more generally, if $D = D_0^2$ is a perfect square, then Q is the product of two linear forms. Then the representation question is much less interesting, and will be left as an exercise. We exclude this case going forward.

Exercise: Let

$$\alpha_Q = \frac{-B + \sqrt{D}}{2A} \tag{6.21}$$

be the root of $Q(x, 1)$ (assuming $A \neq 0$), and suppose $Q' = Q \circ \gamma$. Then $\alpha_{Q'} = \gamma^{-1} \cdot \alpha_Q$, where the action here of γ^{-1} is by fractional linear transformations.

We have seen that if two forms are equivalent, then their discriminants agree. It is then natural to ponder about the converse: does $\mathrm{discr}(Q) = \mathrm{discr}(Q')$ imply that $Q \sim Q'$?

To study this question, let $\mathscr{C}_D$ be the set of all inequivalent, primitive classes having discriminant D,

$$\mathscr{C}_D := \{[Q] : \mathrm{discr}(Q) = D\},$$

and let $h_D := |\mathscr{C}_D|$ be its size; this is called the (narrow) *class number*.

This set $\mathscr{C}_D$ is now called the *class group* (it turns out there is a composition process under which $\mathscr{C}_D$ inherits the structure of an Abelian group, but this fact will not be needed for our investigations; for us, $\mathscr{C}_D$ is just a set). If having the same discriminant implied equivalence, then all class numbers would be 1. This turns out to be false, but actually it is not off by 'very' much, in the following sense: the class number is always finite.

Theorem 6.1.1 (Gauss) *For any non-square integer $D \equiv 0, 1(4)$, we have:*

$$1 \leq h(D) < \infty.$$

It is easy to see that $h(D) \geq 1$. Indeed, if $D \equiv 0(4)$, then $x^2 - \frac{D}{4}y^2$ is primitive of discriminant D; and if $D \equiv 1(4)$, then $x^2 + xy - \frac{D-1}{4}y^2$ works.

We give a very quick sketch of the very well-known Theorem 6.1.1, as it will be relevant to what follows. The proof decomposes according to whether or not the class is definite.

Sketch when $D < 0$. Let α_Q be the root of $Q(x, 1)$ as in (6.21); since Q is definite, $\alpha_Q \in \mathbb{H}$. We have already discussed the standard fundamental domain $\mathfrak{F}$ for $\Gamma\backslash\mathbb{H}$, so we know (from the exercise below (6.21)) that there is a transformation $\gamma \in SL_2(\mathbb{Z})$ taking α_Q to $\mathfrak{F}$. Such a transformation γ is moreover unique (up to technicalities when α_Q is on the boundary of $\mathfrak{F}$), and gives rise to a unique *reduced* representative $Q' = Q \circ \gamma^{-1}$ in the class $[Q]$. It remains to show there are only finitely many such forms having a given discriminant $D < 0$. This is an easy exercise using $|\Re(\alpha_Q)| \leq 1/2$ and $|\alpha_Q| \geq 1$; indeed, one finds that

$$A \leq \sqrt{\frac{|D|}{3}}, \quad \text{and} \quad |B| \leq A,$$

and hence the number of reduced forms is finite. □

Example: Take $D = -5$. This D is not congruent to 0 or 1 mod 4, so instead we consider $D = -20$. Then $A \leq \sqrt{20/3}$; that is, A is at most 2. If $A = 1$, then $B = 0$ or ± 1. The latter case, $B = \pm 1$, gives no integral solutions to $C = (B^2 - D)/(4A)$, but the former gives $C = 5$, corresponding to the reduced form $Q_0 = x^2 + 5y^2$. Next we consider $A = 2$, whence $B \in [-2, 2]$. Only the choices $B = \pm 2$ give integral values for $C = 3$, corresponding to forms $Q_\pm = 2x^2 \pm 2xy + 3y^2$. Actually, this turns out to be a boundary case, and the two forms $Q_\pm$ are equivalent. Hence the class group $\mathscr{C}_{-20} = \{Q_0, Q_+\}$ has class number $h_{-20} = 2$. The fact that this class number is not 1 is well known to be related to the failure of unique factorization in $\mathbb{Z}[\sqrt{-5}]$; e.g. the number 6 factors both as $6 = 2 \cdot 3$ and as $6 = (1 + \sqrt{5}i)(1 - \sqrt{5}i)$.

Sketch when $D > 0$. Again we assume $A > 0$. This case is more subtle, as the root α_Q and its Galois conjugate are real, so cannot be moved to the fundamental domain $\mathfrak{F}$. Instead we notice that, since α_Q is real and quadratic, it has an eventually periodic continued fraction expansion, and transformations $\gamma \cdot \alpha_Q$ simply add (or subtract) letters to (or from) this expansion. In particular, there is a $\gamma \in SL_2(\mathbb{Z})$ which makes the continued fraction exactly periodic; that is, the transformed root $\alpha_{Q.\gamma}$ then satisfies the familiar condition (6.15). (This is our first hint that indefinite forms are related to closed geodesics!) We thus call an indefinite form Q *reduced* if its root α_Q is reduced. As before, it is easy to see using

$$\alpha_Q = \frac{-B+\sqrt{D}}{2A} > 1, \qquad -1 < \overline{\alpha}_Q = \frac{-B-\sqrt{D}}{2A} < 0,$$

that

$$0 < -B < \sqrt{D}, \qquad \frac{1}{2}(\sqrt{D}+B) < A < \frac{1}{2}(\sqrt{D}-B),$$

which forces the class number to be finite. □

Example: Take $D = 7$, or rather $D = 28$, since $7 \not\equiv 0, 1(4)$. Then the possibilities for B range from -1 to -5. To solve $B^2 - D = 4AC$, we must have $B^2 - D \equiv 0(4)$, which leaves only $B = -2$ or $B = -4$. In the former case, the possible positive divisors of $(B^2 - D)/4 = -6$ are $A = 1, 2, 3$, or 6, of which only $A = 2$ and 3 lie in the range $\frac{1}{2}(\sqrt{28}-2) < A < \frac{1}{2}(\sqrt{28}+2)$. These two give rise to the forms

$$Q_1 = 2x^2 - 2xy - 3y^2 \qquad \text{and} \qquad Q_2 = 3x^2 - 2xy - 2y^2. \tag{6.22}$$

The latter case of $B = -4$ leads in a similar way to the two reduced forms $Q_3 = x^2 - 4xy - 3y^2$ and $Q_4 = 3x^2 - 4xy - y^2$.

Note that, unlike the definite case, reduced forms are *not* unique in their class, as any even-length cyclic permutation of the continued fraction expansion of α_Q gives rise to another reduced form. So we cannot conclude from the above computation that $h_{28} \stackrel{?}{=} 4$. Writing α_j for the larger root of Q_j, $j = 1, \ldots, 4$, we compute the continued fractions:

$$\alpha_1 = [\overline{1, 1, 4, 1}],\ \alpha_2 = [\overline{1, 4, 1, 1}],\ \alpha_3 = [\overline{4, 1, 1, 1}],\ \alpha_4 = [\overline{1, 1, 1, 4}]. \tag{6.23}$$

It is then easy to see by inspection that Q_1 and Q_3 are equivalent, as are Q_2 and Q_4, e.g.

$$Q_1 \circ \left[\left(\begin{smallmatrix} 0 & 1 \\ 1 & -1 \end{smallmatrix}\right)\left(\begin{smallmatrix} 0 & 1 \\ 1 & -1 \end{smallmatrix}\right)\right]^{-1} = Q_3.$$

(**Exercise:** Verify this and compute the change of basis matrix to go from Q_2 to Q_4.) Thus $h_{28} = 2$.

Class groups and class numbers are extremely mysterious. A discriminant is defined to be *fundamental* if it is the discriminant of a (quadratic, in our context) field; such Ds are either $\equiv 1 \pmod 4$ and square-free, or divisible by 4 with $D/4$ square-free and $\equiv 2, 3 \pmod 4$. Dirichlet's Class Number Formula (see e.g. [Dav80, IK04]) gives an approach to studying class groups: if D is a fundamental discriminant, then

$$h_D = \sqrt{|D|}\, L(1, \chi_D) \times \begin{cases} 1/(2\pi), & \text{if } D \leq -5, \text{ or} \\ 1/\log \epsilon_D, & \text{if } D > 0. \end{cases} \tag{6.24}$$

Here $L(1, \chi_D)$ is called a 'special L-value', and for $D > 0$, the factor $\epsilon_D \in \mathbb{Q}(\sqrt{D})$ is determined by:

$$\epsilon_D = \frac{t + s\sqrt{D}}{2}, \tag{6.25}$$

where $(T, S) = (t, s)$ is the least solution to the Pellian equation

$$T^2 - S^2 D = 4. \tag{6.26}$$

It follows from (6.26) that ϵ_D is a unit in $\mathbb{Q}(\sqrt{D})$, as its algebraic norm is $\epsilon_D \bar{\epsilon}_D = 1$. (If the Pell equation (6.26) has no solutions with 4 replaced by -4 on the right side, then ϵ_D is the fundamental unit; otherwise it is the square of the latter.) The L-value, which we will not bother to define (as it is not relevant to our discussion), is so fascinating an object about which one could say so much, that we will instead say very little. For example, it is not hard to show that

$$L(1, \chi_D) \leq C \log |D|.$$

Siegel famously proved [Sie35, Lan35] the reverse inequality, that for any $\varepsilon > 0$,

$$L(1, \chi_D) \geq C_\epsilon \cdot |D|^{-\varepsilon},$$

though the constant C_ε is *ineffective*; that is, the proof does not give any means to estimate it for any given ϵ. (These days, we have other less strong but, on the other hand, effective estimates: see [Gol76, GZ86].) Either way, we may think of the L-value as very roughly being of size 1. Then definite class groups are, very roughly, of size

$$h_{-D} \approx \sqrt{|D|},$$

while indefinite ones are of size

$$h_D \approx \sqrt{D}/\log \epsilon_D. \tag{6.27}$$

(We will not give the symbol $\approx$ a precise meaning here.) As a consequence, we obtain Gauss's conjecture, that for definite class numbers, $h_{-D} \to \infty$, as $-D \to -\infty$; see [Deu33, Hei34].

For indefinite forms, the behaviour is even more mysterious, and it is a longstanding conjecture that infinitely often the class number is 1:

$$\liminf h_D \stackrel{?}{=} 1, \tag{6.28}$$

where the limit is over fundamental $D \to +\infty$. (If one does not require fundamental discriminants, this problem was apparently solved long ago by Dirichlet; see [Lag80].) In light of (6.27), this conjecture suggests that the unit ϵ_D, defined in (6.25), should infinitely often be massive, of size about $e^{\sqrt{D}}$. Today, no methods are known to force such large solutions to the Pell equation (6.26), despite rather convincing evidence (see e.g. [CL84]) that this event is far from rare. On the other hand, it is quite easy to make giant class numbers, since one can force very small units; e.g. by taking Ds of the form $D = t^2 - 4$, one sees that $\epsilon_D = \frac{t+\sqrt{D}}{2} \approx \sqrt{D}$, and $h_D \approx \sqrt{D}/\log D$ is as large as possible. Another long-standing conjecture in the indefinite case is that the average class number is roughly bounded, in the crude (more refined conjectures are available) sense that:

$$\sum_{0<D<X} h_D \stackrel{?}{=} X^{1+o(1)}. \tag{6.29}$$

We have insufficient space to delve further into this fascinating story, so we will leave it there.

6.1.3 Duke's Theorem

We now combine the previous two sections, giving a bijection between primitive, indefinite classes $[Q]$ and primitive, oriented, closed geodesics γ on the modular surface. This equivalence was apparently first observed by Fricke and Klein [FK90], and is also discussed in many places, e.g. [Cas78, Sar80, Hej83, IK04, Sar07].

We first attach a form Q to a given hyperbolic matrix $M = \left(\begin{smallmatrix} a & b \\ c & d \end{smallmatrix}\right)$, the passage being through equating αs in (6.7) and (6.21), as follows. Pattern matching (6.7) with (6.21), we obtain a preliminary (possibly imprimitive) set of variables $B_0 = d - a$, $A_0 = c$, $D_0 = \operatorname{tr}^2 M - 4$, leading to

$$C_0 = \frac{B_0^2 - D_0}{4A_0} = -b.$$

Setting $s = \gcd(A_0, B_0, C_0)$, the primitive quadratic form $[A_0, B_0, C_0]/s$ is almost what we want, but does not quite work because M is really in $\mathrm{PSL}_2(\mathbb{Z})$, and, as is, $-M$ could give a different form. To fix this, we set

$$Q = \frac{\operatorname{sgn}(\operatorname{tr} M)}{s}[c, d-a, -b], \tag{6.30}$$

which is now well defined on $\mathrm{PSL}_2(\mathbb{Z})$. The discriminant of this form is

$$D = \frac{\operatorname{tr}^2 M - 4}{\gcd(c, d-a, b)^2}. \tag{6.31}$$

For example, if we take $\widetilde{M} = \left(\begin{smallmatrix} 7 & 5 \\ 4 & 3 \end{smallmatrix}\right)$ in (6.19), then $s = \gcd(4, -4, -5) = 1$ and $\operatorname{sgn}(\operatorname{tr} M) = +1$, so $\widetilde{M}$ corresponds to the binary quadratic form $\widetilde{Q} = 4x^2 - 4xy - 5y^2$ of discriminant 96. Note that if we had done the same operation starting with M in (6.9), the form $Q = -5x^2 - 14xy - 5y^2$, also of discriminant 96, and of course, $Q = \widetilde{Q} \circ \gamma^{-1}$ with change of variables matrix γ given by (6.17). (**Exercise:** Verify all this.)

To invert the map (6.30) given some $Q = [A, B, C]$ with discriminant $D > 0$ and not a perfect square, we seek a matrix $M = \left(\begin{smallmatrix} a & -Cs \\ As & d \end{smallmatrix}\right) \in \mathrm{SL}_2(\mathbb{Z})$ so that $a+d > 0$, say, and $d - a = Bs$. Inserting the last identity into the determinant equation and completing the square gives:

$$1 = a^2 + Bsa + ACs^2 = \frac{1}{4}(2a + Bs)^2 - \frac{1}{4}Ds^2.$$

Multiplying both sides by 4, we come to the familiar Pellian equation (6.26); if (t, s) is a fundamental solution, then $a = (t - Bs)/2$, $d = (t + Bs)/2$, and we have found our desired hyperbolic matrix

$$M = \begin{pmatrix} (t - Bs)/2 & -Cs \\ As & (t + Bs)/2 \end{pmatrix}.$$

That M is primitive follows from the fundamentality of the solution (t, s) (**Exercise**). For example, to turn $Q_1 = 2x^2 - 2xy - 3y^2$ of discriminant $D = 28$ into a closed geodesic, we find the fundamental solution $(t, s) = (16, 3)$ to (6.26), leading to $M_1 = \left(\begin{smallmatrix} 11 & 9 \\ 6 & 5 \end{smallmatrix}\right)$.

Note that the key to finding M above is to solve a Pellian equation, which itself goes through continued fractions; here is a more direct version of this inverse map. We first state the following simple exercise.

Exercise: The matrix

$$M = \begin{pmatrix} a_0 & 1 \\ 1 & 0 \end{pmatrix} \cdot \begin{pmatrix} a_1 & 1 \\ 1 & 0 \end{pmatrix} \cdots \begin{pmatrix} a_\ell & 1 \\ 1 & 0 \end{pmatrix} \tag{6.32}$$

fixes the real quadratic irrational α having continued fraction expansion $\alpha = [\overline{a_0, a_1, \ldots, a_\ell}]$. [Hint: Compare to (6.16).]

Recall from (6.8) that the desired matrix M_1 fixes α_1, where $\alpha_1 = (2+\sqrt{28})/4$ is the root of $Q_1(x, 1)$. As the continued fraction expansion of α_1 is given in (6.23), it is trivial to find the corresponding matrix M_1:

$$M_1 = \begin{pmatrix} 1 & 1 \\ 1 & 0 \end{pmatrix} \cdot \begin{pmatrix} 1 & 1 \\ 1 & 0 \end{pmatrix} \cdot \begin{pmatrix} 4 & 1 \\ 1 & 0 \end{pmatrix} \cdot \begin{pmatrix} 1 & 1 \\ 1 & 0 \end{pmatrix} = \begin{pmatrix} 11 & 9 \\ 6 & 5 \end{pmatrix},$$

again. If the product in (6.32) had odd length, one would obtain a matrix of determinant -1, whose square is the desired element of $\mathrm{SL}_2(\mathbb{Z})$. This corresponds to the situation when $T^2 - Ds^2 = -4$ is solvable; then, writing (t, s) for the least solution, we find that ϵ_D is expressed as $\eta_D^2 = \epsilon_D$, where $\eta_D = (t + s\sqrt{D})/2$ is the fundamental unit.

Definition *The discriminant of a closed geodesic γ on the modular surface, or its corresponding hyperbolic conjugacy class, is defined to be that of its associated equivalence class of binary quadratic forms. Explicitly, the discriminant of $M = \left(\begin{smallmatrix} a & b \\ c & d \end{smallmatrix}\right)$ is $D = (\mathrm{tr}^2 M - 4)/s^2$, where $s = \gcd(c, d-a, b)$.*

The idea in Duke's Theorem is to look at the equidistribution of closed geodesics on the modular surface, but instead of studying them individually, one may, like classes of binary forms, group them by discriminant. First, some examples.

For $D = 1337$ (which is $\equiv 1 \pmod 4$ and square-free, and hence fundamental), we find that the class number is $h_{1337} = 2$ and the class group $\mathscr{C}_{1337}$ is comprised of two classes, represented by $Q_1 = [7, -35, -4]$ and $Q_2 = [4, -35, -7]$. Following the duality above to hyperbolic matrices, we find that Q_1 and Q_2 correspond, respectively, to

$$M_1 = \begin{pmatrix} 2\,676\,336\,167 & 523\,561\,808 \\ 299\,178\,176 & 58\,527\,127 \end{pmatrix}, \quad \text{and} \quad M_2 = M_1^{\top}.$$

The entries are so large because the fundamental solution $(T, S) = (t, s)$ to $T^2 - 1337S^2 = 4$ is:

$$(t, s) = (2\,734\,863\,294,\ 74\,794\,544),$$

which also explains why the class group is so relatively small (cf. the discussion below (6.27)). The larger roots α_1 and α_2 of $Q_1(x, 1)$ and $Q_2(x, 1)$, respectively, have continued fraction expansions:

$$\alpha_1 = [\overline{8, 1, 17, 2, 1, 1, 3, 1, 35, 1, 3, 1, 1, 2, 17, 1, 8, 5}], \tag{6.33}$$

$$\alpha_2 = [\overline{5, 8, 1, 17, 2, 1, 1, 3, 1, 35, 1, 3, 1, 1, 2, 17, 1, 8}].$$

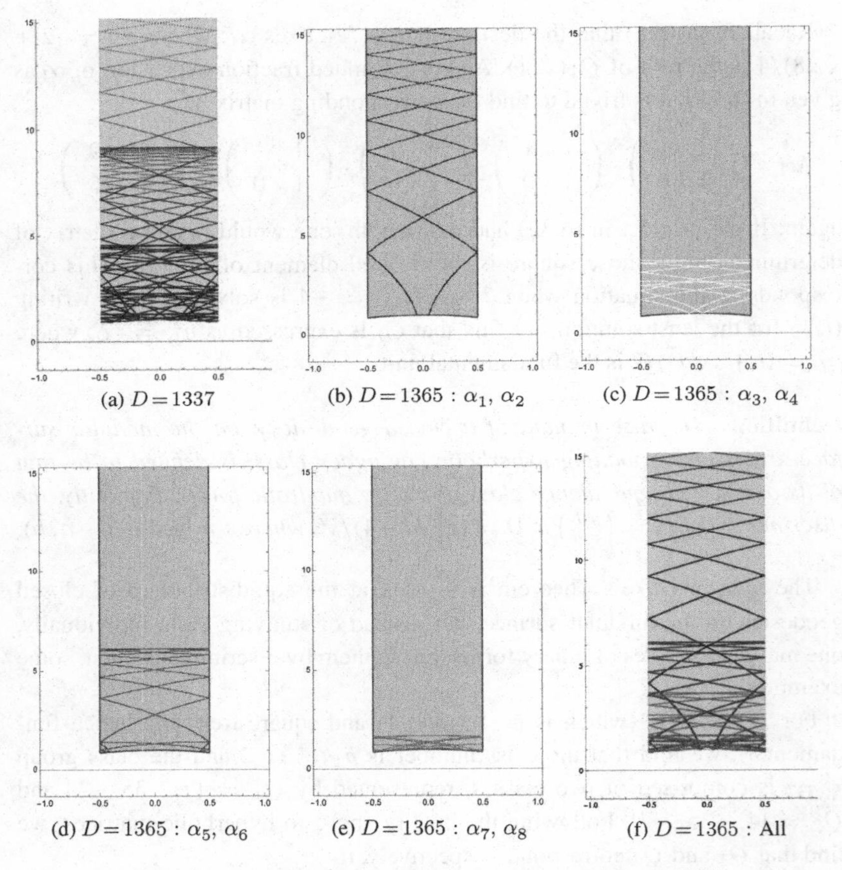

(a) $D = 1337$ (b) $D = 1365 : \alpha_1, \alpha_2$ (c) $D = 1365 : \alpha_3, \alpha_4$

(d) $D = 1365 : \alpha_5, \alpha_6$ (e) $D = 1365 : \alpha_7, \alpha_8$ (f) $D = 1365$: All

Figure 6.4 The geodesics in $\mathscr{C}_D$ corresponding to (6.33) and (6.34).

These have the same even length, and differ by an odd shift (by one), so the two forms Q_1 and Q_2 *are* equivalent under $\mathrm{GL}_2(\mathbb{Z})$-action but not under $\mathrm{SL}_2(\mathbb{Z})$. That is, while the narrow class number of 1337 is 2, the wide class number is 1. The corresponding geodesics are actually the same curve, but with opposite orientation; so they look the same in $\Gamma\backslash\mathbb{H}$ but differ in $T^1(\Gamma\backslash\mathbb{H})$, because the tangent vectors all change direction. See Figure 6.4a for an illustration, which appears as just one curve, since the tangent vectors are not displayed.

For another example, we study the case $D = 1365$. This discriminant is also $\equiv 1 \pmod 4$ and square-free, and therefore fundamental. This time the class number is $h_{1365} = 8$, and we also leave it as an **Exercise** to work out representatives for the classes $[Q_1], \ldots, [Q_8]$, the corresponding hyperbolic matrices $M_1, \ldots, M_8$, and the Pellian solution (t, s). The roots $\alpha_1, \ldots, \alpha_8$ have continued fractions:

$$\alpha_1 = [\overline{1, 35}],\ \alpha_2 = [\overline{35, 1}],\ \alpha_3 = [\overline{5, 7}],\ \alpha_4 = [\overline{7, 5}], \tag{6.34}$$

$$\alpha_5 = [\overline{1, 1, 1, 11}],\ \alpha_6 = [\overline{1, 1, 11, 1}],$$

$$\alpha_7 = [\overline{1, 1, 1, 2, 1, 2}],\ \alpha_8 = [\overline{1, 1, 2, 1, 2, 1}].$$

Each pair $\alpha_{2j-1}, \alpha_{2j}$ is the same geodesic but with opposite orientation; the four distinct geodesics in $\Gamma\backslash\mathbb{H}$ are illustrated in Figures 6.4b–6.4e. Recall from (6.13) that the cutting sequence of the geodesic flow is a symbolic coding of the visual point; then the first pair of αs above corresponds to a geodesic that simply shoots up in the air and falls back down (Figure 6.4b); the next two pairs are mid-level geodesics (Figures 6.4c and 6.4d); and the last pair of αs gives a very low-lying geodesic (Figure 6.4e).

So the behaviour of each geodesic in $\mathscr{C}_{1365}$ is quite different, while there is only one geodesic (up to orientation) in $\mathscr{C}_{1337}$. That said, combining all four geodesics in $\mathscr{C}_{1365}$ into one picture (Figure 6.4f), one finds that, were the geodesics in $\mathscr{C}_{1365}$ all coloured the same, it would be quite difficult to distinguish this image from Figure 6.4a for $\mathscr{C}_{1337}$. Moreover the density of both plots is reminiscent of the invariant measure $dx\,dy/y^2$ on $\mathbb{H}$ from (6.2). That is, these curves, when grouped by discriminant, become 'equidistributed' with respect to the invariant measure as the discriminant grows. Equidistribution here means that the average amount of time this union of curves spends in a given nice region, say $A \subset \Gamma\backslash\mathbb{H}$, becomes proportional to the area of the region. This observation was turned into a beautiful theorem by Duke in [Duk88]. To formulate the statement more precisely, suppose such an A is given, and write $\mathbf{1}_A$ for the indicator function of A in $\Gamma\backslash\mathbb{H}$.

Theorem 6.1.2 (Duke's Theorem) *As $D \to +\infty$ through fundamental discriminants,*

$$\frac{1}{h_D} \sum_{\gamma \in \mathscr{C}_D} \frac{1}{\ell(\gamma)} \int_\gamma \mathbf{1}_A\, ds \longrightarrow \frac{1}{\mathrm{vol}(\Gamma\backslash\mathbb{H})} \int_{\Gamma\backslash\mathbb{H}} \mathbf{1}_A \frac{dx\,dy}{y^2}. \tag{6.35}$$

Here $\ell(\gamma)$ is the length of the closed geodesic γ, and ds is the (hyperbolic) arclength measure.

For simplicity, we have stated (6.35) for the base space $\Gamma\backslash\mathbb{H}$, but a similar result holds for the unit tangent bundle as well. Also, one can prove (6.35) with effective (power savings) rates, and dropping the 'fundamental' condition; see e.g. [CU04]. Again, this is a big theory, of which we have only scratched the surface; the proceeding discussion will suffice for our purposes here.

6.2 Lecture 2: Three Problems in Continued Fractions: ELMV, McMullen, and Zaremba

6.2.1 ELMV

Duke's proof of (6.35) is a tour-de-force of analytic gymnastics, involving a Maass 'theta correspondence' to convert period integrals into Fourier coefficients of half-integral weight forms, and implementing methods pioneered by Iuraniec [Iwa87] (using Kuznestov's formula, sums of Kloosterman sums, and estimates of Bessel-type functions) to give non-trivial estimates of such. Meanwhile, the theorem itself (6.35) seems like a beautifully simple dynamical statement; perhaps there is a more 'ergodic-theoretic' proof? Indeed, one was eventually obtained, after much work, by Einsiedler–Lindenstrauss–Michel–Venkatesh [ELMV12] (which is henceforth referred to as ELMV), with some ideas pre-empted decades earlier by Linnik [Lin68].

ELMV wanted to approach the problem (and higher-rank analogues; see [ELMV09]) from a type of 'measure rigidity' à la Rather's theorems – one must show that the only measure arising as a limit of the measures on the left-hand side of (6.35) is the Haar measure, $dx\, dy/y^2$. This raised the question: does one really need the full *average* over the class group here, or could individual geodesics already equidistribute? Despite some partial progress to the affirmative (see e.g. [HM06, Pop06]), because of the symbolic coding of the geodesic flow, closed geodesics can be made to have arbitrary behaviour simply by choosing the partial quotients in the visual point and working backwards. So, certainly not all long closed geodesics equidistribute. But perhaps if one restricts oneself only to *fundamental* closed geodesics – that is, ones whose corresponding discriminant is fundamental – the equidistribution will be restored? Even then, it is easy to produce examples of non-equidistributing sequences of closed geodesics, which, for example, have limit measure dy/y supported on the imaginary axis (see [Sar07, p. 233]). But what if we do not allow the 'mass' to escape into the cusp of $\Gamma\backslash\mathbb{H}$? Can we find closed fundamental geodesics which stay away from the cusp? Among these would certainly be some interesting limiting measures, which would demonstrate the difficulty of Duke's Theorem. Such considerations naturally led ELMV around 2004 to propose the following:

Problem: (ELMV) Does there exist a compact subset $\mathcal{Y} \subset \mathcal{X} = T^1(\Gamma\backslash\mathbb{H})$ of the unit tangent bundle of the modular surface which contains infinitely many fundamental closed geodesics?

The answer to this question turns out to be YES, as resolved (in a nearly best-possible quantitative sense) by Bourgain and the author in [BK15a]. We will say more about the proof in the last lecture, but we first move on to two other (seemingly unrelated) problems.

6.2.2 McMullen's (Classical) Arithmetic Chaos Conjecture

Here is a more recent problem posed by McMullen [McM09, McM12], which he calls 'Arithmetic Chaos' (not to be confused with Arithmetic Quantum Chaos, for which see e.g. [Sar11]). The problem's statement begins in a similar way to ELMV, asking for closed geodesics on $\mathcal{X}$ contained in a fixed compact set $\mathcal{Y}$, but the source is quite different.

McMullen was studying questions around the theme of Margulis's conjectures on the rigidity of higher-rank torus actions, and observed that a very interesting problem in rank 1 had been overlooked.[3] The statement is the following.

Conjecture 6.2.1 (Arithmetic Chaos) *There is a compact subset $\mathcal{Y}$ of $\mathcal{X}$ such that, for all real quadratic fields K, the set of closed geodesics defined over K and lying in $\mathcal{Y}$ has positive entropy.*

Before explaining the meaning of the above words, we reformulate the conjecture as a simple statement about continued fractions.

Conjecture 6.2.2 (Arithmetic Chaos II) *There is an $A < \infty$ such that, for any real quadratic field K, the set*

$$\{[\overline{a_0, a_1, \ldots, a_\ell}] \in K : \text{ all } a_j \leq A\} \tag{6.36}$$

has exponential growth (as $\ell \to \infty$).

The reformulation is quite simple. Recall yet again that the geodesic flow is a symbolic coding of the continued fraction expansion of the visual point. Thus, going high in the cusp means having large partial quotients, and vice

[3] See [McM09, Conj. 6.1] and for a more precise statement which implies Arithmetic Chaos, and moreover predicts that the entropy mentioned below can be made arbitrarily close to the natural limit.

versa. So a geodesic which is 'low-lying' in some compact set $\mathcal{Y}$ – that is, avoiding the cusp – is one whose visual point has only small partial quotients. Now, since closed geodesics correspond to classes of binary forms with roots that are quadratic irrationals, the visual points automatically lie in some real quadratic field; in fact, it is easy to see that they lie in $K = \mathbb{Q}(\sqrt{D})$, where D is the discriminant of the geodesic (i.e. that of the class). This explains the appearance of real quadratic fields in both versions of the conjecture, as well as how being 'low-lying' in Conjecture 6.2.1 corresponds to having small partial quotients in Conjecture 6.2.2. Without defining entropy, let us simply say that this condition in the first version corresponds in the second to the exponential growth of the set in (6.36).

In the third lecture, below, we will present a certain 'Local–Global Conjecture' which would easily imply McMullen's in the strongest from: that is, with $A = 2$ in Conjecture 6.2.2. The same conjecture also has as an immediate consequence the aforementioned resolution of the ELMV Problem, as well as Zaremba's Conjecture (to be described below). Unlike the latter two problems, it seems McMullen's problem requires the full force of this Local–Global Conjecture; embarrassingly, the only meagre progress made so far is numerical, as we now describe.

In McMullen's lecture [McM12], he gives numerical evidence for Conjecture 6.2.2 with $A = 2$, taking the case $K = \mathbb{Q}(\sqrt{5})$: he is able to find the following (primitive, modulo cyclic permutations and reversing of partial quotients) continued fractions:

$$[\overline{1}] = \frac{1+\sqrt{5}}{2}, \quad [\overline{1,1,1,1,1,1,2,1,1,2,2,1,1,1,1,2,2}] = \frac{554+421\sqrt{5}}{923}.$$

McMullen presents these two surds as evidence of the conjectured exponential growth.

Using the Local–Global Conjecture as a guide, the author found the following further examples:

$$[\overline{1,1,1,2,1,2,2,2,2,1,1,2,2,1,2,1,1,2,2,1}] = \frac{90603+105937\sqrt{5}}{207538},$$

$$[\overline{2,1,1,2,1,1,1,1,2,2,1,1,1,1,1,1,1,1,1,2}] = \frac{12824+7728\sqrt{5}}{11667}.$$

Most recently, Laurent Bartholdi and Dylan Thurston (private communication) have pushed these numerics even further, finding the following further examples:

$$[\overline{1, 1, 1, 1, 1, 1, 2, 2, 2, 1, 1, 1, 1, 1, 2, 1, 1, 1, 2, 2, 2, 1, 2, 2, 1, 1, 1, 2}],$$

$$[\overline{1, 1, 1, 1, 1, 2, 2, 1, 1, 1, 2, 1, 1, 1, 1, 1, 2, 2, 2, 1, 2, 1, 1, 1, 1, 2, 2, 2}],$$

$$[\overline{1, 1, 1, 2, 1, 1, 2, 1, 2, 2, 2, 2, 2, 2, 1, 2, 1,}$$
$$\overline{1, 2, 1, 2, 1, 1, 2, 1, 2, 2, 2, 2, 2, 2, 1, 2, 2}],$$

$$[\overline{1, 1, 1, 1, 1, 2, 2, 1, 1, 1, 2, 2, 1, 2, 1, 2, 2, 1,}$$
$$\overline{2, 2, 2, 1, 1, 2, 2, 2, 2, 2, 1, 2, 2, 1, 2, 1, 2, 2}],$$

$$[\overline{1, 1, 1, 1, 2, 1, 1, 2, 2, 1, 1, 1, 2, 2, 1, 1, 1, 2,}$$
$$\overline{2, 2, 1, 2, 1, 1, 2, 2, 2, 1, 1, 1, 2, 1, 1, 1, 2, 2}],$$

$$[\overline{1, 1, 2, 1, 2, 1, 1, 2, 2, 2, 1, 1, 2, 1, 2, 2, 2, 2,}$$
$$\overline{1, 1, 2, 2, 1, 2, 1, 1, 2, 2, 2, 1, 1, 2, 1, 2, 2, 2}],$$

$$[\overline{1, 1, 1, 1, 1, 1, 1, 1, 2, 1, 1, 1, 1, 1, 2, 2, 2, 2, 2, 2, 1,}$$
$$\overline{2, 2, 1, 2, 2, 2, 1, 1, 2, 1, 2, 1, 2, 2, 2, 2, 1, 2, 1, 1, 2}],$$

$$[\overline{1, 1, 1, 1, 1, 1, 1, 1, 1, 2, 1, 1, 2, 1, 1, 1, 1, 1, 2, 1, 2,}$$
$$\overline{2, 2, 2, 2, 1, 1, 2, 1, 2, 2, 2, 2, 2, 1, 2, 1, 2, 2, 1, 2, 2, 2}],$$

$$[\overline{1, 1, 1, 2, 1, 1, 2, 1, 2, 2, 1, 1, 1, 2, 1, 2, 1, 1, 2, 1, 1, 1, 1,}$$
$$\overline{2, 2, 1, 1, 1, 2, 2, 2, 2, 2, 1, 1, 1, 2, 2, 2, 2, 1, 1, 1, 2, 2}]$$

all in $\mathbb{Q}(\sqrt{5})$. One may now argue whether this list (of 13 distinct surds in all) is yet indicative of exponential growth.

6.2.3 Zaremba's Conjecture

Our final problem originates in questions about pseudorandom numbers and numerical integration. A detailed discussion of these questions is given in [Kon13, §2], so we will not repeat it here. The statement of the conjecture, understandable by Euclid, is as follows.[4]

[4] See [McM09, §6] for McMullen's connection of Zaremba's Conjecture to Arithmetic Chaos.

Conjecture 6.2.3 (Zaremba [Zar72]) *There is some $A < \infty$ such that, for every integer $d \geq 1$, there is a coprime integer $b \in (0, d)$ such that the reduced rational b/d has the (finite) continued fraction expansion*

$$\frac{b}{d} = [0, a_1, \ldots, a_\ell],$$

with all partial quotients a_j bounded by A.

Progress on the advertised Local–Global Conjecture allowed Bourgain and the author to resolve a density version of Zaremba's Conjecture:

Theorem 6.2.4 ([BK14]) *There is an $A < \infty$ such that the proportion of $d < N$ for which Zaremba's conjecture holds approaches 1 as $N \to \infty$.*

In the original paper, $A = 50$ was sufficient, and this has since been reduced to $A = 5$ in [Hua15, FK14]. Most recently, Zaremba's conjecture has found application to counterexamples to Lusztig's conjecture on modular representations, via the groundbreaking work of Geordie Williamson: see [Wil15].

6.3 Lecture 3: The Thin Orbits Perspective

All three problems discussed in Lecture 2 are collected here under a common umbrella as a 'Local–Global' problem for certain 'thin' orbits. We discuss recent joint work with Jean Bourgain which settles the first problem (ELMV), and makes some progress towards the last (Zaremba).

To motivate the discussion, recall from the exercise below (6.32) and the expression (6.7) that the quadratic surd

$$\alpha = [\overline{a_0, a_1, \ldots, a_\ell}]$$

is fixed by the matrix

$$M = \begin{pmatrix} a_0 & 1 \\ 1 & 0 \end{pmatrix} \cdot \begin{pmatrix} a_1 & 1 \\ 1 & 0 \end{pmatrix} \cdots \begin{pmatrix} a_\ell & 1 \\ 1 & 0 \end{pmatrix},$$

with

$$\alpha \in \mathbb{Q}(\sqrt{\operatorname{tr}^2 M - 4}).$$

This tells us that studying traces of such matrices M, we might learn something about both ELMV and McMullen's conjectures. For example:

Exercise: If $\mathrm{tr}^2 M - 4$ is square-free, then the corresponding closed geodesic is fundamental.

Yet another elementary exercise is that, if $b/d = [0, a_1, \ldots, a_\ell]$, then

$$\begin{pmatrix} a_1 & 1 \\ 1 & 0 \end{pmatrix} \cdots \begin{pmatrix} a_\ell & 1 \\ 1 & 0 \end{pmatrix} = \begin{pmatrix} d & b \\ * & * \end{pmatrix},$$

so studying the top-left entries of matrices of the above form tells us about Zaremba's conjecture.

In all the above problems, the partial quotients a_j are bounded by some absolute constant A. Thus we should study the semigroup generated by matrices of the form $\left(\begin{smallmatrix} a & 1 \\ 1 & 0 \end{smallmatrix}\right)$, with $a \le A$. Actually, we will want all elements in $\mathrm{SL}_2(\mathbb{Z})$ (whereas the generators have determinant -1), so we define the key semigroup

$$\Gamma_A := \left\langle \begin{pmatrix} a & 1 \\ 1 & 0 \end{pmatrix} : a \le A \right\rangle^+ \cap \mathrm{SL}_2,$$

of even length words in the generators; here the superscript '+' denotes generation as a semigroup.

We first claim that this semigroup Γ_A is *thin*. We give the definition by example. As soon as $A \ge 2$, the Zariski closure of Γ_A is SL_2; that is, the zero set of all polynomials vanishing on all of Γ_A is that of the single polynomial $P(a, b, c, d) = ad - bc - 1$. The integer points, $\mathrm{SL}_2(\mathbb{Z})$, of the Zariski closure grow like:

$$\#(\mathrm{SL}_2(\mathbb{Z}) \cap B_X) \asymp X^2,$$

where B_X is a ball about the origin of radius X in one's favourite fixed Archimedean norm. On the other hand, an old result of Hensley [Hen89] gives that

$$\#(\Gamma_A \cap B_X) \asymp X^{2\delta_A}. \tag{6.37}$$

Here δ_A is the Hausdorff dimension of the 'limit set' $\mathfrak{C}_A$ of Γ_A, defined as follows:

$$\mathfrak{C}_A := \{[0; a_1, a_2, \ldots] : \forall j,\ a_j \le A\}.$$

The dimensions of these Cantor-like sets have been studied for a long time [Goo41, JP01], e.g.

$$\delta_2 \approx 0.5312, \qquad \delta_3 \approx 0.7056, \qquad \delta_4 \approx 0.7889, \tag{6.38}$$

and as $A \to \infty$, Hensley [Hen92] showed that

$$\delta_A = 1 - \frac{6}{\pi^2 A} + o\left(\frac{1}{A}\right).$$

The point is that all of these dimensions δ_A are strictly less than 1, so it follows from (6.37) that

$$\#(\Gamma_A \cap B_X) = o\left(\#(\mathrm{SL}_2(\mathbb{Z}) \cap B_X)\right),$$

as $X \to \infty$. This is the defining feature of a so-called *thin integer set* (defined this way in [Kon14, p. 954]) – it has Archimedean zero density in the integer points of its Zariski closure. When the set in question is actually a sub*group* of a linear group, this definition agrees with the 'other' definition of thinness, namely the infinitude of a corresponding co-volume, or index; see e.g. [Sar14].

The ELMV, McMullen and Zaremba problems do not just study the semigroup Γ_A itself; they all study the image of Γ_A under some linear map F on $M_{2\times 2}(\mathbb{Z})$ taking integer values on Γ_A. For example, in Zaremba's conjecture, one takes $F : \left(\begin{smallmatrix} a & b \\ c & d \end{smallmatrix}\right) \mapsto a$, which picks off the top-left entry. For ELMV and Arithmetic Chaos, one studies the trace, $F : \left(\begin{smallmatrix} a & b \\ c & d \end{smallmatrix}\right) \mapsto a + d$. Then the key object of interest is $F(\Gamma_A) \subset \mathbb{Z}$, and even more precisely, the multiplicity with which an integer n is represented in $F(\Gamma_A)$. We define the multiplicity as:

$$\mathrm{mult}(n) := \#\{\gamma \in \Gamma_A : F(\gamma) = n\}.$$

A priori this count may be infinite (e.g. if F is constant), so it is useful to also define a quantity guaranteed to be finite, by truncating:

$$\mathrm{mult}_X(n) := \#\{\gamma \in \Gamma_A \cap B_X : F(\gamma) = n\}.$$

Since F is linear, the image of an Archimedean ball B_X will also be of order X, so one may naively expect the multiplicity of some $n \asymp X$ to be of order

$$\frac{1}{X} \cdot \#(\Gamma_A \cap B_X) \asymp X^{2\delta_A - 1}. \tag{6.39}$$

It is easy to see that such a prediction is too primitive; e.g. if $F : \left(\begin{smallmatrix} a & b \\ c & d \end{smallmatrix}\right) \mapsto 2a$, then all odd integers are missing in the image.

Given Γ_A and F, we define an integer $n \in \mathbb{Z}$ to be *admissible* if it passes all 'local obstructions':

$$n \in F(\Gamma_A) \pmod{q},$$

for every integer $q \geq 1$. While this condition may seem difficult to verify (for instance, it asks about infinitely many congruences!), it turns out that, thanks to the theory of Strong Approximation, it is very easy to check in practice; see e.g. [Kon13, §2.2] for a discussion.

The Local–Global Conjecture, formulated by Bourgain and the author, states that, once these local obstructions are passed, the naive heuristic (6.39) holds.

Conjecture 6.3.1 (The Local–Global Conjecture) *Assume that $A \geq 2$, so that Γ_A is Zariski dense in SL_2, and that its image under the linear map F is infinite; equivalently the Zariski closure of $F(\Gamma_A)$ is the affine line. For a growing parameter X and an integer $n \asymp X$ which is admissible, we have*

$$\mathrm{mult}_X(n) = X^{2\delta_A - 1 - o(1)}. \tag{6.40}$$

Notice that the dimensions in (6.38) all exceed $1/2$, whence the exponents $2\delta_A - 1$ in (6.40) are all positive. So, for large X, these multiplicities are non-zero; that is, large numbers which pass all local obstructions should be 'globally' represented in $F(\Gamma_A)$.

We leave it as a pleasant **Exercise** to prove McMullen's Conjecture 6.2.2 from (6.40); see also [BK13, Lemma 1.16].

The progress leading to Theorem 6.2.4 uses the fact that the Zaremba map $F : \left(\begin{smallmatrix} a & b \\ c & d \end{smallmatrix}\right) \mapsto a$ is of 'bilinear type', in that F can be written as $F(M) = \langle v_1 \cdot M, v_2 \rangle$, where $v_1 = v_2 = (1, 0)$. A similar result can be proved (see [BK10]) whenever F is of this form; equivalently (**Exercise**) whenever $\det F = \alpha\delta - \beta\gamma = 0$, where $F : \left(\begin{smallmatrix} a & b \\ c & d \end{smallmatrix}\right) \mapsto \alpha a + \beta b + \gamma c + \delta d$.

Quite recently, a more precise formulation of Zaremba's conjecture, based on the Local–Global Conjecture, and the execution of certain Hardy–Littlewood-type estimations and identities for the 'singular series' was given by Cohen [Coh15], a Rutgers summer REU student.

Conjecture 6.3.2 *In the Zaremba setting of $F : \left(\begin{smallmatrix} a & b \\ c & d \end{smallmatrix}\right) \mapsto a$, we have*

$$\mathrm{mult}(n) \sim 2\delta_A \frac{\#(\Gamma_A \cap B_n)}{n} \times \frac{\pi^2}{6} \times \prod_{p \mid n} \left(1 - \frac{1}{p}\right).$$

A plot of the left-hand side divided by the right-hand side is given in Figure 6.5; the data is rather convincing in support of the refined conjecture, which hopefully also serves as evidence for the more general Local–Global Conjecture.

6.3.1 Tools: Expansion and Beyond

We give here but a hint of some of the methods developed to prove the results in [BK14, BK15a]. A key initial ingredient is what we shall refer to broadly as 'expansion', or 'SuperApproximation'. This has also been discussed rather extensively in numerous surveys; see e.g. [Sar04, Lub12], and [Kon13, §5.2], [Kon14, §3.3]. To go 'Beyond Expansion', one needs to

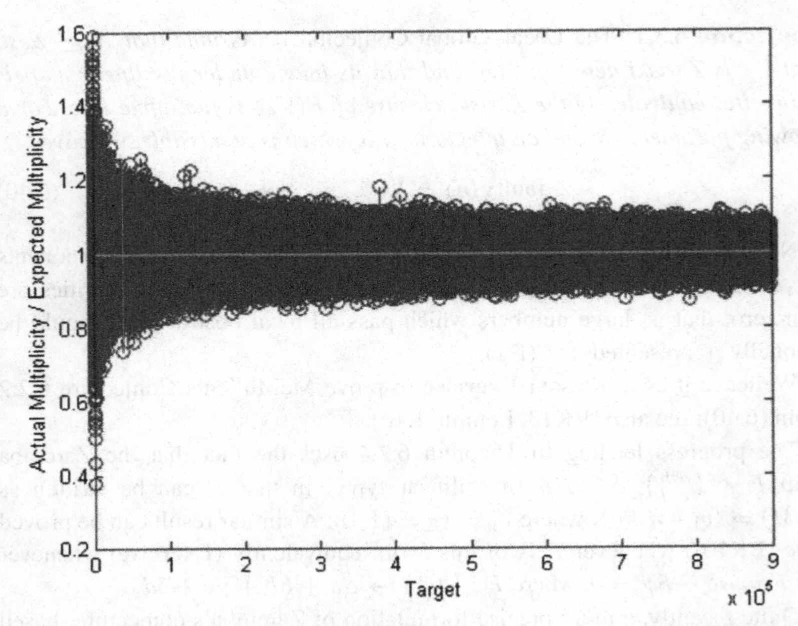

Figure 6.5 Numerical verification of Conjecture 6.3.2 for $A = 5$. Image by P. Cohen in [Coh15].

develop more sophisticated tools outside the scope of this survey; a hint is given in [Kon14, §3.7], and the interested reader is invited to peruse the original papers [BK15a, BK15b, BK16].

Acknowledgements The author is grateful to the organisers for the opportunity to collect these topics, and to Jean Bourgain for the collaborative work described here. Thanks also to Valentin Blomer and Curt McMullen for comments on an earlier draft.

Alex Kontorovich
Department of Mathematics, Rutgers University
110 Frelinghuysen Rd, Piscataway, NJ 08854, USA.
email: `alex.kontorovich@rutgers.edu`
`math.rutgers.edu/~alexk`

References

[BK10] J. Bourgain and A. Kontorovich. On representations of integers in thin subgroups of SL(2, **Z**). *GAFA*, 20(5):1144–1174, 2010.

[BK13] J. Bourgain and A. Kontorovich. Beyond expansion II: Traces of thin semigroups, 2013. Arxiv posting, `arXiv:1310.7190`.

[BK14] J. Bourgain and A. Kontorovich. On Zaremba's conjecture. *Annals Math.*, 180(1):137–196, 2014.

[BK15a] J. Bourgain and A. Kontorovich. Beyond expansion II: Low-lying fundamental geodesics, 2015. To appear, *JEMS* `arXiv:1406.1366`.

[BK15b] J. Bourgain and A. Kontorovich. The Affine Sieve Beyond Expansion I: Thin Hypotenuses. *Int. Math. Res. Not. IMRN*, (19):9175–9205, 2015.

[BK16] J. Bourgain and A. Kontorovich. Beyond expansion III: Reciprocal geodesics, 2016. Preprint.

[BM00] M. B. Bekka and M. Mayer. *Ergodic Theory and Topological Dynamics of Group Actions on Homogeneous Spaces*, volume 269 of *London Mathematical Society Lecture Note Series*. Cambridge University Press, 2000.

[Cas78] J. W. S. Cassels. *Rational Quadratic Forms*. Number 13 in London Mathematical Society Monographs. London–New York–San Francisco: Academic Press, 1978.

[CL84] H. Cohen and H. W. Lenstra, Jr. Heuristics on class groups of number fields. In *Number theory, Noordwijkerhout 1983 (Noordwijkerhout, 1983)*, volume 1068 of *Lecture Notes in Math.*, pages 33–62. Berlin: Springer, 1984.

[Coh15] P. Cohen. A heuristic for multiplicity computation for zarembas conjecture, 2015. Rutgers University Summer 2015 REU paper.

[Cox13] D. A. Cox. *Primes of the Form* $x^2 + ny^2$. Pure and Applied Mathematics (Hoboken). Hoboken, NJ: John Wiley & Sons, Inc., second edition, 2013. Fermat, class field theory, and complex multiplication.

[CU04] L. Clozel and E. Ullmo. Equidistribution des pointes de hecke. In *Contribution to Automorphic Forms, Geometry and Number Theory*, pages 193–254. Johns Hopkins University Press, 2004.

[Dav80] H. Daveport. *Multiplicative Number Theory*, volume 74 of *Grad. Texts Math.* New York: Springer-Verlag, 1980.

[Deu33] M. Deuring. Imaginäre quadratische Zahlkörper mit der Klassenzahl 1. *Math. Z.*, 37(1):405–415, 1933.

[Duk88] W. Duke. Hyperbolic distribution problems and half-integral weight Maass forms. *Invent. Math.*, 92(1):73–90, 1988.

[ELMV09] M. Einsiedler, E. Lindenstrauss, P. Michel and A. Venkatesh. Distribution of periodic torus orbits on homogeneous spaces. *Duke Math. J.*, 148(1):119–174, 2009.

[ELMV12] M. Einsiedler, E. Lindenstrauss, P. Michel and A. Venkatesh. The distribution of closed geodesics on the modular surface, and Duke's theorem. *L'Enseignement Mathematique*, 58:249–313, 2012.

[EW11] M. Einsiedler and T. Ward. *Ergodic theory with a view towards number theory*, volume 259 of *Graduate Texts in Mathematics*. London: Springer-Verlag London, Ltd, 2011.

[FK90] R. Fricke and F. Klein. *Vorlesungen über die Theorie der elliptischen Modulfunktionen*. 2 Bände, Teubner, Leipzig, 1890.

[FK14] D. A. Frolenkov and I. D. Kan. A strengthening of a theorem of Bourgain–Kontorovich II. *Mosc. J. Comb. Number Theory*, 4(1):78–117, 2014.

[GF51] I. M. Gel'fand and S. V. Fomin. Unitary representations of Lie groups and geodesic flows on surfaces of constant negative curvature. *Doklady Akad. Nauk SSSR (N.S.)*, 76:771–774, 1951.

[Gol76] D. M. Goldfeld. The class number of quadratic fields and the conjectures of Birch and Swinnerton–Dyer. *Ann. Scuola Norm. Sup. Pisa Cl. Sci. (4)*, 3(4):624–663, 1976.

[Goo41] I. J. Good. The fractional dimensional theory of continued fractions. *Proc. Cambridge Philos. Soc.*, 37:199–228, 1941.

[GZ86] B. H. Gross and D. B. Zagier. Heegner points and derivatives of L-series. *Invent. Math.*, 84(2):225–320, 1986.

[Hei34] H. Heilbronn. On the class number in imaginary quadratic fields. *Quarterly J. of Math.*, 5:150–160, 1934.

[Hej83] D. A. Hejhal. *The Selberg trace formula for* PSL(2, **R**). *Vol. 2*, volume 1001 of *Lecture Notes in Mathematics*. Berlin: Springer-Verlag, 1983.

[Hen89] D. Hensley. The distribution of badly approximable numbers and continuants with bounded digits. In *Théorie des nombres (Quebec, PQ, 1987)*, pages 371–385. Berlin: de Gruyter, 1989.

[Hen92] D. Hensley. Continued fraction Cantor sets, Hausdorff dimension, and functional analysis. *J. Number Theory*, 40(3):336–358, 1992.

[HM06] G. Harcos and P. Michel. The subconvexity problem for Rankin-Selberg L-functions and equidistribution of Heegner points. II. *Invent. Math.*, 163(3):581–655, 2006.

[Hua15] ShinnYih Huang. An improvement to Zaremba's conjecture. *Geom. Funct. Anal.*, 25(3):860–914, 2015.

[Hum16] G. Humbert. Sur les fractions continues ordinaires et les formes quadratiques binaires indéfinies. *Journal de mathématiques pures et appliquées 7e série*, 2:104–154, 1916.

[IK04] H. Iwaniec and E. Kowalski. *Analytic number theory*, volume 53 of *American Mathematical Society Colloquium Publications*. Providence, RI: American Mathematical Society, 2004.

[Iwa87] H. Iwaniec. Fourier coefficients of modular forms of half-integral weight. *Invent. Math.*, 87:385–401, 1987.

[JP01] O. Jenkinson and M. Pollicott. Computing the dimension of dynamically defined sets: E_2 and bounded continued fractions. *Ergodic Theory Dynam. Systems*, 21(5):1429–1445, 2001.

[Kon13] A. Kontorovich. From Apollonius to Zaremba: local–global phenomena in thin orbits. *Bull. Amer. Math. Soc. (N.S.)*, 50(2):187–228, 2013.

[Kon14] A. Kontorovich. Levels of distribution and the affine sieve. *Ann. Fac. Sci. Toulouse Math. (6)*, 23(5):933–966, 2014.

[Lag80] J. C. Lagarias. On the computational complexity of determining the solvability or unsolvability of the equation $x^2 - dy^2 = -1$. *Trans. AMS*, 260(2):485–508, 1980.

[Lan35] E. Landau. Bemerkungen zum Heilbronnschen Satz. *Acta Arith*, pages 1–18, 1935.

[Lin68] Yu. V. Linnik. *Ergodic properties of algebraic fields*. Translated from the Russian by M. S. Keane. Ergebnisse der Mathematik und ihrer Grenzgebiete, Band 45. New York, NY: Springer-Verlag, 1968.

[Lub12] A. Lubotzky. Expander graphs in pure and applied mathematics. *Bull. Amer. Math. Soc.*, 49:113–162, 2012.

[McM09] C. T. McMullen. Uniformly Diophantine numbers in a fixed real quadratic field. *Compos. Math.*, 145(4):827–844, 2009.

[McM12] C.T. McMullen. Dynamics of units and packing constants of ideals, 2012. Online lecture notes, www.math.harvard.edu/ ctm/expositions/home/text/-papers/cf/slides/slides.pdf.

[Pop06] A. Popa. Central values of Rankin L-series over real quadratic fields. *Compos. Math.*, 142:811–866, 2006.

[Sar80] P. C. Sarnak. Prime Geodesic Theorems. Ann Arbor, MI: ProQuest LLC, 1980. Thesis (Ph.D.)–Stanford University.

[Sar04] P. Sarnak. What is... an expander? *Notices Amer. Math. Soc.*, 51(7):762–763, 2004.

[Sar07] P. Sarnak. Reciprocal geodesics. In *Analytic number theory*, volume 7 of *Clay Math. Proc.*, pages 217–237. Amer. Math. Soc., Providence, RI, 2007.

[Sar11] P. Sarnak. Recent progress on the quantum unique ergodicity conjecture. *Bull. Amer. Math. Soc. (N.S.)*, 48(2):211–228, 2011.

[Sar14] P. Sarnak. Notes on thin matrix groups. In *Thin Groups and Superstrong Approximation*, volume 61 of *Mathematical Sciences Research Institute Publications*, pages 343–362. Cambridge University Press, 2014.

[Ser85] C. Series. The modular surface and continued fractions. *J. London Math. Soc. (2)*, 31(1):69–80, 1985.

[Sie35] C. L. Siegel. Über die analytische Theorie der quadratischen Formen. *Ann. of Math. (2)*, 36(3):527–606, 1935.

[Wil15] G. Williamson. Schubert calculus and torsion explosion, 2015. `arXiv:1309.5055v2` with Appendix by A. Kontorovich, P. McNamara, G. Williamson.

[Zar72] S. K. Zaremba. La méthode des 'bons treillis' pour le calcul des intégrales multiples. In *Applications of number theory to numerical analysis (Proc. Sympos., Univ. Montreal, Montreal, Que., 1971)*, pages 39–119. New York: Academic Press, 1972.

Index

Printed in the United States
By Bookmasters